Telecommunications Switching, Traffic and Networks

J. E. Flood

Prentice Hall
New York   London   Toronto   Sydney   Tokyo   Singapore

First published 1995 by
Harvester Wheatsheaf,
An imprint of
Pearson Education Limited
Edinburgh Gate
Harlow
Essex CM20 2JE, England

© Pearson Education Limited 1999

All rights reserved. No part of this publication may be
reproduced, stored in a retrieval system, or
transmitted, in any form, or by any means, electronic,
mechanical, photocopying, recording or otherwise,
without prior permission, in writing, from the
publisher.

Transferred to digital print on demand 2001

Printed and bound by Antony Rowe Ltd, Eastbourne

Library of Congress Cataloging-in-Publication Data

Flood, J.E. (John Edward), 1925-
   Telecommunications switching, traffic and networks / J.E. Flood.
      p. cm.
   Includes bibliographical references and index.
   ISBN 0-13-033309-3
   1. Telecommunication—Switching systems.
2. Telecommunications–traffic.   3. Telecommunication systems.
I. Title.
TK5103.8.F56   1995
621.382—dc20   94–29046
                                                                           CIP

British Library Cataloguing in Publication Data

A catalogue record for this book is available from the British
Library

ISBN 0-13-033309-3

---

6 7 8 02 01 00 99

Contents

Preface	xi
List of abbreviations	xiii

1 Introduction ... 1
   1.1 The development of telecommunications ... 1
   1.2 Network structures ... 2
   1.3 Network services ... 8
   1.4 Terminology ... 9
   1.5 Regulation ... 10
   1.6 Standards ... 11
   1.7 The OSI reference model for open systems interconnection ... 12
   1.8 Conclusion ... 14
   References ... 14

2 Telecommunications transmission ... 16
   2.1 Introduction ... 16
   2.2 Power levels ... 17
   2.3 Four-wire circuits ... 20
      2.3.1 Principle of operation ... 20
      2.3.2 Echoes ... 21
      2.3.3 Stability ... 23
   2.4 Digital transmission ... 25
      2.4.1 Bandwidth and equalization ... 25
      2.4.2 Noise and jitter ... 25
   2.5 Frequency-division multiplexing ... 27
   2.6 Time-division multiplexing ... 29
      2.6.1 Principles ... 29
      2.6.2 The PCM primary multiplex group ... 33

[vi] *Contents*

	2.6.3 The plesiochronous digital hierarchy	34
	2.6.4 The synchronous digital hierarchy	37
2.7	Transmission performance	39
	2.7.1 Telephony	39
	2.7.2 Digital transmission	41
2.8	Transmission systems	43
Notes		45
Problems		45
References		47

3 Evolution of switching systems — 49

3.1	Introduction	49
3.2	Message switching	49
3.3	Circuit switching	51
3.4	Manual systems	51
3.5	Functions of a switching system	56
3.6	The Strowger step-by-step system	57
3.7	Register–translator–senders	62
3.8	Distribution frames	64
3.9	Crossbar systems	66
3.10	A general trunking	73
3.11	Electronic switching	76
3.12	Reed–electronic systems	78
3.13	Digital switching systems	80
Problems		84
References		86

4 Telecommunications traffic — 87

4.1	Introduction	87
4.2	The unit of traffic	88
4.3	Congestion	90
4.4	Traffic measurement	92
4.5	A mathematical model	92
4.6	Lost-call systems	96
	4.6.1 Theory	96
	4.6.2 Traffic performance	99
	4.6.3 Loss systems in tandem	103
	4.6.4 Use of traffic tables	103
4.7	Queuing systems	105
	4.7.1 The second Erlang distribution	105
	4.7.2 Probability of delay	107
	4.7.3 Finite queue capacity	109
	4.7.4 Some other useful results	109
	4.7.5 System with a single server	111

Telec... Swit... Netw...

	4.7.6 Queues in tandem	112
	4.7.7 Delay tables	112
	4.7.8 Applications of delay formulae	113
4.8	Simulation	113
Notes		114
Problems		115
References		116

5 Switching networks 117

5.1	Introduction	117
5.2	Single-stage networks	117
5.3	Gradings	119
	5.3.1 Principle	119
	5.3.2 Design of progressive gradings	121
	5.3.3 Other forms of grading	123
	5.3.4 Traffic capacity of gradings	124
	5.3.5 Applications of gradings	126
5.4	Link systems	126
	5.4.1 General	126
	5.4.2 Two-stage networks	129
	5.4.3 Three-stage networks	132
	5.4.4 Four-stage networks	137
	5.4.5 Discussion	137
5.5	Grades of service of link systems	138
	5.5.1 General	138
	5.5.2 Two-stage networks	139
	5.5.3 Three-stage networks	140
	5.5.4 Four-stage networks	141
5.6	Application of graph theory to link systems	142
5.7	Use of expansion	144
5.8	Call packing	146
5.9	Rearrangeable networks	146
5.10	Strict-sense nonblocking networks	147
5.11	Sectionalized switching networks	151
Note		153
Problems		153
References		154

6 Time-division switching 156

6.1	Introduction	156
6.2	Space and time switching	158
	6.2.1 General	158
	6.2.2 Space switches	158
	6.2.3 Time switches	159

[viii] *Contents*

    6.3 Time-division switching networks   161
        6.3.1 Basic networks   161
        6.3.2 Bidirectional paths   162
        6.3.3 More complex switching networks   163
        6.3.4 Concentrators   164
        6.3.5 PBX switches   166
        6.3.6 Digital cross-connect units   166
    6.4 Grades of service of time-division switching networks   167
    6.5 Nonblocking networks   169
    6.6 Synchronization   170
        6.6.1 Frame alignment   170
        6.6.2 Synchronization networks   171
    Problems   173
    References   175

7 Control of switching systems   176

    7.1 Introduction   176
    7.2 Call-processing functions   177
        7.2.1 Sequence of operations   177
        7.2.2 Signal exchanges   181
        7.2.3 State transition diagrams   183
    7.3 Common control   186
    7.4 Reliability, availability and security   191
    7.5 Stored-program control   193
        7.5.1 Processor architecture   193
        7.5.2 Distributed processing   196
        7.5.3 Software   197
        7.5.4 Overload control   199
    Note   200
    Problems   200
    References   202

8 Signalling   204

    8.1 Introduction   204
    8.2 Customer line signalling   205
    8.3 Audio-frequency junctions and trunk circuits   207
    8.4 FDM carrier systems   210
        8.4.1 Outband signalling   210
        8.4.2 Inband (VF) signalling   211
    8.5 PCM signalling   213
    8.6 Inter-register signalling   214
    8.7 Common-channel signalling principles   218
        8.7.1 General   218

	8.7.2 Signalling networks	219
8.8	CCITT signalling system no. 6	220
8.9	CCITT signalling system no. 7	221
	8.9.1 General	221
	8.9.2 The high-level data-link control protocol	223
	8.9.3 Signal units	223
	8.9.4 The signalling information field	225
8.10	Digital customer line signalling	226
Problems		227
References		229

9 Packet switching 230

9.1	Introduction	230
9.2	Statistical multiplexing	232
9.3	Local-area and wide-area networks	235
	9.3.1 Bus networks	235
	9.3.2 Ring networks	239
	9.3.3 Comparison of bus and ring networks	241
	9.3.4 Optical-fiber networks	241
9.4	Large-scale networks	242
	9.4.1 General	242
	9.4.2 Datagrams and virtual circuits	242
	9.4.3 Routing	243
	9.4.4 Flow control	243
	9.4.5 Standards	243
	9.4.6 Frame relay	245
9.5	Broadband networks	246
	9.5.1 General	246
	9.5.2 The asynchronous transfer mode	247
	9.5.3 ATM switches	248
Notes		250
Problems		250
References		252

10 Networks 254

10.1	Introduction	254
10.2	Analog networks	255
10.3	Integrated digital networks	258
10.4	Integrated services digital networks	262
10.5	Cellular radio networks	267
10.6	Intelligent networks	269
10.7	Private networks	272
10.8	Numbering	273

[x] *Contents*

	10.8.1 National schemes	273
	10.8.2 International numbering	275
	10.8.3 Numbering plans for the ISDN era	276
	10.8.4 Public data networks	277
10.9	Charging	277
10.10	Routing	280
	10.10.1 General	280
	10.10.2 Automatic alternative routing	283
10.11	Network management	286
10.12	Conclusion	287
Problems		288
References		291

Appendix 1   Basic probability theory   293

A1.1	Definitions	293
	A1.1.1 Probability	293
	A1.1.2 Conditional probability	293
	A1.1.3 Joint probability	293
	A1.1.4 Mutually exclusive events	293
	A1.1.5 Complementary events	293
A1.2	Discrete probability distributions	293
	A1.2.1 Mean and variance	293
	A1.2.2 Mean and variance of sum or difference of independent random variables	294
	A1.2.3 Bernoulli or binomial distribution	294
	A1.2.4 Poisson distribution	294
A1.3	Continuous probability distributions	296
	A1.3.1 General	296
	A1.3.2 Negative exponential distribution	296
	A1.3.3 Gaussian or normal distribution	297
References		297

Appendix 2   Non-Poissonian traffic   298

A2.1	Smooth and peaky traffic	298
A2.2	Time congestion and call congestion	299
References		300

Appendix 3   Method of Jacobaeus for determining grade of service of a link system   301

References	301

Answers to problems   302

Index   304

Preface

The communication of information over distance is essential to civilization. This communication is increasingly provided by electronic means, in order to transport large amounts of information very rapidly. Thus, telecommunications services are extensively used in business, in social life and for entertainment. The importance of telecommunications is reflected by the inclusion of Communication Engineering as a subject in Electrical and Electronic Engineering courses.

Most users of telecommunications obtain the services they require from a telecommunications network provided by a public operator. A telecommunications network conveys traffic by means of transmission links connected by switching systems. However, conventional first degree and diploma courses in Communication Engineering, and most undergraduate textbooks, are mainly devoted to the technology of information transmission. Networks, traffic and switching have largely been ignored.

There are signs of change. Some recent undergraduate textbooks on telecommunications contain a chapter on networks and their components. There are also advanced texts, suitable for postgraduate students and for practising engineers, on networks, traffic, switching systems and signalling; however, these are too specialized for undergraduate courses. It is hoped that this book will help to fill the gap between the detailed treatments in the advanced texts and the brief treatments of these subjects in existing undergraduate textbooks. The book contains much that the author learned while working in industry, so it is hoped that it will be of interest to practising engineers as well as to students.

The book deals with switching, signalling and traffic in the context of telecommunications networks. The first chapter provides an introduction to telecommunications networks and the final chapter treats them in more detail, showing the application of technology described in preceding chapters. Both circuit switching and packet switching are treated in some depth. The reader is led through the evolution of switching systems, from manual and electromechanical systems to stored-program-controlled digital systems and future broadband systems. This approach enables fundamental ideas to be introduced progressively and explained by means of real examples. Both channel-associated and common-channel signalling systems are also described.

The theory of telecommunications traffic engineering is developed from first principles, using simple probability theory, and examples of lost-call systems and queuing systems are considered. Telecommunications transmission is not a principal subject of the book. However, a review of transmission fundamentals is included, since an appreciation of them is necessary for an understanding of networks. The book can be studied by anyone with a basic level of mathematics that includes elementary calculus. A summary of basic probability theory is included as an appendix.

As a result of teaching these topics in courses at Aston University and elsewhere, I have had many enjoyable discussions with students, academic colleagues and friends in industry. I am grateful to everyone for helping to clarify my ideas. I particularly wish to thank Mr R.W. Whorwood and Mr A.W. Joel Jr for their useful comments on the manuscript while I was writing the book.

The final form of this book arose from a course given to final year students while a visiting professor at Auckland University in New Zealand. Thanks are due to the University of Auckland for granting a University Foundation Award and to Professor A. G. Williamson, who invited me to give the course. It is hoped that the book which has resulted will also be found useful by students elsewhere.

Acknowledgements are also made to the following organizations for permission to publish information which they provided: Bell Telephone Laboratories, British Telecom, GPT Ltd, Northern Telecom (Europe) and the Institution of Electrical Engineers.

J. E. Flood
Aston University
Birmingham, UK

List of abbreviations

AAR	automatic alternative routing
ACP	action control point
A/D	analog-to-digital
ADPCM	adaptive differential pulse-code modulation
AFNOR	Association Française de Normalisation
AMPS	Advanced Mobile Phone System
ANI	automatic number indication
ANSI	American National Standards Institute
ARPANET	Advanced Research Projects Agency Network
AT	access tandem
ATD	asynchronous time division
ATM	asynchronous transfer mode
AT&T	American Telephone and Telegraph Company
Bellcore	Bell Companies Research and Engineering Organization
BHCA	busy-hour call attempts
BIB	backward indicator bit
B-ISDN	broadband ISDN
BORSCHT	battery feed, over-voltage protection, ringing, supervisory signalling, coding, hybrid and testing
BRL	balance return loss
BSI	British Standards Institution
BSN	backward sequence number
BT	British Telecom
CCIR	Comité Consultatif International des Radiocommunications
CCITT	Comité Consultatif International Télégraphique et Téléphonique
CCS	common-channel signalling
CCS	hundreds of call seconds

List of abbreviations

CHILL	CCITT high-level language
CIC	carrier identification code
CLI	calling line identification
Codec	coder/decoder
COS	class of service
C/R	command/response bit
CRC	cyclic redundancy check
CSH	called subscriber held
CSMA/CD	carrier-sense multiple access with collision detection
CT	centre de transit
D/A	digital-to-analog
dB	decibels
dBm	decibels relative to 1 mW
dBm0	corresponding dBm at zero reference point
dBr	dB relative to level at zero reference point
dBW	dB relative to 1 W
DCC	data country code
DCCE	digital cell centre exchange
DCE	data-circuit terminating equipment
DCS	digital crossconnect system
DDD	direct distance dialling
DDF	digital distribution frame
DDI	direct dialling-in
DDSN	digital derived-services network
DDSSC	digital derived-services switching centre
DIN	Deutsches Institut für Normung
DMSU	digital main switching unit
DN	directory number
DNHR	dynamic non-hierarchical routing
DNIC	data network identification code
DPNSS	digital private-network signalling system
DQDB	distributed-queue dual bus
DUP	data-user part
DSC	district switching centre
DSS	digital switching subsystem
DSSS	digital subscribers' switching subsystem
DTE	data terminal equipment
DTMF	dual-tone multi-frequency signalling
E	erlang (the unit of traffic intensity)
$E_{1,N}(A)$	loss probability for lost-call system of N trunks offered A erlangs
$E_{2,N}(A)$	delay probability for queueing system of N servers offered A erlangs
EN	equipment number

ET	exchange termination
ETSI	European Telecommunications Standards Institute
EU	European Union
FAS	flexible access system
FAW	frame-alignment word
Fax	facsimile
FCC	Federal Communications Commission
FDDI	fiber-optic distributed data interface
FDM	frequency-division multiplex(ing)
FEXT	far-end crosstalk
FIB	forward indicator bit
FIFO	first-in first-out
FISU	fill-in signal unit
FLMTS	Future Land Mobile Telecommunications System
FPS	fast packet switching
FSN	forward sequence number
GOS	grade of service
GSM	Groupe Spéciale Mobiles
GSC	group switching centre
HDLC	high-level data-link control
IAM	initial address message
IBCN	integrated broadband communications network
IBM	International Business Machines Company
IDF	intermediate distribution frame
IDN	integrated digital network
IEEE	Institute of Electrical and Electronic Engineers (USA)
IFRB	International Frequency Registration Board
IN	intelligent network
IP	intelligent peripheral
ISD	international subscriber dialling
ISDN	integrated-services digital network
ISO	International Standards Organization
ISP	intermediate-service part
ISPBX	integrated-services PBX
ISUP	ISDN-user part
ITU	International Telecommunications Union
ITU-R	ITU-Radiocommunication
ITU-T	ITU – Telecommunications
IXC	inter-exchange carrier

List of abbreviations

LAN	local-area network
LAP	link-access protocol
LAP-B	link-access protocol – balanced
LAP-D	link-access protocol – D channel
LATA	local-access and transport area
LDDC	long-distance DC signalling
LE	local exchange
LEC	local-exchange carrier
LI	length indicator
LPC	linear predictive coding
LSSU	link-status signal unit
LT	line termination
MAN	metropolitan-area network
MDF	main distribution frame
MF	multifrequency
Modem	modulator/demodulator
Muldex	multiplexer/demultiplexer
Mux	multiplexer
MSC	main switching centre
MSU	message signal unit
MTBF	mean time between failures
MTP	message-transfer part
MTTF	mean time to failure
MTTR	mean time to repair
MTS	message transfer system
N	neper
NAMTS	Nippon Automatic Mobile Telephone System
NCP	network control point
NCTE	network channel terminating equipment
NDC	network destination code
NEXT	near-end crosstalk
N-ISDN	narrow-band ISDN
NMC	network management centre
NMT	Nordic Mobile Telephone Service
NRM	normal response mode
NT	network termination
NTN	network terminal number
O&M	operations and maintenance
OFTEL	Office of Telecommunications
OLR	overall loudness rating
OMC	operations and maintenance centre

ONA	Open Network Access
ONP	Open Network Provision
OSI	open systems interconnection
PAD	packet assembler/disassembler
P&T	Posts and Telegraphs
PBX	private branch exchange
PC	primary centre
PCN	personal communication network
PDH	plesiochronous digital hierarchy
PDN	public data network
PG	permanent glow
PLE	principal local exchange
POP	point of presence
POTS	plain ordinary telephone service
PSPDN	packet-switched public data network
PSS	packet-switched service
PSTN	public switched telephone network
PTO	public telecommunications operator
PTT	posts, telegraphs and telephones
PVC	permanent virtual call
RAM	random-access memory
RBS	radio base station
RCU	remote concentrator unit
RMS	root mean square
RSU	remote switching unit
RTNR	Real-time network routing
SCCP	signalling-connection control part
SCP	service control point
SD	space division
SDH	synchronous digital hierarchy
SDL	specification and description language
SIB	service-independent building block
SIF	signalling-information field
SIO	service-information octet
SLEE	service logic execution environment
SLIC	subscriber's line interface circuit
SLP	service logic programs
SMS	service management system
SMX	synchronous multiplexer
SOH	section overhead
SONET	synchronous optical network

[xviii]   *List of abbreviations*

SPC	stored-program control
SSP	service switching point
Statmux	statistical multiplexer
STD	state transition diagram
STD	subscriber trunk dialling
STM	synchronous transport module
STP	signal-transfer point
S–T–S	space–time–space
SU	signal unit
TA	terminal adapter
TACS	Total Access Communication System
TC	transaction capabilities
TCAP	transaction-capabilities application part
TD	time division
TDF	trunk distribution frame
TDM	time-division multiplex(ing)
TE	terminal equipment
TRG	transmission relay group
T–S–T	time–space–time
TU	tributary unit
TUP	telephone-user part
TV	television
UMTS	Universal Mobile Telecommunications Service
VANS	value-added network service
VC	virtual container
VDU	visual-display unit
VF	voice frequency
VPN	virtual private network
VT	virtual tributary
WAN	wide-area network
WARC	World Administrative Radio Conference

Author's note

Some chapters are supplemented by notes at the end. These are indicated in the text by superscript numbers. References are indicated by numbers in square brackets.

CHAPTER 1

Introduction

1.1 The development of telecommunications

Most human activities depend on using information. This comes in a variety of forms, including human speech, written and printed documents and computer data. The information can be processed, stored and transported, and technologies have been developed to perform all these functions.[1] One of the most important means of transporting information is by converting it into electrical signals and transmitting these over a distance, i.e. telecommunications.

Electrical communications began with the invention of the telegraph independently by Wheatstone and Morse in 1837. Telegraph systems consisted mainly of separate point-to-point lines, sending information in one direction at a time. The advent of telephony made it necessary for lines to be connected together to permit two-way conversations. Alexander Graham Bell invented the telephone in 1876 and the first telephone exchange, at New Haven, Connecticut, was opened in 1878. In the same year, Bell wrote:[2]

> It is conceivable that cables of telephone wires could be laid underground, or suspended overhead, communicating by branch wires with private dwellings, country houses, shops, manufactories, etc., etc., uniting them through the main cable with a central office where the wires could be connected as desired, establishing direct communication between any two places in the city. Such a plan, although impracticable at the present moment, will, I firmly believe, be the outcome of the introduction of the telephone to the public. Not only so, but I believe, in the future, wires will unite the head offices of the Telephone Company in the different cities, and a man in one part of the country may communicate by word of mouth with another in a distant place.

Bell clearly conceived the idea of a national telecommunications network, in a form similar to that which exists today. The term 'central office', which Bell introduced, is still used in North America to describe a telephone exchange.

Telecommunications networks have grown up in all the countries of the world

[2] *Introduction*

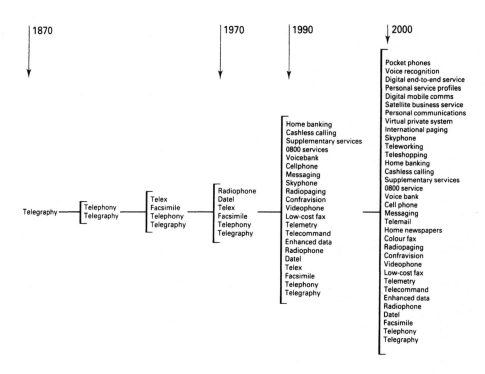

Figure 1.1 Growth of telecommunications services (past, present and future).

and have been joined up by an international network, which links more than 800 million telephones in over 200 countries. These networks now provide many different services, including telegraphy, telephony, data communications and television transmission.[3–5] The amount of traffic carried and the variety of services provided have grown steadily and may be expected to grow further in the future, as illustrated in Figure 1.1.

The business of telecommunications involves many participants. These include: the users, the public telecommunications operators (PTO), providers of services that involve telecommunications, the manufacturers of equipment and components (both hardware and software), financial investors and governments. Since the users must pay charges to cover the cost of providing the network, they are usually called *subscribers* or *customers*.

1.2 Network structures

If communication is required between n users' stations, it could be provided by a network consisting of a line from each station to every other, as shown in Figure 1.2(a).

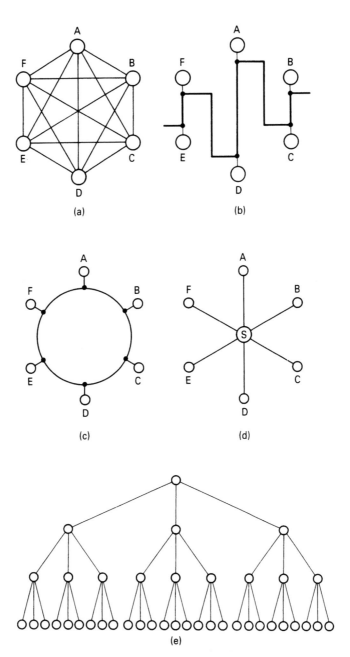

Figure 1.2 Network configurations. (a) Mesh. (b) Bus. (c) Ring. (d) Star. (e) Tree.

Introduction

This is called a *mesh network*. Each station needs lines to $n-1$ others. Thus, if the line from A to B can also convey calls from B to A, the total number of lines required is $N = \frac{1}{2}n(n-1)$. If $n \gg 1$, then N is approximately proportional to n^2. This arrangement is practicable if n is small and the lines are short. It has been used, for example, for small systems serving a number of telephones in the same office. However, as n increases and the lines become longer, the arrangement becomes much too expensive. For example, a system serving 10 000 users' stations would need nearly 50 million lines.

Instead of each station being connected to every other, they can all be connected to a single line, forming a *bus* as shown in Figure 1.2 (b), or a *ring* as shown in Figure 1.2(c). These networks are useless for normal telephony, since only one conversation at a time can take place. However, bus and ring networks can be used for data communication by transmitting data over the common circuit at a much higher rate than it is generated by the individual terminals. When the common circuit is already in use, a terminal that needs to send a message stores it until the circuit becomes free. These configurations are used for *local-area networks* (LANs) for data transmission over short distances, as described in Chapter 9.

For telephony, two-way communication is required, on demand, between any pair of stations and it must be possible for many conversations to take place at the same time. These requirements can be met by providing a line from each user's station to a central switching centre (e.g. a telephone exchange), which connects the lines together as required. This network configuration, which is shown in Figure 1.2(d), is called a *star network*. The number of lines is reduced from $N = \frac{1}{2}n(n-1)$ to only $N = n$. If N is large, the cost of providing the switching centre is far outweighed by the saving in line costs.

As the area covered by a star network and the number of stations served by it grow, line costs increase. It then becomes economic to divide the network into several smaller networks, each served by its own exchange, as shown in Figure 1.3. The average length of a customer's line, and thus the total line cost, decreases with the number of exchanges, but the cost of providing the exchanges increases. Thus, as shown in Figure 1.4, there is an optimum number of exchanges for which the total cost is a minimum.

If an area is served by several exchanges, customers on each exchange will wish to

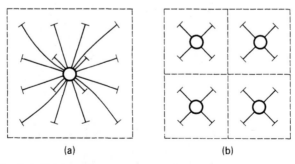

Figure 1.3 Subdivision of an exchange area. (a) Area with a single exchange. (b) Area with several exchanges.

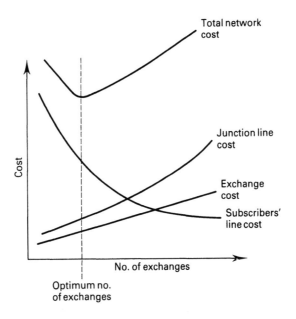

Figure 1.4 Variation of network cost with number of exchanges.

converse with customers on other exchanges. It is therefore necessary to provide circuits between the exchanges. These are called *junction circuits* and they form a *junction network*. If junctions are provided between all exchanges, the junction network has a mesh configuration, as shown in Figure 1. 2(a). However, if the cost of the junction circuits is high, it will be uneconomic to connect all the exchanges directly and cheaper to make connections between the customers' local exchanges via a central switching centre called a *tandem exchange*. The junction network then has a star configuration, as shown in Figure 1.2(d).

In practice, direct junctions between two local exchanges prove economic when there is a high community of interest between their customers (resulting in a high traffic load) or when the distance between them is short (resulting in low transmission costs). Conversely, indirect routing via a tandem exchange is cheaper when the traffic is small or the distance large. Consequently, a multi-exchange area usually has direct junctions between some exchanges, but traffic between the others is routed via the tandem exchange. The network of the area, as shown in Figure 1.5, is then a mixture of a star network, joining all local exchanges to the tandem exchange, and mesh networks connecting some of the local exchanges together.

Customers wish to communicate with people in other parts of the country in addition to those in their own area. The different areas of the country are therefore interconnected by long-distance circuits, which form a *trunk network* or a *toll network*. Since all local exchanges in an area have junctions to the tandem exchange, this provides convenient access to the trunk network. However, in large cities, the

[6] *Introduction*

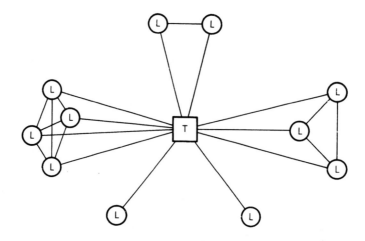

Figure 1.5 Multi-exchange area. L = local exchange, T = tandem exchange.

long-distance traffic is sufficient for the local-tandem switching and the trunk switching functions to be performed by separate exchanges.

Just as it is usually uneconomic for all the local exchanges in an area to be fully interconnected, it is often uneconomic for all the trunk exchanges of a country to be fully interconnected. Consequently, routings between different areas are provided by tandem connections via *trunk transit exchanges*. In a large national network, even these may not be fully interconnected and one or more higher levels of switching centre are introduced. This produces a concatenation of star networks, resulting in a tree configuration, as shown in Figure 1.2(e). However, direct routes are provided where the traffic is high and transmission costs are low. Thus, the 'backbone' tree is complemented by lateral routes between some exchanges at the same level, as shown in Figure 1.6.

In a network of the kind shown in Figure 1.6, when there is a direct route between two exchanges at the same level, there is also a possible alternative route between them via an exchange at the next higher level. Thus, if the direct circuit is not available (e.g. because of a cable breakdown), it is possible to divert traffic to the indirect route. In older switching systems, such changes must be made by manual rearrangements. However, modern switching systems provide *automatic alternative routing* (AAR).

With AAR, if an originating exchange is unable to find a free circuit on the direct route to a destination exchange, it automatically routes the call through the higher-level exchange. This takes place not only when there are no direct circuits available because of a breakdown but also when they are all busy. Thus, tandem connections augment the the number of circuits available to carry peak traffic and fewer circuits are needed on the direct route. In a modern network, use of AAR improves the resilience of the network to withstand both breakdowns and traffic overloads.

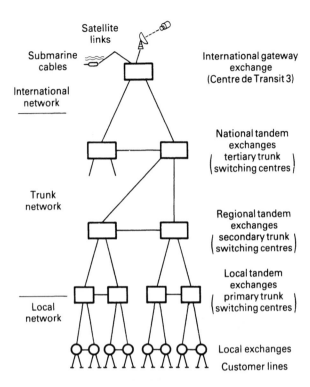

Figure 1.6 National telecommunications network. (Primary centre, secondary centre, tertiary centre and Centre de Transit 3 are the names used internationally for these switching centres.)

A national public switched telecommunications network (PSTN), as shown in Figure 1.6, consists of the following hierarchy:

1. Local networks, which connect customers' stations to their local exchanges. (These are also called subscribers' distribution networks, customer access networks or the customer loop).
2. Junction networks, which interconnect a group of local exchanges serving an area and a tandem or trunk exchange.
3. The trunk network or toll network, which provides long-distance circuits between local areas throughout the country.

The totality of (2) and (3) is sometimes called the *core network*, the inner core consisting of the trunk network and the outer one consisting of the junction networks.

Above this hierarchy, there is the international network, which provides circuits linking the national networks of different countries. The national network is connected to the international network by one or more *international gateway exchanges*.

Below the hierarchy of the national public network, some customers have

[8] *Introduction*

internal lines serving extension telephones. These are connected to one another and to lines from the public exchange by a *private branch exchange* (PBX). For data communications, they may have a LAN which is also connected to a public data network. Large companies also have private networks (usually employing circuits leased from the public telecommunications operator) which link their PBXs in different parts of the country, or even across several countries.

A telecommunications network contains a large number of transmission links joining different locations, which are known as the *nodes* of the network. Thus, each customer's terminal is a node. Switching centres form other nodes. At some nodes, certain circuits are not switched but their transmission paths are joined semi-permanently. Customers require connection to nodes where there are telephone operators to assist them in making calls and to public emergency services (e.g. police, fire and ambulance services). They also wish to obtain connections to commercial providers of 'value-added' network services (VANS), such as voice mail boxes, stock-market prices and sports results. Consequently, a telecommunications network may be considered to be the totality of the transmission links and the nodes, which are of the following types:

- Customer nodes
- Switching nodes
- Transmission nodes
- Service nodes

In order to set up a connection to the required destination, and clear it down when no longer required, the customer must send information to the exchange. For a connection that passes through several exchanges, such information must be sent between all exchanges on the route. This interchange of information is called *signalling*. A telecommunications network may therefore be considered as a system consisting of the following interacting subsystems:

- Transmission systems
- Switching systems
- Signalling systems

These systems are described more fully in later chapters.

1.3 Network services

The customers of a public telecommunications operator (PTO) require many different services, which appear to require different networks. Examples include:

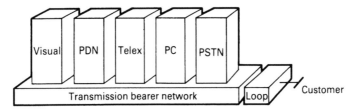

Figure 1.7 Relationship of service and bearer networks. PC = private circuits, PDN = public data network, PSTN = public switched telephone network.

- The public switched telephone network (PSTN)
- The public switched telegraph network (Telex)
- Private networks for voice and data (using circuits leased from the PTO)
- Cellular radio networks providing mobile communications
- Public data networks (PDN), usually employing packet switching
- Special service networks, introduced to meet specialized demands from customers

These services may use separate switching centres and the private circuits use transmission links semi-permanently connected together at the network nodes without switching. However, as shown in Figure 1.7, the different services all use a common *transmission bearer network* consisting of the junction and trunk circuits. Customers are connected to this at their local exchange via the local access network.

The services provided over telecommunication networks can thus be divided into two categories:

1. Teleservices, in which provision of the service depends on particular terminal apparatus (e.g. a telephone or teleprinter).
2. Bearer services, which present the customer with transmission capacity that can be used for any desired function (e.g. private circuits).

1.4 Terminology

Different names for the various networks and their switching centres are used not only in different languages but also in different English-speaking countries. For example, a switching centre is called an exchange in the UK, but a central office in North America. (In the USA, the term 'exchange' is used for a geographical area for which specific filed tariffs are applicable.) An exchange on which customers' lines terminate is a local exchange in the UK but an end office or class 5 office in North America. An exchange that switches long-distance traffic is called a trunk exchange in the UK but is a toll office in North America. Circuits between local exchanges, or between a local exchange and a tandem or trunk exchange are called junctions in the UK but are trunks in North America. In the UK, the term 'trunk' is used for a circuit between switches in an

[10] *Introduction*

Table 1.1 Comparison of nomenclature

North American	British
Customer's loop	Local network
	Access network
Central office	Exchange
End office	Local exchange
Class 5 office	
Inter-office trunk	Junction
Junctor	Trunk
Toll office	Trunk exchange
Toll network	Trunk network

exchange, but this is called a junctor in North America. These differences in terminology are summarised in Table 1.1.

This book mainly uses European terminology and the author hopes that North American readers will kindly tolerate this. However, many examples of North American practice and standards are described in the following chapters.

In the UK the customer access network of a local exchange is usually called the local network. However, in some countries the term 'local-area network' is used to describe the complete network of a multi-exchange area, including both the customer-access networks and the junction network. The term 'local-area network', or LAN, is also used to denote a private data network confined to the premises of its owner.

Internationally, trunk exchanges are called primary centres, secondary centres and tertiary centres, as shown in Figure 1.6. The primary centre is at the lowest level of the trunk hierarchy and interfaces with local exchanges. However, in North America, the highest-level toll office is a class 1 office and the lowest is a class 4 office.

In ITU terminology, an international gateway exchange is called a centre de transit 3 (CT3). International exchanges CT2 and CT1 connect only international circuits. Exchanges CT2 switch traffic between regional groups of countries and CT1 exchanges switch traffic between continents.

1.5 Regulation

The business of operating telecommunications networks has tended to be a monopoly. It is enormously expensive to dig up streets, install conduits or ducts and lay cables throughout a country. This high cost is a barrier against competitors entering the market. It is obviously desirable that the power of a monopoly should be limited to protect its customers from exploitation. Different countries have used different methods to regulate the telecommunications business.[6]

In most countries the telecommunications monopoly has been controlled by means of state ownership. Typically, the telecommunications operator has been a Posts and Telegraphs Department (P&T) of the government or a public corporation.

In some countries the telecommunications operating companies are privately owned and efforts have been made to introduce competition. In the USA the customer can only obtain local service from the regional Bell operating company or independent company serving the area, but can choose which of several competing long-distance carriers to use. Tariffs are regulated by the Federal Communications Commission (FCC) for long-distance traffic and by the Public Utilities Commissions of individual states for local service. In Britain, both British Telecom and Mercury Communications provide local and trunk services. In addition, cable television companies have been licensed to provide telephone service to their customers. Also, three competing cellular mobile radio companies have been established. The Office of Telecommunications (OFTEL) was set up as the government's regulatory body.

Similar changes have occurred in other parts of the world, but operation of the telecommunication network is still a state monopoly in most countries. However, in the European Union, the EU Commission has issued an Open Network Provision (ONP) Directive, which requires the telecommunications operators of member states to allow other service providers fair and equal access to leased lines in their networks. A similar requirement, known as Open Network Access (ONA), has been introduced in the USA.

1.6 Standards

Successful planning and operation of international telecommunications depends on cooperation between all the countries involved. The standardization which has made an effective international network possible is carried out through the International Telecommunications Union (ITU). This was founded in 1865 as the International Telegraph Union and is the oldest of the specialized agencies of the United Nations.[7]

The work of the ITU is carried out through two main bodies:

1. The ITU Telecommunications Sector (ITU-T), which was formerly the Comité Consultatif Télégraphique et Téléphonique (CCITT). Its duties include the study of technical questions, operating methods and tariffs for telephony, telegraphy and data communications.
2. The ITU Radiocommunication Sector (ITU-R), which was formerly the Comité Consultatif International des Radiocommunications (CCIR). It studies all technical and operating questions relating to radio communications, including point-to-point communications, mobile services and broadcasting. Associated with it is the International Frequency Registration Board (IFRB), which regulates the assignment of radio frequencies to prevent interference between different transmissions.

The ITU-R and ITU-T are composed of representatives of governments, operating administrations and industrial organizations. Both have a large number of active study groups. The recommendations of the study groups are reported to plenary sessions which meet every few years. The results of the plenary sessions are published in a series of volumes which provide authoritative records of the state of the art.

In theory, these bodies issue recommendations and these apply only to international communications. However, an international call must pass over parts of the national networks of two countries in addition to the international circuits concerned. Consequently, national standards are inevitably affected. For example, an international telephone connection could not meet the transmission requirements of the ITU-T if these were violated by the part of the national network between either the calling or called customer and the international gateway exchange. Thus, in practice, PTOs must take account of ITU-T recommendations in planning their networks and manufacturers must produce equipment that meets ITU-T specifications.

In addition, there is the International Standards Organization (ISO). This produces standards in many fields, including information technology. Of particular importance to telecommunications is the ISO Reference Model for Open Systems Interconnection, which is described below. There is also the European Telecommunications Institute (ETSI). In the USA, standards are produced by the American National Standards Institute (ANSI) and the Institute of Electrical and Electronic Engineers (IEEE). Other national standards bodies include the Association Française de Normalisation (AFNOR), the British Standards Institution (BSI) and the Deutsches Institut für Normung (DIN).

The standards of individual large companies can also be influential. For example, other computer companies manufacture IBM- compatible equipment. In the USA, the Bell Companies Research and Engineering organization (Bellcore) produces standards to facilitate communications between the different Bell regional operating companies.

1.7 The ISO reference model for open systems interconnection

For successful data communication across a network, appropriate operating procedures must be established. They must be specified in detail and strictly adhered to by the sending data terminal, the receiving terminal and any intervening switching centres. These procedures are called *protocols*.

Many LANs interconnect data terminals from the same manufacturer and operate using proprietary protocols. However, as data communications developed, the need arose for communication between computers and terminals from different manufacturers. This led to the concept of *open systems interconnection* (OSI), to enable networks to be machine-independent.

Development of the necessary specifications and protocols for OSI was undertaken by the International Standards Organization (ISO). The ISO standards are based on a seven-layer protocol known as the *ISO Reference Model for OSI*. The principle of this model[8] is shown in Figure 1.8. Each layer is a service user to the layer

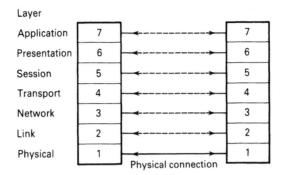

Figure 1.8 The OSI seven-layer model for open systems interconnection.

below and a service provider to the layer above. Also, each layer is specified independently of the other layers; however, it has a defined interface with the layer above and the layer below.

Thus, as far as users are concerned, communication appears to take place across each layer, as shown by the broken lines in Figure 1.8. In fact, however, each data exchange passes down to the bottom layer (the physical layer) at the sending terminal, crosses the network to the receiving terminal and then passes up again.
The layers of the OSI model are as follows:

Layer 1: *The physical layer*   This defines an interface in terms of the connections, voltage levels and data rate, in order for signals to be transmitted bit by bit.

Layer 2: *The link layer*   This provides error detection and correction for a link to ensure that the exchange of data is reliable. It may require the data stream to be divided into blocks, called *packets*, for inserting error-checking bits or for synchronization. However, transparency is preserved for the data bits in these blocks.

Layer 3: *The network layer*   This is concerned with the operation of the network between the terminals. It is responsible for establishing the correct connections between the appropriate network nodes.

Layer 4: *The transport layer*   This is responsible for establishing a network-independent communication path suitable for the particular terminal equipments (e.g. providing the appropriate data rate and end-to-end error control). It thus relieves the user from being concerned with such details.

Layer 5: *The session layer*   This is concerned with setting up and maintaining an operational session between terminals. For example, 'signing on' at the commencement of a task and 'signing off' at its end.

Layer 6: *The presentation layer*   This is concerned with the format of the data presented, in order to overcome differences in representation of the information as supplied to one terminal and required at the other. Its purpose is to make communication over the network machine-independent.

Layer 7: The application layer This defines the nature of the task to be performed. It supplies the user with the applications programs needed. Examples include electronic mail, word processing, banking transactions, etc.

It is useful to analyze communication systems in terms of the OSI model.[9–11] Systems are now being specified and designed with layered protocols conforming to the OSI model. An example is the CCITT No.7 Signalling System described in Section 8.9.

The designers of telecommunication networks are mainly concerned with layers 1 to 3. The higher layers are the concerns of the designers of software for particular applications of the network. In the case of telephony, layer 3 only requires calling and clearing signals (off-hook and on-hook), ringing, tones and addressing (dialling). Protocols for the higher layers can be developed *ad hoc* by the users as a conversation proceeds. In contrast, for machine communication, these must all be worked out in detail and programmed in advance.

1.8 Conclusion

An introduction has been given to the structure of telecommunications networks against the background of the telecommunications business and network services. The systems which form the components from which networks are built and the engineering of them to carry traffic are discussed in later chapters.

During recent years, rapid developments in technology have greatly changed the components of telecommunication networks and the structure and operation of networks themselves. Transmission systems and switching systems have become digital. A network that uses digital exchanges and digital transmission throughout is called an *integrated digital network* (IDN). Extension of digital transmission from the local exchange to the customer enables a wide variety of services (including both telephony and high-speed data transmission) to be provided through the same network. Such a network is called an *integrated-services digital network* (ISDN). The effects of these developments are discussed in the final chapter.

References

[1] Pinkerton, J.(1990), *Understanding Information Technology*, Ellis Horwood, Chichester.
[2] Field, K. (1878), *The History of Bell's Telephone*, Bradbury and Agnew.
[3] Flood, J. E. (1975), *Telecommunication networks*, Peter Peregrinus, Stevenage.
[4] Clark, M. P. (1991), *Networks and Telecommunications: design and operation*, Wiley, Chichester.
[5] van Duuren, J., Kastelein, P. and Schoute, F. C. (1992), *Telecommunication Networks and Services*, Addison-Wesley, Reading, MA.
[6] Littlechild, S. C. (1979), *Elements of Telecommunications Economics* Peter Peregrinus, Stevenage.

[7] Michaelis, A. R. (1965), *From Semaphore to Satellite*, International Telecommunications Union Centenary Volume, ITU.
[8] International Standards Organization: Specification ISO 7498 (1984), *Information Processing Systems: Open systems interconnection basic reference model*.
[9] Schwartz, M. (1987), *Telecommunication Networks: protocols, modelling and analysis*, Addison-Wesley, Reading, MA.
[10] Waband, J. (1991), *Communication Networks*, Aksand Associates.
[11] Halsall, F. (1992), *Data Communications, Computer Networks and Open Systems*, 3rd edition, Addison-Wesley, Reading, MA.

CHAPTER 2

Telecommunications transmission

2.1 Introduction

Although transmission is not a principal subject in this book, an appreciation of it is necessary for an understanding of networks. Some basic principles of transmission are therefore reviewed in this chapter.

Transmission systems provide circuits between the nodes of a telecommunications network. If a circuit uses a separate transmision path for each direction, these are called *channels*. In general, a complete channel passes through sending equipment at a *terminal station*, a *transmission link*, which may contain *repeaters* at *intermediate stations*, and receiving equipment at another terminal station.[1] Present-day transmission systems[1-4] range in complexity from simple unamplified audio-frequency circuits to satellite communication systems.

Both channels and the signals that they convey can be classified in two broad classes: *analog* and *digital*. An analog signal is a continuous function of time; at any instant it may have any value between limits set by the maximum power that can be transmitted. Speech signals are an obvious example. A digital signal can only have discrete values. The most common digital signal is a binary signal having only two values (e.g. 'mark' and space or '0' and '1'). Telegraph signals and binary-coded data from a computer are thus digital signals.

A signal consisting of a sinusoidal waveform is completely predictable, so it can convey no information. An analog signal must therefore contain a range of frequencies; this is known as its *bandwidth*. For a digital signal, the number of signal elements transmitted per second is called the signalling rate in *bauds*. For a non-redundant binary code, the rate of transmission of information in binary digits (bits) per second equals the signalling rate in bauds. For a redundant code, the bit rate is less than the number of bauds. If a multi-level code is used (e.g. ternary or quaternary), each element conveys more than one bit of information, so the bit rate is greater than the number of bauds.

To transmit an analog signal without distortion, the channel must be a linear system. A digital channel does not need to be linear. An example is a telegraph channel, whose output signal is provided by the operation of a relay. It does not follow that analog signals must always be transmitted over an analog channel and digital signals over a digital channel. Data communication and voice-frequency telegraphy over telephone lines are examples of digital signals sent over analog circuits. Analog signals may be sent over digital channels by using analog-to-digital conversion. An example is the use of pulse-code modulation[5,6] (PCM) to transmit speech.

An advantage of digital transmission over analog transmission is its relative immunity to noise and interference. For example, error-free transmission of a binary signal requires detection only of the presence or absence of each pulse and this can be done in the presence of a high level of noise. It is also possible to use *regeneration*. Provided that a received signal is not so corrupted that it is detected erroneously, it can cause the generation of an almost-perfect signal for retransmission. Thus, the transmission performance of digital circuits can be almost independent of their length, whereas analog signals deteriorate progresssively with distance.

If a link has a greater bandwidth than the signals to be transmitted, it can be used to provide a plurality of channels. At the sending terminal, the signals of different channels are combined to form a composite signal of wider bandwidth. At the receiving terminal, the signals are separated and retransmitted over different channels. This process is known as *multiplexing* and *demultiplexing*. The separate channels that enter and leave the terminal stations are called *baseband channels* and the transmission link which carries the multiplex signal is called a *broadband channel* or a *bearer channel*. The combination of a multiplexer and demultiplexer at a terminal station is often called a *muldex*, which is sometimes abbreviated to *mux*.

The principal methods of multiplexing are *frequency-division multiplexing* (FDM) and *time-division multiplexing* (TDM). In FDM transmission, each baseband channel uses the bearer channel for all of the time, but it is allocated only a fraction of the bandwidth. In TDM transmission, each baseband channel uses the entire bandwidth of the bearer, but only for a fraction of the time. These methods are discussed in Sections 2.5 and 2.6.

2.2 Power levels

A wide range of power levels is encountered in telecommunication-transmission systems. It is therefore convenient to use a logarithmic unit for powers. This is the *decibel* (dB), which is defined as follows:

If the output power P_2 is greater than the input power P_1, then the gain G in decibels is

$$G = 10 \log_{10} \frac{P_2}{P_1} \text{ dB} \tag{2.1a}$$

If, however, $P_2 < P_1$, then the loss or attenuation in decibels is

$$L = 10 \log_{10} \frac{P_1}{P_2} \text{ dB} \tag{2.1b}$$

If the input and output circuits have the same impedance, then $P_2/P_1 = (V_2/V_1)^2 = (I_2/I_1)^2$ and

$$G = 20 \log_{10} \frac{V_2}{V_1} = 20 \log_{10} \frac{I_2}{I_1} \text{ dB} \qquad (2.2)$$

In some countries, the unit employed is the *neper* (N), defined as follows:

$$\text{Gain in nepers} = \log_e \frac{I_2}{I_1} \text{ N} \qquad (2.3)$$

Thus, if the input and output circuits have the same impedance, a gain of 1 N corresponds to 8.69 dB.

A logarithmic unit of power is convenient when a number of circuits having gain or loss are connected in tandem. The overall gain or loss of a number of circuits in tandem is simply the algebraic sum of their individual gains and losses measured in decibels or nepers.

If a passive network, such as an attenuator pad or a filter, is inserted in a circuit between its generator and load, the increase in the total loss of the circuit is called the *insertion loss* of the network. If an active network, such as an amplifier, is inserted, the power received by the load may increase. There is thus an *insertion gain*.

The decibel, as defined above, is a unit of relative power level. To measure absolute power level in decibels, it is necessary to specify a *reference level*. This is usually taken to be 1 mW and the symbol dBm is used to indicate power levels relative to 1 mW. For example, 1 W = + 30 dBm and 1 μW = − 30 dBm. Sometimes (e.g. in satellite systems) the reference level is taken to be 1 W. The symbol used is then dBW.

Since a transmission system contains gains and losses, a signal will have different levels at different points in the system. It is therefore convenient to express levels at different points in the system in relation to a chosen point called the *zero reference point*. The *relative level* of a signal at any other point in the system with respect to its level at the reference point is denoted by dBr. This is, of course, equal to the algebraic sums of the gains and losses between that point and the reference point, as shown in Figure 2.1. For a four-wire circuit (see Section 2.3), the zero-level reference point is usually taken to be the two-wire input to the hybrid transformer.

It is often convenient to express a signal level in terms of the corresponding level at the reference point; this is denoted by dBmO. Consequently,

$$\text{dBmO} = \text{dBm} - \text{dBr}$$

For example, if a signal has an absolute level of − 6 dBm at a point where the relative level is − 10 dBr, the signal level referred back to the zero reference point is + 4 dBmO.

Example 2.1

An amplifier has an input resistance of 600Ω and a resistive load of 75Ω. When it has an r.m.s. input voltage of 100 mV, the r.m.s. output current is 20 mA. Find the gain in dB.

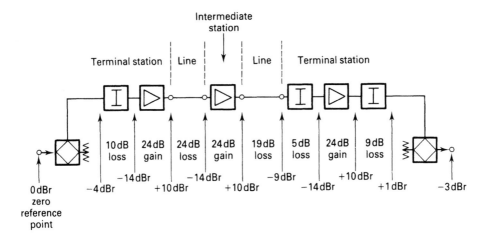

Figure 2.1 Example of relative levels in an analog transmission system.

Input power is $P_1 = (100 \times 10^{-3})^2/600$ W $= 16.7$ μW.
Output power is $P_2 = (20 \times 10^{-3})^2 \times 75$ W $= 30$ mW.

Gain is $\dfrac{P_2}{P_1} = \dfrac{30 \times 10^{-3}}{16.67 \times 10^{-6}} = 1.80 \times 10^3$

$10 \log_{10} 1.8 + 10 \log_{10} 10^3 = 2.6 + 30 = \underline{32.6 \text{ dB}}$.

Example 2.2

1. Express the following power ratios in decibels:
   (i) 1.0, (ii) 2.0, (iii) 4.0, (iv) 10, (v) 20, (vi) 100.

   (i) $10 \log_{10} 1.0 = 0$ dB.
   (ii) $10 \log_{10} 2 = 10 \times 0.30 = 3$ dB.
   (iii) $10 \log_{10}(2 \times 2) = 3 + 3 = 6$ dB.
   (iv) $10 \log_{10} 10 = 10 \times 1.0 = 10$ dB.
   (v) $10 \log_{10} (2 \times 10) = 3 + 10 = 13$ dB.
   (vi) $10 \log_{10} 100 = 10 \times 2 = 20$ dB.

2. What power ratios correspond to the following?
   (i) 0 dB, (ii) 0.1 dB, (iii) 1.0 dB, (iv) 30 dB, (v) 33 dB.

   (i) $10^0 = 1$.
   (ii) $10^{0.01} = 1.023$.
   (iii) $10^{0.1} = 1.26$.
   (iv) $10^3 = 1000$. (v) $10^3 \times 10^{0.3} = 1000 \times 2.0 = 2000$.

Example 2.3

3. Express the following power levels in dBm and dBW.

(i) 1 mW, (ii) 1 W, (iii) 2 mW, (iv) 100 mW.

(i) 1 mW = 0 dBm = − 30 dBW.
(ii) 1 W = 0 dBW = + 30 dBm.
(iii) 2 mW = 0 dBm + 3 dB = + 3 dBm = − 30 dBW + 3 dB = − 27 dBW.
(iv) 100 mW = 0 dBm + 20 dB = + 20 dBm = − 30 dBW + 20 dB = − 10 dBW.

2.3 Four-wire circuits

2.3.1 Principle of operation

It is frequently necessary to use amplifiers to compensate for the attenuation of a transmission path. Since most amplifiers are unidirectional, it is usually necessary to provide separate channels for the 'go' and 'return' directions of transmission. The term *four-wire circuit* is then used, although the go and return paths may be provided by channels in a multiplex transmission system, as shown in Figure 2.2, instead of on physical cable pairs.

At each end, the four-wire circuit must be connected to a two-wire line leading to

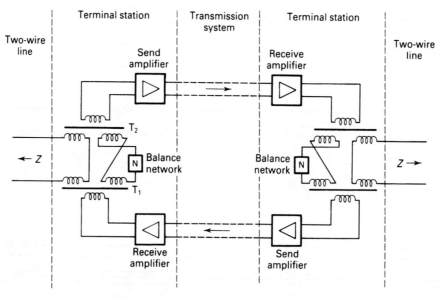

Figure 2.2 Four-wire circuit.

a telephone. If both paths of the four-wire circuit were connected directly to the two-wire circuit at each end, a signal could circulate round the complete loop thus created. This would result in continuous oscillation, known as *singing*, unless the sum of the gains in the two directions were less than zero. To avoid this, the two-wire line at each end is connected to the four-wire circuit by a *four-wire/two-wire terminating set*. This contains a *hybrid transformer*[2] (consisting of two cross-connected transformers, as shown in Figure 2.2) and a *line-balance network* whose impedance is similar to that of the two-wire circuit over the required frequency band.

The output signal from the 'receive' amplifier causes equal voltages to be induced in the secondary windings of transformer T_1. If the impedances of the two-wire line and the line balance are equal, then equal currents flow in the primary windings of transformer T_2. These windings are connected in antiphase; thus, no EMF is induced in the secondary winding of T_2 and no signal is applied to the input of the 'send' amplifier. It should be noted that the output power from the receive amplifier divides equally between the two-wire line and the line balance. When a signal is applied from the two-wire line, the cross-connection between the transformer windings results in zero current in the line-balance impedance. The power thus divides equally between the input of the send amplifier and the output of the receive amplifier, where it has no effect. The price for avoiding singing is thus 3 dB loss in each direction of transmission, together with any loss in the transformers (typically, 0.5 to 1 dB).

The impedance of the two-wire line varies with frequency. To achieve correct operation of the two-wire/four-wire termination, it would be necessary to design a complex balance network to match it closely over the frequency band. This would be expensive, if practicable. Moreover, when the four-wire line is connected to the two-wire line by switches in an electromechanical telephone exchange, it is not known in advance to which two-wire line the four-wire line will be connected. Consequently, a simple *compromise balance* is usually employed, e.g. a 600Ω or 900Ω resistor. Thus, a small fraction of the power received from the receive side of the four-wire circuit will pass through the hybrid transformer and be retransmitted in the other direction. The effects of this are considered in Sections 2.3.2 and 2.3.3.

2.3.2 Echoes

In a four-wire circuit an imperfect line balance causes part of the signal energy transmitted in one direction to return in the other. The signal reflected to the speaker's end of the circuit is called *talker echo* and that at the listener's end is called *listener echo*. The paths traversed by these echoes are shown in Figure 2.3.

The attenuation between the two-wire line and the four-wire line and between the four-wire line and the two-wire line has been shown in Section 2.3.1 to be 3 dB. Thus, the total attenuation from one two-wire circuit to the other is

$$L_2 = 6 - G_4 \text{ dB} \tag{2.4}$$

where G_4 is the net gain of one side of the four-wire circuit (i.e. total amplifier gain minus total line loss).

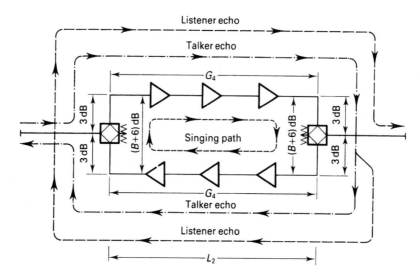

Figure 2.3 Echo and singing paths in a four-wire circuit.

The attenuation through the hybrid transformer from one side of the four-wire circuit to the other is called the *transhybrid loss*. It can be shown[2] that this loss is $6 + B$ dB, where

$$B = 20 \log_{10} \left| \frac{N + Z}{N - Z} \right| \text{ dB} \tag{2.5}$$

Z is the impedance of the two-wire line, N is the impedance of the balance network.

The loss B represents that part of the transhybrid loss which is due to the impedance mismatch between the two-wire line and the balance network. It is known as the *balance-return loss* (BRL).

The attenuation L_t of the echo that reaches the talker's two-wire line, round the path shown in Figure 2.3, is

$$L_t = 3 - G_4 + (B + 6) - G_4 + 3 \text{ dB} = 2L_2 + B \text{ dB}$$

This echo is delayed by a time

$$D_t = 2T_4$$

where T_4 is the delay of the four-wire circuit (between its two-wire terminations). The attenuation L_1 of the echo that reaches the listener's two-wire line (relative to the signal received directly) is

$$L_1 = (B + 6) - G_4 + (B + 6) - G_4 \text{ dB} = 2L_2 + 2B \text{ dB}$$

and this is delayed by a time $2T_4$ relative to the signal received directly.

The effect of an echo is different for the speaker and the listener. For the speaker it interrupts his or her conversation and for the listener it reduces the intelligibility of the received speech. Talker echo is usually the more troublesome because it is louder (by an amount equal to the BRL). The annoying effect of echo increases with its magnitude and the delay. The longer the circuit, the greater is the echo attenuation L_t required. This can be achieved by making the overall loss L_2 increase with the length of the circuit.

There is a limit to which the loss of connections can be increased to control echo. This is usually reached when the *round-trip delay*, $2T_4$, is about 40 ms. This delay is exceeded on very long transcontinental and intercontinental circuits, so it is impossible to obtain both an adequately low transmission loss and an adequately high echo attenuation. On such circuits, it is necessary to control echo by fitting devices called *echo suppressors* or *echo cancellers*.

An echo suppressor consists of a voice-operated attenuator, which is fitted on one path of the four-wire circuit operated by speech signals on the other path. Whenever speech is being transmitted in one direction, transmission in the opposite direction is attenuated, thus interrupting the echo path. There is one such suppressor (called a 'half echo suppressor') at each end of the circuit.

A number of difficulties arise with simple echo suppressors of this type. In a very-long-distance switched connection it is possible to have a number of circuits fitted with echo suppressors connected in tandem; if these operate independently, 'lock-out' conditions can arise. It is therefore necessary to disable the echo suppressors on intermediate links in the connection. It is also necessary to disable echo suppressors during data transmission, since data-transmission systems often use a return channel to request retransmission of blocks of information when errors are detected. More-sophisticated echo suppressors have been designed[7] both to provide these facilities and to cater for the very long propagation times (250 ms each way) encountered on synchronous satellite links.

Echo cancellers are now also used.[8] The echo is cancelled by subtracting a replica of it. This replica is synthesized by means of a filter, controlled by a feedback loop, which adapts to the transmission characteristic of the echo path and tracks any variations in it that may occur during a conversation.

2.3.3 Stability

If the balance-return losses of the terminations of a four-wire circuit are sufficiently small and the gains of its amplifiers are sufficiently high, the net gain round the loop may exceed zero and singing will occur. The net loss L_s of the singing path shown in Figure 2.3 is

$$L_s = 2(B + 6 - G_4) \text{ dB} \qquad (2.6)$$

Substituting from equation (2.4) into equation (2.6) gives

$$L_s = 2(B + L_2) \text{ dB} \qquad (2.7)$$

Thus, the loss of the singing path equals the sum of the two-wire to two-wire losses in the two directions of transmission and the BRLs at each end. The necessary condition for stability is $L_s > 0$. This requires that $L_2 + B > 0$, i.e.

$$G_2 < B \text{ (where } G_2 = -L_2) \tag{2.8}$$

The gain G_2 which can be obtained over a four-wire circuit is thus limited by the BRL. Equation (2.5) shows that if $N = Z$, the balance-return loss is infinite. In the limiting cases where either Z or N is zero or infinite, the balance-return loss is zero. The loss between the return and go channels is then only 6 dB (plus any loss due to transformer inefficiencies). Four-wire circuits are usually set up to be unconditionally stable, i.e. to be stable even when the two-wire lines at each end are open-circuited or short-circuited ($B = 0$). This requires operation with an overall net loss ($G_2 < 0$).

In practice, the attenuation of the singing path is deliberately made greater than zero. This provides a safety margin and avoids the attenuation distortion caused by echoes when the circuit is operating close to its singing point. The *singing point* of a circuit is defined as the maximum gain S that can be obtained (from two-wire to two-wire line) without producing singing. Thus, from equation (2.7), $S = B$, i.e. the singing point is given by the BRL (or the average of the two BRLs if these are different at the two ends of the circuit). The *stability margin* is defined as the maximum amount of additional gain, M, that can be introduced (equally and simultaneously) in each direction of transmission without causing singing, i.e. $L_s - 2M = 0$. Hence, from equation (2.7),

$$M = B + L_2 \text{ dB} \tag{2.9}$$

Thus, the stability margin is the sum of the two-wire to two-wire loss and the BRL.

In practice, a stability margin of 3 dB is found to be adequate[3] (i.e. $L_s = 6$ dB). If the circuit is to cater for zero BRL, the overall loss from two-wire circuit to two-wire circuit is then 3 dB.

In setting up long-distance switched connections, it is often necessary to connect a number of four-wire circuits in tandem. It is advantageous to eliminate terminating sets from the interfaces between the four-wire lines rather than interconnect them on a two-wire basis. The complete connection therefore consists of a number of four-wire circuits in tandem with a four-wire/two-wire termination at each end of the connection. It is necessary to ensure that this complete circuit has adequate stability.

Since the standard deviation of G_4 increases with the number of circuits in tandem, so must the overall loss. A simple rule that has been adopted by operating administrations in some countries is:

$$L_2 = 4.0 + 0.5n \text{ dB}$$

where L_2 is the overall loss (two-wire to two-wire) and n is the number of four-wire circuits in tandem on the switched connection.

Example 2.4

A four-wire circuit has an overall loss (two-wire to two-wire) of 1 dB and the balance-return loss at each end is 6 dB. Find:

1. The singing point;
2. The stability margin;
3. The attenuation of talker and listener echo.

1. From equation (2.7), $L_s = 0$ when $L_2 = -B$
   $\therefore$ Singing point is $+6$ dB.
2. From equation (2.9): $M = B + L_2 = 6 + 1 = 7$ dB.
3. $L_t = 2L_2 + B = 2 + 6 = 8$ dB.
   $L_l = 2L_2 + 2B = 2 + 12 = 14$ dB.

2.4 Digital transmission

2.4.1 Bandwidth and equalization

The minimum bandwidth needed to transmit a digital signal at B bauds was shown by Nyquist[9] to be $W_{min} = \frac{1}{2}B$. If the signal is sent through an ideal low-pass network with this cut-off frequency, it is possible to detect every pulse without error, i.e. there is no *inter-symbol interference*. In practice, it is not possible to obtain a channel with an ideal low-pass characteristic. However, Nyquist also showed[9] that zero inter-symbol interference can be obtained if the gain of the channel changes from unity to zero over a band of frequencies with a gain/frequency response that is skew-symmetrical about $f = \frac{1}{2}B$. The transfer characteristic of the channel should therefore be equalized so that the output signal has such a spectrum. A commonly used signal is that having a 'raised-cosine' spectrum which rolls off from unity to zero over the frequency band up to $f = B$. It thus occupies twice the minimum bandwidth.

Digital transmission systems can use gain and phase equalization to obtain negligible inter-symbol interference. However, 'time-domain' equalizers can be designed on a basis of impulse response. A common form of time-domain equalizer is the transversal equalizer.[4] If the characteristics of the transmission path change with time, such an equalizer can be made to adjust itself automatically. An equalizer which does this during the course of normal operation is called an *adaptive equalizer*.[10]

2.4.2 Noise and jitter

The principal advantage of PCM and other forms of digital transmission is that it is possible to obtain satisfactory transmission in the presence of severe crosstalk and noise. In the case of binary transmission, it is necessary to detect only the presence or absence of each pulse. Consider an idealized train of unipolar pulses as shown in Figure 2.4(a). The receiver compares the signal voltage, v_s, with a threshold voltage of $\frac{1}{2}V$. If a noise voltage, v_n, is added, an error occurs if $|v_n| > \frac{1}{2}V$. If v_n has a Gaussian

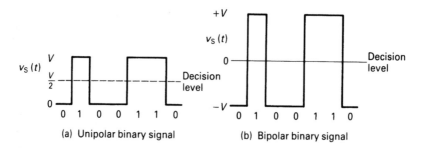

Figure 2.4 Detection of digital signals (a) Unipolar binary signal. (b) Bipolar binary signal.

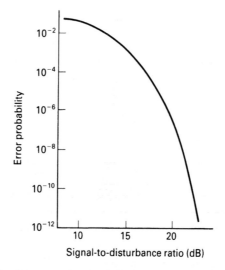

Figure 2.5 Error rate for transmission of unipolar binary signal.

probability-density distribution, the probability of an error can be calculated from a normal probability table.

If a bipolar signal is used, as shown in Figure 2.4(b), an error occurs if $|v_n| > V$. Consequently, the same error rate is obtained with a 3 dB lower signal/noise ratio.[4] Alternatively, a much lower error rate can be obtained for the same signal/noise ratio.

The variation of error probability with signal/noise ratio for a unipolar binary signal disturbed by Gaussian noise is shown in Figure 2.5. For example, if the signal/noise ratio is 20 dB, less than one digit per million is in error. For telephone transmission, an error rate of 1 in 10^3 is intolerable, but an error rate of 1 in 10^5 is acceptable. Lower error rates are required for data transmission. If the error rate of a link is inadequate, it is necessary to use an error-detecting or error-correcting code for the data.[10,11]

In digital transmission it is possible to use regenerative repeaters instead of analog amplifiers. A regenerative repeater,[1,11] samples the received waveform at intervals corresponding to the digit rate. If the received voltage at the sampling instant exceeds a threshold voltage, this triggers a pulse generator which sends a pulse to the next section of line. If bipolar pulses are used, the received voltage is compared against both a positive and a negative threshold in order to send output pulses of either polarity.

The instants at which pulses are retransmitted by a regenerative repeater are determined by a local oscillator synchronized to the digit rate, which must be extracted from the received waveform. Variations in the extracted frequency can cause a periodic variation of the times of the regenerated pulses, which is known as *jitter*.[12–14] The tolerance to jitter of any subsequent equipment in a link must therefore exceed the amount of jitter produced by preceding equipment. There can also be a long-term variation in the times of the regenerated pulses due to changes in propagation time. This is known as *wander*.

2.5 Frequency-division multiplexing

In frequency-division multiplex (FDM) transmission a number of baseband channels are sent over a common wideband transmission path by using each channel to modulate a different carrier frequency. Systems using this process are called *multichannel carrier systems*.

A *channel translating equipment*, or *channelling equipment*, for multiplexing 12 telephone channels is shown in Figure 2.6(a). At the sending end, each incoming baseband signal ($0 < f_m < F_m$) from an audio-frequency circuit is applied to a balanced modulator supplied with the appropriate carrier (f_c). The output of this modulator is a double-sideband suppressed-carrier signal, $f_c \pm f_m$. This signal is applied to a band-pass filter which suppresses the upper sideband ($f_c + f_m$) and transmits the lower sideband ($f_c - f_m$). The outputs of these filters are commoned to give a composite output signal containing the signal of each telephone channel translated to a different portion of the frequency spectrum, as shown in Figure 2.6(c).

At the receiving end, the incoming signal is applied to a bank of band-pass filters, each of which selects the frequency band containing the signal of one channel. This signal is applied to a modulator supplied with the appropriate carrier (f_c) and the output of this modulator consists of the baseband signal and unwanted high-frequency components. The unwanted components are suppressed by a low-pass filter and the baseband signal is transmitted to the audio-frequency circuit at the correct level by means of an amplifier.

Suppressed-carrier modulation is used to minimize the total power to be handled by amplifiers in the wide-band transmission system. The use of single-sideband modulation maximizes the number of channels that can be transmitted in the bandwidth available. To avoid interchannel crosstalk the sidebands of adjacent

[28] Telecommunications transmission

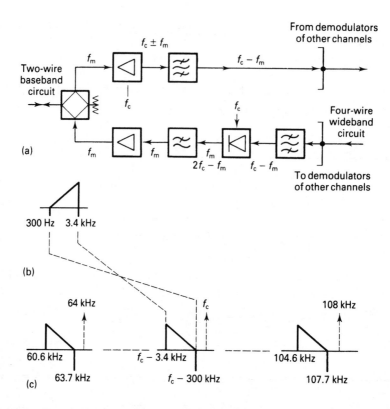

Figure 2.6 Principle of frequency-division multiplexing. (a) Channel translating equipment. (b) Frequency band of baseband signal. (c) Frequency band of wideband signal (CCITT basic group).

channels obviously must not overlap, so the spacing between carrier frequencies is determined by the highest frequency F_m of the baseband signal. Practical bandpass filters cannot have a perfectly sharp cut-off, so it is necessary to leave a small guardband between the frequency bands of adjacent channels. Figure 2.6(c) shows the standard basic group of 12 channels (CCITT basic group B). The carrier spacing is 4 kHz; thus, 12 channels occupy the band from 60 to 108 kHz. Each channel has a baseband from 300 Hz to 3.4 kHz. The frequency guardband between adjacent channels is only 900 Hz, so crystal filters are used to obtain the necessary sharp transitions between pass and stop bands.

For transmission over a balanced-pair cable, the basic 12-channel group modulates a carrier of 120 kHz in order to produce a lower sideband in the frequency band from 12 kHz to 60 kHz. The basic group is also used as a building block from which larger systems are constructed,[15] as shown in Figure 2.7.

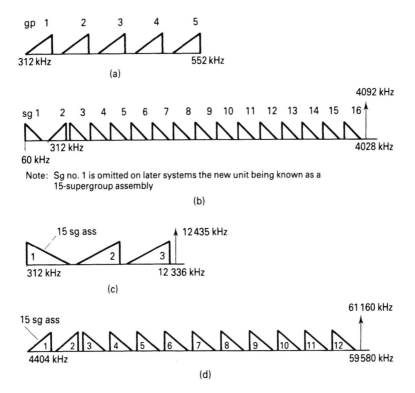

Figure 2.7 Hierarchy of FDM channel assemblies. (a) Basic supergroup. (b) Assembly of 15 + 1 supergroups (as used in 4 MHz system). (c) Assembly of three hypergroups (as used in 12 MHz system). (d) Assembly of 12 hypergroups (as used in 60 MHz system).

2.6 Time-division multiplexing

2.6.1 Principles

A basic time-division multiplex (TDM) system is shown in Figure 2.8(a). Each baseband channel is connected to the transmission path by a sampling gate which is opened for short intervals by means of a train of pulses. In this way, samples of the baseband signal are sent at regular intervals by means of amplitude-modulated pulses. Pulses with the same repetition frequency (f_r) but staggered in time, as shown in Figure 2.8(b), are applied to the sending gates of the other channels. Thus, the common transmission path receives interleaved trains of pulses modulated by the signals of different channels. At the receiving terminal, gates are opened by pulses coincident with those received from the transmission path, so that the demodulator of each channel is connected to the transmission path for its allotted intervals.

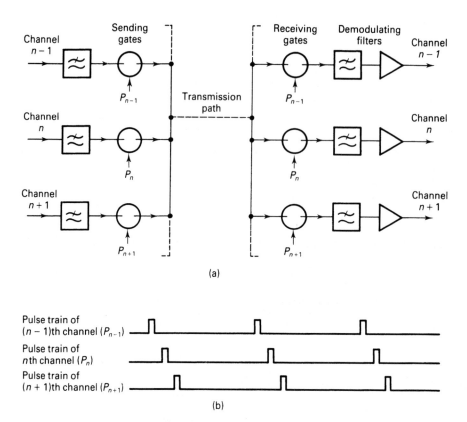

Figure 2.8 Principle of time-division multiplex transmission. (a) Elementary TDM system (one direction of transmission only). (b) Channel pulse trains.

The pulse-amplitude-modulated signal can be correctly demodulated by a low-pass filter with a cut-off frequency $\tfrac{1}{2}f_r$ provided that f_r is greater than twice the highest baseband frequency.[4,16] The input signal is therefore band limited by the input filter shown in Figure 2.8(a). To accommodate telephone channels with a band from 300 Hz to 3.4 kHz using inexpensive low-pass filters, the internationally agreed sampling frequency is 8 kHz.

The pulse generator at the receiving terminal must be synchronized with that at the sending terminal. A distinctive synchronizing pulse signal is therefore sent in each repetition period in addition to the channel pulses. The complete waveform transmitted during each repetition period thus contains a *time-slot* for each channel and one for the sync signal and it is called a *frame*.

The elementary TDM system shown in Figure 2.8 uses pulse-amplitude modulation. Pulse-length modulation and pulse-position modulation can also be employed.[16] However, these methods are not used for line transmission, because attenuation and delay distortion cause dispersion of the transmitted pulses. They

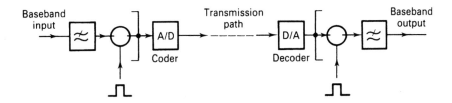

Figure 2.9 PCM system (one direction of transmission only is shown).

spread in time and interfere with the pulses of adjacent channels, thus causing inter-channel crosstalk. To overcome this problem, pulse-code modulation (PCM) is used.

In pulse-code modulation, each analog sample is applied to an analog-to-digital (A/D) converter, which produces a group of pulses that represents its voltage in a binary code. At the receiving end, a digital-to-analog (D/A) converter performs the inverse process. Since the coder used for A/D conversion and the decoder used for D/A conversion are required to perform their operations within the duration of the time-slot of one channel, they can be common to all the channels of a TDM system, as shown in Figure 2.9.

The group of bits (i.e binary pulses) representing one sample is called a *word* or a *byte*. An 8-bit byte is often called an *octet*. For telephony, sampling is carried out at 8 kHz and 8-bit encoding is used. Thus, a telephone channel requires binary digits to be sent at the rate of $8 \times 8 = 64$ kilobauds. Since the minimum bandwidth required is half the pulse rate, a bandwidth of at least 32 kHz is needed to transmit a single telephone channel. The advantages of digital transmission are won at the expense of a much greater bandwidth requirement.

PCM transmission introduces an impairment, known as *quantizing distortion* which is not present in analog transmission. This arises because the system can only transmit a finite number of sample values. For example, if the PCM system uses an 8-bit code, it can only transmit $2^8 = 256$ different sample values. There is thus a small difference between the output signal and the input signal. This process of *quantizing*, which is inherent in analog/digital conversion, thus introduces nonlinear distortion. If the amplitude of the input signal is large compared with a quantizing step, the errors in successive samples are nearly random. The spectrum of the distortion products thus approximates to that of Gaussian noise and the effect is often called *quantizing noise*.

If the coder uses quantizing steps of uniform size, as shown in Figure 2.10(a), then a large-amplitude signal is represented by a large number of steps and is reproduced with little distortion. However, a small-amplitude signal will range over few steps and a large percentage distortion will be present in the output signal. Thus, the quantizing signal/noise ratio varies with the level of the input signal. The effect of quantizing noise can be reduced by using smaller quantizing steps for small input voltages and larger steps for large input voltages, as shown in Figure 2.10(b). This process, including the corresponding non-uniform decoding process, is called *instantaneous companding*. By this means, the quantizing signal/noise ratio can be made nearly constant over a range

[32] *Telecommunications transmission*

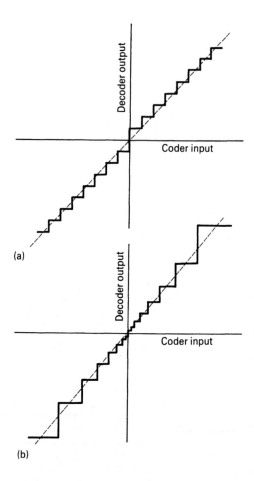

Figure 2.10 Quantization characteristics in PCM. (a) Uniform quantization. (b) Nonuniform quantization.

of input levels of about 30 dB. Two such non-uniform quantizing laws have been standardized by the CCITT. These are known as the *A law* and the *mu law* respectively.[5,12,17]

In PCM, each speech sample is coded independently. However, since samples very rarely change between their minimum and maximum voltages from one sample to the next, reductions can be made in the output digit rate if use is made of the *a priori* knowledge that the source of the signal is the human voice. Adaptive differential pulse-code modulation (ADPCM)[17] uses only 32 kbit/s. More complex coding methods, such as linear predictive coding (LPC),[18] enable digit rates as low as 4.8 kbit/s to be obtained. However, the signal processing involved introduces a delay (typically 20 ms) into the coding process. Such delays are undesirable, as explained in

Section 2.3.2, so 64 kbit/s PCM remains the standard coding technique used in digital networks.

2.6.2 The PCM primary multiplex group

PCM systems were first developed for telephone transmission over cables originally designed for audio-frequency transmission. It was found that these are satisfactory, using suitable bipolar coding,[11] for transmitting up to 2 Mbit/s. Consequently, telephone channels are combined by time-division multiplexing to form an assembly of 24 or 30 channels. This is known as the *primary multiplex group*. It is also used as a building block for assembling larger numbers of channels in higher-order multiplex systems, as described in Sections 2.6.3 and 2.6.4.

The operation of a primary multiplexer is shown in Figures 2.11 and 2.12. The length of the frame is 125 μs, corresponding to the sampling interval. It contains one speech sample from each channel, together with additional digits used for synchronization and signalling. Two frame structures are widely used: the European 30-channel system and the DS1 24-channel system used in North America and Japan. Both systems employ 8-bit coding: however, the 30-channel system uses A-law companding and the 24-channel system uses the mu law.

As shown in Figure 2.11, the frame of the 30-channel system is divided into 32 time-slots, each of 8 digits. Thus, the total bit rate is 8 kHz × 8 × 32 = 2.048 Mbit/s. Time-slots 1 to 15 and 17 to 31 are each allotted to a speech channel. Time-slot 0 is used for frame alignment. Time-slot 16 is used for signalling, as described in Section 8.5.

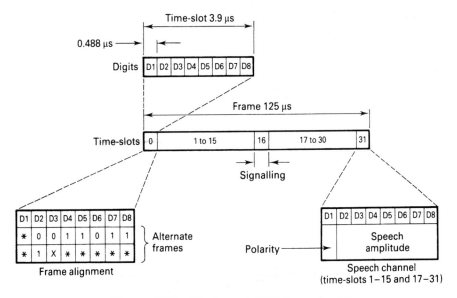

Figure 2.11 30-channel PCM frame format.

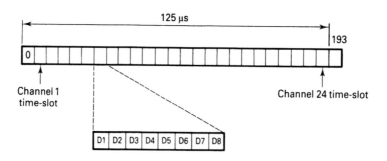

Figure 2.12 24-channel PCM frame format.

The frame format of the 24-channel system is shown in Figure 2.12. The basic frame consists of 193 bits; thus, the digit rate is 193 × 8 kbit/s = 1.544 Mbit/s. The first bit is used for framing and is called the F bit; the others form 24 8-bit time-slots for speech channels. On odd-numbered frames, the F bit takes on the alternating pattern '1,0,1,0,...', which is the pattern for frame alignment. This is a distributed frame-alignment signal as opposed to the block alignment signal used in the 30-channel system. The even frames carry the pattern '0,0,1,1,1,0,...', which defines a 12-frame multiframe. On frames 6 and 12 of the multiframe, bit D8 of each channel time-slot is used for signalling for that channel. This process of bit stealing causes a small degradation in quantizing distortion, which is none the less considered acceptable.

2.6.3 The plesiochronous digital hierarchy

The primary multiplex group of 24 or 30 shannels is used as a building block for larger numbers of channels in higher-order multiplex systems. At each level in the hierarchy, several bit streams, known as *tributaries*, are combined by a multiplexer. The output from a multiplexer may serve as a tributary to a multiplexer at the next higher level in the hierarchy, or it may be sent directly over a line or radio link.

In a transmission network which has not been designed for synchronous operation, the inputs to a digital multiplexer will not generally be exactly synchronous. Although they have the same nominal bit rate, they commonly originate from different crystal oscillators and can vary within the clock tolerance. They are said to be *plesiochronous*. The first generation of higher-order digital multiplex systems, described in this section, was designed for this situation. They form the *plesiochronous digital hierarchy* (PDH). More recently, the introduction of integrated digital networks has resulted in the transmission systems being fully synchronized and this has led to the emergence of a new *synchronous digital hierarchy* (SDH), which is described in Section 2.6.4.

If the inputs to a multiplexer are synchronous, i.e. they have the same bit rate and are in phase, they can be interleaved by taking a bit or a group of bits from each in turn. This can be done by a switch that samples each input under the control of the multiplex clock. As shown in Figure 2.13, there are two main methods of interleaving digital

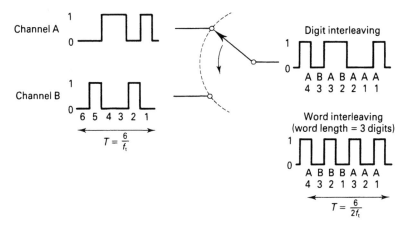

Figure 2.13 Interleaving digital signals. (a) Bit interleaving. (b) Word interleaving.

signals: *bit interleaving* and *word interleaving*. In bit interleaving, one bit is taken from each tributary in turn. If there are N input signals, each with a rate of f_t bit/s, then the combined rate will be Nf_t bit/s and each element of the combined signal will have a duration equal to $1/N$ of an input digit. In word interleaving, groups of bits are taken from each tributary in turn and this involves the use of storage at each input to hold the bits waiting to be sampled. Since bit interleaving is simpler, it was chosen for the PDH. Later, word interleaving was chosen for the SDH.

There are three incompatible sets of standards for plesiochronous digital multiplexing, centred on Europe, North America and Japan. The European standards are based on the 30-channel primary multiplex and the North American and Japanese standards on the 24-channel primary multiplex. The European and North American hierarchies are shown in Figures 2.14 and 2.15.

These systems all use bit interleaving. The frame length is the same as for the primary multiplex, i.e. 125 μs, since this is determined by the basic channel sampling rate of 8 kHz. However, when N tributaries are combined, the number of digits contained in the higher-order frame is greater than N times the number of digits in the

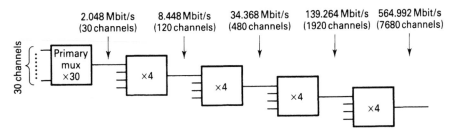

Figure 2.14 European plesiochronous digital hierarchy.

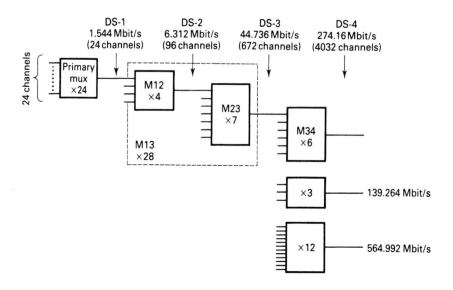

Figure 2.15 North American plesiochronous digital hierarchy.

tributary frame. This is because it is necessary to add extra 'overhead' digits for two reasons.

The first reason is frame alignment. A higher-order demultiplexer must recognize the start of each frame in order to route subsequent received digits to the correct outgoing tributaries, just as a primary demultiplexer must route received digits to the correct outgoing channels. The same technique is employed. A unique code is sent as a frame-alignment word (FAW), which is recognized by the demultiplexer and used to maintain its operation in synchronism with the incoming signal. The European hierarchy uses a block FAW at the start of each frame, but the other hierarchies use distributed FAWs.

The second reason for adding extra digits to the frame is to perform the process known as *justification*. This process is to enable the multiplexer and demultiplexer to maintain correct operation, although the input signals of the tributaries entering the multiplexer may drift relative to each other. If an input tributary is slow, a dummy digit (i.e a justification digit) is added to maintain the correct output digit rate. If the input tributary speeds up, no justification digit is added. These justification digits must be removed by the demultiplexer, in order to send the correct sequence of signal digits to the output tributary. Consequently, further additional digits, called *justification service digits*, must be added to the frame for the multiplexer to signal to the demultiplexer whether a justification digit has been used for each tributary.

The term 'justification' originated in the printing trade. Since different lines of print on a page contain unequal numbers of letters, the printer inserts additional spaces to ensure that all the lines on a page are of equal length. Word processors can also perform justification.

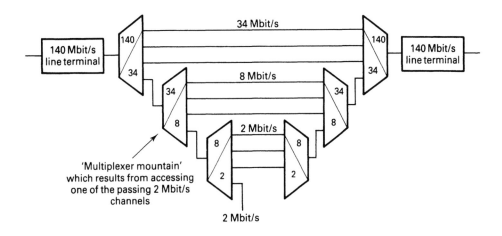

Figure 2.16 The PDH multiplex mountain.

When bit interleaving is used, bits for a particular channel occur in different bytes of a higher-order frame. In order to separate one channel from the aggregate bit steam, a total demultiplexing process is required. This results in the 'multiplexing mountain' shown in Figure 2.16. The new synchronous digital hierarchy, described in the next section, employs byte interleaving. This enables *drop-and-insert* or *add/drop muldexes* to insert or remove lower-order assemblies, down to a primary group, with relative ease.

2.6.4 The synchronous digital hierarchy

Networks are becoming fully digital, operating synchronously, using high-capacity optical-fiber transmission systems and time-division switching. It is advantageous for the multiplexers used in these networks to be compatible with the switches used at the network nodes, i.e. they should be synchronous rather than plesiochronous. In 1990, the CCITT defined a new multiplex hierarchy, known as the *synchronous digital hierarchy* (SDH). In the USA this is called the *synchronous optical network* (SONET), since the muldexes use optical interfaces. The SDH uses a digit rate of 155.52 Mbit/s and multiples of this by factors of 4n, e.g. 622.08 Mbit/s and 2488.32 Mbit/s, giving the hierarchy shown in Figure 2.17. Any of the existing CCITT plesiochronous rates up to 140 Mbit/s can be multiplexed into the SDH common transport rate of 155.52 Mbit/s. The SDH also includes management channels, which have a standard format for network-management messages.[20]

The basic SDH signal, called the *synchronous transport module at level* 1 (STM-1) is shown in Figure 2.18(a). This has nine equal segments, with 'overhead' bytes at the start of each. The remaining bytes contain a mixture of traffic and overheads, depending on the type of traffic carried. The total length is 2430 bytes, with each

[38] *Telecommunications transmission*

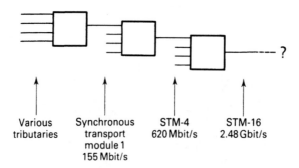

Figure 2.17 The synchronous digital hierarchy.

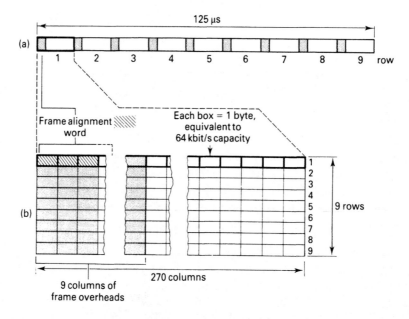

Figure 2.18 SDH frame structure (STM-1). (a) Outline frame structure. (b) Frame structure shown in rows and columns.

overhead using nine bytes. Thus, the overall bit rate is 155 520 kbit/s, which is usually called '155 Mbit/s'.

This frame is usually represented as nine rows and 270 columns of 8-bit bytes, as shown in Figure 2.18(b). The first nine columns are for *section overheads* (SOH), such as frame alignment, error monitoring and data. The remaining 261 columns comprise the *payload*, into which a variety of signals can be mapped.

Each tributary to the multiplex has its own payload area, known as a *tributary unit* (TU). In North America, a TU is called a *virtual tributary* (VT). Each column contains 9 bytes (one from each row), with each byte having 64 kbit/s capacity. Three columns (i.e. 27 bytes) can hold a 1.5 Mbit/s PCM signal, with 24 channels and some overheads. Four columns (i.e. 36 bytes) can hold a 2 Mbit/s PCM signal with 32 time-slots. The STM-1 frame can also hold payloads at the European rates of 8, 34 and 140 Mbit/s and the North American rates of 6 and 45 Mbit/s.

In the multiplexing process, the bytes from a tributary are assembled into a *container* and a *path overhead* is added to form a *virtual container* (VC). In North America, the VC is known as a *virtual tributary synchronous payload envelope*. The VC travels through the network as a complete package until it is demultiplexed. Since the VC may not be fully synchronized with the STM-1 frame, its start point is indicated by a *pointer*. The VC together with its pointer constitute the TU; thus, it is the TU that is locked to the STM-1 frame. The pointers occupy fixed places in the frame and their numerical values show where the VCs start to enable demultiplexing to be done.

The STM-1 frame is used in this way to transport signals from tributaries which use conventional TDM. An alternative method is used to transport signals in the *asynchronous transfer mode* (ATM), which is described in Section 9.5.2. Messages are transported in *cells* of 53 bytes, each cell being identified by a combination of digits known as a *header*.

Because the SDH provides interfaces for network-management messages in a standard format, it can lead to a managed transmission bearer network, in which transport capacity can be allocated flexibly to various services.[20] The network can be reconfigured under software control from remote terminals.

The ability of the SDH to provide add/drop multiplexers can lead to novel network structures.[20] Figure 2.19 shows four remote switching units (RSU) connected to a principal local exchange (PLE) in a ring configuration. There are two alternative routes between each pair of exchanges and the synchronous multiplexers (SMX) can be arranged to reroute traffic in the event of a failure, without any higher-level network-management intervention.

2.7 Transmission performance

2.7.1 Telephony

In a telephone connection the complete path includes the air path from the talker's mouth to a telephone transmitter and from a telephone receiver to the listener's ear, in addition to the telephones and the switched connection between them. The overall attenuation of such a path is expressed in terms of its *overall loudness rating* (OLR) in decibels.[21,22] This is measured by comparing the perceived loudness of the received sound with that from a standard speech path, called the Intermediate Reference System, which has been defined by the CCITT.

Subjective experiments have been carried out to determine users' opinions on a variety of typical telephone connections. In this way, it has been possible to plot curves of *percentage difficulty* (i.e the percentage of connections considered unsatisfactory)

[40] *Telecommunications transmission*

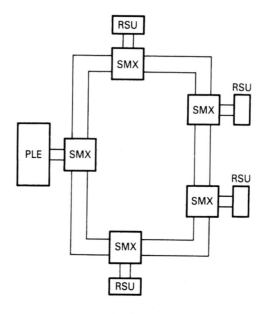

Figure 2.19 Local exchanges connected by synchronous ring. SMX = synchronous mux. PLE = principal local exchange. RSU = remote switching unit.

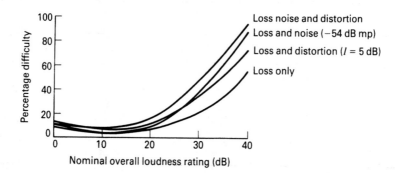

Figure 2.20 Variation of percentage difficulty with nominal OLR for combinations of noise power and attenuation distortion.

against the overall loudness ratings of the connections. Some results[23] are shown in Figure 2.20. It shows that there is a preferred range of loudness ratings from about + 5 dB to + 15 dB.

In an analog network, four-wire connections have an overall nominal loss to ensure stability when component links have losses less than their nominal values, as described in Section 2.3.3. When these component losses are greater than the nominal values, the overall loss of the connection is increased. For a large number of analog links in tandem, it can be very large and the worst-case connection can have an extremely high OLR. For example, the limiting connection in the former UK analog network had an OLR of 29 dB, which did not give acceptable performance. Fortunately, this was encountered by only a small proportion of calls, because it required a combination of links that were all adverse.

In an integrated digital network (IDN), all the inter-exchange links have zero attenuation with zero tolerance. Consequently, the variation in OLR is due only to customers' lines and telephones and A/D conversion in local exchanges. Thus, this variation is much less than in an analog network and it is possible to ensure that all connections have an OLR within the preferred range.

In analog transmission, noise tends to increase with the length of the transmission circuits and with the number of multiplexing stages employed. In digital telephone transmission, the use of regenerators ensures that noise introduced into a connection is negligible, regardless of the length of the circuit. Thus, the noise is substantially only that due to quantizing distortion introduced by coding. In an integrated digital network, coding and decoding take place only at the originating and terminating exchanges. Consequently, noise is negligible for all connections.

In four-wire circuits, propagation delay results in echo, as explained in Section 2.3.2. As the delay of the echo path increases, the loss of the circuit should increase to avoid the troublesome subjective effects of echo. The CCITT recommends that echo suppresssors or echo cancellers should be fitted if the one-way delay of the echo path exceeds 25 ms. This delay can be exceeded on international and transcontinental circuits. In particular, transmission to and from a geostationary satellite results in a one-way delay of about 250 ms. Echo control is thus essential.

2.7.2 Digital transmission

Digital transmission links introduce a number of impairments, including bit errors, slip (i.e. loss of consecutive digits due to failure of synchronisation), short breaks, jitter and wander.[21] These are usually unnoticeable for speech transmission, but they can cause unacceptable errors in data transmission.

The CCITT has specified error parameters and has set objectives for them as shown in Table 2.1. Parameters (a) and (b) are most relevant to telephony and parameter (c) is more relevant to data services.

Example 2.5

A public telecommunications operator has an analog national network as shown in Figure 1.6. All trunk circuits are four-wire circuits, but junction circuits between local exchanges (LE) and primary centres (PC) are two-wire unamplified circuits having a maximum attenuation of 3 dB. The switching systems introduce negligible attenuation. Overall loudness ratings (OLR) for calls between customers on the same LE range from − 1 dB to + 13 dB.

Find the minimum and maximum nominal attenuations between local exchanges

Table 2.1 CCITT objectives for error performance of an international hypothetical reference connection at 64 kbit/s

Performance classification	Objective
(a) Degraded minutes	< 10% of 1 min intervals to have a bit error ratio $> 1 \times 10^{-6}$
(b) Severely errored seconds	< 0.2% of 1 s intervals to have a bit error ratio $> 1 \times 10^{-3}$
(c) Errored seconds	< 8% of 1 s intervals to have any errors. (i.e. 92% to be error free)

for trunk calls and the corresponding OLRs if:

1. The trunk exchanges use two-wire switching and, to obtain unconditional stability, the four-wire circuits have a two-wire to two-wire attenuation of 3 dB.
2. The trunk exchanges use four-wire switching and the four-wire circuits are lined up so that the overall loss (two-wire to two-wire) of a connection through the trunk network is

$$L_2 = 4 + 0.5n \text{ dB}$$

where n is the number of links in tandem.

1. Minimum loss occurs when there is only one trunk link and the junctions between LEs and PCs are very short.

    ∴ minimum loss $= 0 + 3 + 0 = 3$ dB.

    Maximum loss occurs when there are five trunk links in tandem and the LE–PC junctions are of maximum length.

    ∴ maximum loss $= 3 + 5 \times 3 + 3 = 21$ dB.

    The minimum OLR is $3 - 1 = 2$ dB.
    The maximum OLR is $21 + 13 = 34$ dB.

2. Minimum loss $= 0 + 4.5 + 0 = 4.5$ dB.
    Maximum loss $= 3 + (4 + 2.5) + 3 = 12.5$ dB.
    The minimum OLR is $4.5 - 1 = 3.5$ dB.
    The maximum OLR is $12.5 + 13 = 25.5$ dB.

Example 2.6

The above network is being converted to an integrated digital network. The trunk network is already fully digital, but some local exchanges still have analog equipment with unamplified two-wire junctions to their PCs. The overall loss (two-wire to two-wire) between digital exchanges is 2 dB and OLRs for own-exchange calls vary between + 1 dB (for short customers' lines) and + 15 dB (for long customers' lines).

Find the minimum and maximum attenuation between local exchanges for trunk calls and the corresponding OLRs, both at present and when conversion to an IDN is completed.

Minimum attenuation occurs when both LEs are digital.

$\therefore$ minimum loss = 2 dB.

The corresponding minimum OLR is +1 dB.

Maximum attenuation occurs when both LEs are analog and have maximum-length junctions to their PCs.

$\therefore$ maximum loss = 3 + 2 + 3 = 8 dB.

The corresponding maximum OLR is 8 + 13 = 21 dB.

When conversion to an IDN is complete, all calls will have an attenuation (two-wire to two-wire) between LEs of 2 dB and OLRs will range from +1 dB to +15 dB.

2.8 Transmission systems

The circuits comprising a national telecommunications network fall into the following categories:

- Customers' lines
- Junction circuits
- Trunk circuits.

A wide range of transmission systems exists to meet these needs.[1–4]

Customers' lines are the most numerous and therefore need to be provided as economically as possible. Traditionally, local networks have consisted of balanced-pair audio cables, using the smallest wire gauge consistent with meeting the overall transmission plan of the national network and providing a sufficiently low resistance for loop/disconnect DC signalling.[24] In rural areas, where lines are long and expensive, a single line has sometimes been shared by several customers. For very long customers' lines, analog carrier systems have sometimes been used. Frequency-division multiplexing enables several customers to be served by a four-wire circuit. This is preferable to a multi-party line, since privacy is obtained. These systems are said to provide *pair gain*.

The advent of *integrated-services digital networks* (ISDN) has required digital transmission over customers' lines. For *basic-rate access*, 144 kbit/s are used to provide two B channels at 64 kbit/s (for speech and data) and a D channel (for signalling) at 16 kbit/s in each direction over a single two-wire line. For *primary-rate access*, a 24-channel or 30-channel PCM system is used on two cable pairs. PCM is also sometimes used to provide access for ordinary telephone customers via multiplexers, thus obtaining pair gain. The multiplexers may be connected to the local exchange over metallic pairs or optical fibers. Increasingly, optical fibers are being used to provide access for large business customers who require many circuits. The penetration of optical fibers into customers' access networks may be expected to increase as the technology develops.[24]

Radio links are sometimes used to provide access to isolated remote customers and radio is, of course, used to provide access for mobile customers in cellular radio systems.[25] In future, microcellular radio systems may be used as an alternative to cable distribution in access networks.[24]

Junction networks have also, traditionally, used audio cables. Since the number of circuits is less than in customer-access networks, it is economic for these cables to have larger conductors in order to reduce attenuation. Loading coils have also been used to reduce attenuation.[2,4] FDM carrier systems have been used on junction routes requiring very large numbers of circuits. The first application of PCM primary multiplex systems was to increase the number of circuits provided on junction cables.[26] PCM is now employed extensively and higher-order multiplexing is used on routes requiring large numbers of circuits.

The high cost of audio transmission for long-distance circuits led to the introduction of FDM carrier systems in trunk networks. Initially, 4-channel, 12-channel and 24-channel systems were used on open-wire lines and balanced-pair cables. Later, high-capacity FDM systems operating over coaxial cables and microwave radio links have been employed.[15,26,27]

The success of PCM primary multiplex systems in junction applications led to the development of higher-order multiplex systems for providing large numbers of circuits over long distances. These digital systems use coaxial cables,[26] microwave radio links[27] and optical-fiber cables.[28,29]

Digital transmission has now made analog carrier systems obsolete. Digital transmision on optical fibers is now being introduced for most applications, ranging from short-haul systems to inter-continental submarine cables. Continuing research may be expected to produce further significant advances in optical-fiber transmission systems and extend their applications still further.[30,31]

Satellite communication systems[32–34] were first introduced to provide large numbers of telephone channels across oceans. However, their inherent large propagation delay (of about 0.25 second) makes them less suitable than submarine cables for this application and optical-fiber submarine cables can now convey such large volumes of traffic at much lower cost.

In future, satellite links are likely to be used in applications where they are uniquely suitable. They will still be used for lower-capacity routes, where links to many different destinations can share a common satellite transponder. In addition, applications will increase in the areas of broadcasting, mobile communications (to ships and aircraft), business services for individual companies and personal communications. Digital techniques will be used extensively and the trend will be to smaller earth stations,[35] but more powerful satellites having regenerative transponders with on-board signal processing and scanning spot-beam antennas.[32]

Notes

1. In practice, a building that houses any kind of transmission equipment is usually called a repeater station.
2. The anti-sidetone transformer (induction coil) in a telephone set also acts as a hybrid transformer, connecting the two-wire customer's line to the four-wire circuit consisting of the microphone and receiver.
3. If the standard deviation of G_4 is 1.0 dB and a normal probability distribution is assumed, then the probability of the gain exceeding the 3 dB stability margin is 1 in 1000.

Problems

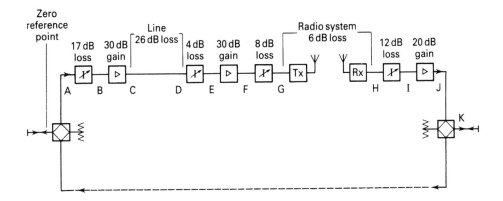

1. On the above diagram of a transmission link, mark at points A to K:
   (a) The relative levels (dBr)
   (b) The power levels (dBm) when the input level is -2 dBm
   (c) The referred levels (dBm0) when the signal power at F is 20 mW.

2. A four-wire circuit has a round-trip delay of 20 ms. The propagation time for the two-wire circuit connected to each end is 1 ms, and its attenuation is 6 dB, The balance-return loss is 3 dB and the stability margin is 3 dB. Determine:
   (a) Attenuation of talker echo
   (b) Delay of talker echo
   (c) Attenuation of listener echo

3. In a national telecommunications network, all trunk circuits are four-wire and trunk transit switching centres use four-wire switching. The four-wire/two-wire terminating sets are thus located at the primary centres (which are the interfaces between the four-wire trunk network and two-wire junction networks).

Assume that balance-return losses have a nominal value of 6 dB with a standard deviation of 2.5 dB. Assume that, for each four-wire circuit, the gains in each direction are correlated and have standard deviations of 1 dB. Both the circuit gains and the balance-return losses may be assumed to have normal distributions.

Determine the minimum overall loss

(two-wire to two-wire) to ensure that the probability of obtaining a stability margin less than 3 dB is less than 0.1% for a connection containing n links in tandem, for $n = 1$ to $n = 8$. Hence, show that a suitable planning rule for the network would be:
$$L_2 = 4 + 0.5n \text{ dB}$$
where L_2 is the overall loss (two-wire to two-wire) of a connection made up of n four-wire circuits in tandem.

4. For a long analog transmission link, it might appear possible to compensate for the line attenuation by three methods:
(a) To amplify the signal only at the sending end
(b) To amplify the signal only at the receiving end
(c) To amplify the signal at various points along the line.
Explain why method (c) is universally employed.

A 10 mW audio signal is to be transmitted over 100 km of line having an attenuation of 2 dB/km to produce a signal at the receiving end also of 10 mW. The noise level on the line is -138 dBm. Why would methods (a) and (b) be impracticable?

If method (c) is used and an amplifier is available having 50 dB gain and a noise factor of 7 dB, determine the output signal-to-noise ratio in dB.

5. A newspaper of 30 pages contains 310 Mbits of data. Its contents are to be sent from the editorial office to the printing plant over a data link. How long does this take at the following data rates?
(a) 9.6 kbit/s,
(b) 64 kbit/s,
(c) 2 Mbit/s,
(d) 140 Mbit/s.

6. The distance between London and Birmingham in England is 100 miles and the distance from Boston to Los Angeles in the USA is 3200 miles. (1 mile = 1.6 km). Optical-fiber transmission systems have been installed between these cities and their propagation velocity is 2×10^5 km/s.

If the links operate at 140 Mbit/s and are fully loaded, how many bits are in transit between London and Birmingham and between Boston and Los Angeles?

7. When fully loaded, how many telephone channels can be conveyed by each of the systems below?
(a) The following FDM carrier coaxial line systems:
4 MHz, 12 MHz, 60 MHz.
(b) The following European PDH systems:
2 Mbit/s, 8 Mbit/s, 34 Mbit/s, 140 Mbit/s, 565 Mbit/s.
(c) The following North American PDH systems:
1.5 Mbit/s, 6.3 Mbit/s, 45 Mbit/s, 140 Mbit/s, 275 Mbit/s, 565 Mbit/s.
(d) The following SDH synchronous transport modules:
STM-1, STM-4, STM-16.

8. In a data communications system, messages are transmitted in packets. Each packet contains checking digits, which enable errors to be detected. When the receiving data terminal detects an error, it sends a message back to the sending terminal requesting retransmission. The sending terminal then sends the packet again.
(a) If bit errors are independent random events, find the proportion of packets received in error when:
(i) The packet length is 50 octets and the bit error rate is 10^{-6}
(ii) The packet length is 50 octets and the bit error rate is 10^{-4}
(iii) The packet length is 500 octets and the bit error rate is 10^{-6}
(iv) The packet length is 500 octets and the bit error rate is 10^{-4}.
(b) It may be assumed that the transmission delay and the time taken to detect an error and initiate retransmission are negligible compared with the time taken to send a packet. Find, in each case, the average percentage increase in time taken to send packets, compared with the time taken when there are no errors.

9. (a) Ruritania has an analog national telecommunications network as shown in Figure 1.6. Each local exchange (LE) obtains access to the trunk network via its primary centre (PC).

The PCs use two-wire switching and the trunk exchanges use four-wire switching. Junctions between local exchanges and between LEs and PCs use two-wire unamplified circuits and their maximum attenuation is 3 dB. All circuits in the trunk network are four-wire circuits having a nominal attenuation of 0.5 dB with a tolerance of ± 1 dB. Switches in the exchanges cause negligible attenuation and the hybrid transformers in the PCs may be assumed to be perfect. The balance return losses are 3 dB ± 1 dB.

(i) Find the stability margin obtained under worst-case conditions for a trunk connection between primary centres.

(ii) Find the maximum and minimum attenuation between local exchanges for trunk calls in Ruritania.

(b) It is proposed to replace the analog network of Ruritania with an integrated digital network (IDN). Suggest a transmission plan for the new network, stating the reasons for any choices you make.

10. An integrated digital network has the following parameters:

Two-wire to two-wire loss between local exchanges: 2 dB
Balance return loss at local exchange: ≥ 12 dB
Attenuation of customers' lines: ≤ 10 dB
End-to-end propagation delay: ≤ 15 ms

Find for the worst-case conditions:
(a) Overall attenuation of a connection between customers' telephones
(b) Attenuation and delay of talker echo.

References

[1] Bell Laboratories Staff (1980), *Transmission Systems for Communications*, 5th edn, Bell Telephone Laboratories.
[2] Hills, M.T. and Evans, B.G. (1973), *Transmission Systems*, Allen and Unwin, London.
[3] Freeman, R.L. (1981), *Telecommunications Transmission Handbook*, 2nd edn, Wiley, New York.
[4] Flood, J.E. and Cochrane, P. (eds.) (1991), *Transmission Systems*, Peter Peregrinus, Stevenage.
[5] Cattermole, K.W. (1969), *Principles of Pulse Code Modulation*, Iliffe, London.
[6] Bellamy, J. (1991), *Digital Telephony*, 2nd edn, Wiley, Chichester.
[7] Shanks, P.H. (1968), 'A new echo suppressor for long-distance communications', *Post. Off. Elec. Eng. Jour.*, **60**, 288–92.
[8] Sonhi, M.M. (1967), 'An adaptive echo canceller', *Bell Syst. Tech. Jour.*, **46**, 497–511.
[9] Nyquist, H. (1928), 'Certain topics in telegraph transmission theory', *Trans. AIEE*, **47**, 617.
[10] Lucky, R.W., Saltz, J. and Weldon, E.J. (1968), *Principles of Data Communication*, McGraw-Hill, New York.
[11] Dorward, R.M. (1991), 'Digital transmission principles', Chapter 7 in Flood, J.E. and Cochrane, P. (eds), *Transmission Systems*, Peter Peregrinus, Stevenage.
[12] Bylanski, P. and Ingram, D.G.W. (1980), *Digital Transmission Systems*, Peter Peregrinus, Stevenage, 2nd edn.
[13] Byrne, C.J., Karafin, B.J. and Robinson, D.B. (1963) 'Systematic jitter in a chain of digital regenerators', *Bell Syst. Tech. Jour.*, **43**, 2679–714.
[14] Kearsey, B.N. and McLintock, R.N. (1984), 'Jitter in digital telecommunication networks', *Brit. Telecom. Eng. Jour.*, **3**, 108–16.
[15] Kingdom, D.J. (1991), 'Frequency-division multiplexing', Chapter 5 in Flood, J.E. and Cochrane, P. (eds), *Transmission Systems*, Peter Peregrinus, Stevenage.

[16] Black, H.S. (1953), *Modulation Theory*, Van Nostrand, New York.
[17] Wallace, A.D., Humphrey, L.D. and Sexton, M.J. (1991), 'Analogue to digital conversion and the primary multiplex group', Chapter 6 in Flood, J.E. and Cochrane, P. (eds), *Transmission Systems*, Peter Peregrinus, Stevenage.
[18] Jayant, N.S. and Noll, P. (1989), *Digital Coding of Waveforms*, Prentice Hall, Englewood Cliffs, NJ.
[19] Ferguson, S.P. (1991), 'Plesiochronous higher-order multiplexing', Chapter 8 in Flood, J.E. and Cochrane, P. (eds), *Transmission Systems*, Peter Peregrinus, Stevenage.
[20] Sexton, M.J. and Ferguson, S.P. (1991), 'Synchronous higher-order multiplexing', *ibid.*, Chapter 9.
[21] McLintock, R.W. (1991), 'Transmission performance', *ibid.*, Chapter 4.
[22] Richards, D.L. (1973), *Telecommunications by Speech*, Butterworths, London.
[23] Fry, R.A. (1975), 'Transmission standards and planning', Chapter 7 in Flood, J.E. (ed.), *Telecommunication Networks*, Peter Peregrinus, Stevenage.
[24] Adams, P.F., Rosher, P.A. and Cochrane, P. (1991), 'Customer access', Chapter 15 in Flood, J.E. and Cochrane, P. (eds), *Transmission Systems*, Peter Peregrinus, Stevenage.
[25] Macario, R.C.V. (ed.) (1991), *Personal and Mobile Radio Systems*, Peter Peregrinus, Stevenage.
[26] Howard, P.J. and Catchpole, R.J. (1991), 'Line systems', Chapter 10 in Flood, J.E. and Cochrane, P. (eds), *Transmission Systems*, Peter Peregrinus, Stevenage.
[27] de Belin, M.J. (1991), 'Microwave radio links', *ibid.*, Chapter 12.
[28] Senior, J. M. (1991), *Optical Fiber Communication*, 2nd edn Prentice Hall, Englewood Cliffs, NJ.
[29] Bickers, L. (1991), 'Optical-fibre transmission systems', Chapter 11 in Flood, J.E. and Cochrane, P. (eds), *Transmission Systems*, Peter Peregrinus, Stevenage.
[30] Cochrane, P. (1991), 'Future trends', *ibid.*, Chapter 17.
[31] Midwinter, J.E. and Guo, Y.L. (1992), *Optoelectronics and Lightwave Technology*, Wiley, Chichester.
[32] Nouri, M. (1991), 'Satellite communication', Chapter 13 in Flood, J.E. and Cochrane, P. (eds), *Transmission Systems*, Peter Peregrinus, Stevenage.
[33] Dalgleish, D.I. (1989), *An Introduction to Satellite Communications*, Peter Peregrinus, Stevenage.
[34] Evans, B.G. (ed.) (1991), *Satellite Communication Systems*, 2nd edn, Peter Peregrinus, Stevenage.
[35] Everett, J. (ed.) (1992), *VSATS: very small aperture terminals*, Peter Peregrinus, Stevenage.

CHAPTER 3

Evolution of switching systems

3.1 Introduction

Switching systems, and associated signalling systems, are essential to the operation of telecommunications networks. The functions performed by a switching system, or a subsystem of it, in order to provide customers with services are called *facilities*. Over the years, the design of switching systems has become ever more sophisticated, in order to provide additional facilities which enable networks to provide more services to customers and to facilitate operation and maintenance.

In spite of the complexity of modern switching systems, there are basic functions which must be performed by all switching systems. These can be demonstrated more clearly for simple early systems[1] than for complex modern ones.[2] This chapter will therefore adopt a historical approach. Each system feature will be explained in the context of the system which first implemented it. For example, the manual switchboard provides an excellent demonstration of features now found in modern stored-program-controlled switching systems.

3.2 Message switching

In the early days of telegraphy a customer might wish to send a message from town A to town B although there was no telegraph circuit between A and B. However, if there was a circuit between A and C and one between C and B, this could be achieved by the process known as *message switching*. The operator at A sent the message to C, where it was written down by the receiving operator. This operator recognized the address of the message as being at B and then retransmitted the message over the circuit to B. This manual process is shown in Figure 3.1(a).

Subsequent technical developments enabled improvements to be made in message switching.[3,4] First, the message received at C was automatically recorded on punched tape and subsequently torn off the receiver by the operator, who read the address from the tape. The message was then retransmitted automatically from the

[50] *Evolution of switching systems*

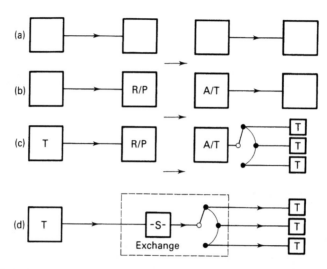

Figure 3.1 Evolution of message switching. (a) Manual transfer of hard copy. (b) Manual transfer of paper tape. (c) Manual transfer of paper tape with automatic route selection (torn-tape relay system). (d) Automatic message-switching system. T = teleprinter, R/P = reperforator, A/T = automatic transfer, S = store.

same tape, as shown in Figure 3.1(b). This was known as a *torn-tape relay system*. Later, the outgoing route was also selected automatically, as shown in Figure 3.1(c). Finally, the paper tape was eliminated by storing the messages electronically and analyzing their addresses by electronic logic, as shown in Figure 3.1(d). A modern message-switching centre is thus a special-purpose electronic computer. In fact, message switching was the first area in telecommunications to adopt stored-program control (SPC).

In a message-switching centre an incoming message is not lost when the required outgoing route is busy. It is stored in a queue with any other messages for the same route and retransmitted when the required circuit becomes free. Message switching is thus an example of a *delay system* or a *queuing system*. The theory of traffic in such systems is explained in Chapter 4.

Message switching is still used for telegraph traffic and a modified form of it, known as *packet switching*, is used extensively for data communications. A data-communications network may need to handle a wide variety of traffic. Some messages may be very short. For example, the user of a visual display unit (VDU) may press only a few keys to access a mainframe computer and expects a quick response. In contrast, the transfer of data files between computers results in very long messages. The VDU operator will not obtain the desired quick response to a message if it has to wait for the completion of a large file transfer. This problem is solved by dividing long messages into smaller units, known as *packets*. A packet switch sends each of these as a separate message. Thus, packets of different messages are interleaved on an outgoing circuit and a short message (e.g. a single packet) does not wait for the transmission of a long message to be completed. Packet switching is discussed in Chapter 9.

3.3 Circuit switching

Invention of the telephone introduced a new requirement: simultaneous both-way communication in real time. Message switching could not meet this requirement because of its inherent delays. It became necessary to connect the circuit of a calling telephone to that of the called telephone on demand and to maintain this connection for the duration of the call. This is called *circuit switching* and the remainder of this chapter will be devoted to circuit-switching systems.

It is inherent in circuit switching that, if the required outgoing cicuit from a switch is already engaged on another call, the new call offered to it cannot be connected. The call cannot be stored, as in message switching; it is lost. Circuit switching is thus an example of a *lost-call* system. The theory of traffic in lost-call systems is also explained in Chapter 4.

3.4 Manual systems

The earliest form of switchboard[5] had incoming circuits connected to vertical metal bars and outgoing links connected to horizontal metal bars, as shown in Figure 3.2. The operator makes a connection by inserting a brass peg where the appropriate vertical and horizontal bars crossed, i.e. at a *crosspoint*. This was the forerunner of the crosspoint matrices used in modern switching systems.

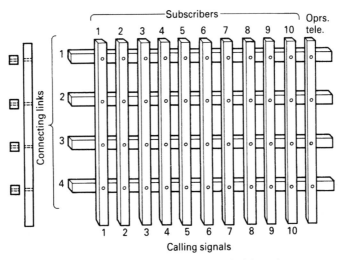

Figure 3.2 Early crossbar switchboard.

[52] Evolution of switching systems

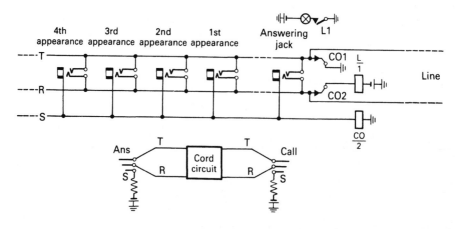

Figure 3.3 Multiple switchboard.

If all customers' lines are connected to vertical bars, the operator makes a connection from a calling line j to a called line k by choosing a horizontal link z and inserting pegs at the crosspoints with coordinates (j,z) and (k,z). Thus, the connection is made through two stages of crosspoint switching and an intermediate link. Such *link systems*, with two or more switching stages, are used in modern telephone exchanges.

The need for larger exchanges, with many operators to handle the traffic, led to the cord type of switchboard.[6–8] As shown in Figure 3.3, each operator answers calls from one group of customers. When one of these calls, the operator responds to a lamp signal by inserting a plug into the corresponding answering jack and operating a key to connect the headset to the cord circuit attached to that plug. The operator obtains the number of the called line by speaking to the caller and then completes the connection, if the line is free, by inserting the other plug of the cord circuit into a jack associated with that line. As shown in Figure 3.3, there are a number of such jacks at intervals along the switchboard, so that each operator can obtain access to every line.

Having made a connection to the called line, the operator alerts the called customer by operating a key in the cord circuit to connect ringing (i.e. a low-frequency alternating current) to the line. The operator is informed by a lamp signal when the called customer answers and disconnects the ringing. The operator then monitors the connection to detect lamp signals from the customers which indicate the end of their conversation and clears down the connection by removing the two plugs from the jacks. This monitoring process is called *supervision*.

The manual-exchange example demonstrates the following features that are also present in automatic switching systems:

- Central-battery operation
- Loop/disconnect signalling
- The multiple
- Busy testing

- Concentration
- Metering and ticketing
- Classes of service
- Common control
- Scanning
- Stored-program control
- Common-channel signalling.

Central-battery operation means that current for the customers' telephones is supplied by a large secondary battery in the exchange, instead of by small primary batteries at the telephones. When the handset of a telephone is in the *on-hook* state the line is disconnected and no current flows. When the handset is in the *off-hook* state the loop of the line is completed and current flows. This provides *loop/disconnect signalling* from the telephone to the exchange.

A customer calls the exchange simply by lifting the handset. Current flows in the line and operates the *line relay* of the customer's line circuit in the exchange (e.g. relay L in Figure 3.3). In the case of the manual switchboard this gives a 'call request' signal to the operator by lighting a lamp. When the operator makes connection to the line, this operates the *cut-off relay* (relay CO in Figure 3.3), which extinguishes the calling lamp. When the called customer responds to ringing by lifting the handset, loop current flows to give the operator an 'answer' signal. At the end of the conversation the telephones of both customers return to the on-hook condition. This interrupts the loop currents and so provides 'clear' signals from lamps in the cord circuit, which instruct the operator to clear down the connection.

So that any operator can make connections easily to all customers' lines, a jack for every line appears in a field of rows and columns and this is repeated at intervals around the switchboard. As shown in Figure 3.3, corresponding contacts of all the jacks for a subscriber are wired together. This arrangement is called a *multiple*. An operator at one end of the switchboard wishing to make a connection to a particular line cannot see whether that line already has a connection made by an operator at the other end of the switchboard. It would be both time wasting and intrusive for the operator to plug into the line and listen to find if a conversation is in progress. Instead, the operator performs a busy test before making the connection.

Each connection through the switchboard has three wires, although the external lines have only two. Each plug has three concentric contacts, known as the tip (T), ring (R) and sleeve (S), the tip being the inner and the sleeve being the outer contact. The wires connected to the tip and the ring of a plug provide the speech path (and are connected to the positive and negative sides of the exchange battery). The wire connected to the sleeve is known as the *private wire* (P wire), since it does not extend beyond the exchange, and this wire provides the busy test. When a connection has been made to a line, the potential of the sleeve wire in the cord circuit is raised by the voltage drop across the cut-off relay (CO). To make the busy test, the operator touches the tip of the plug on the sleeve of the jack. If the circuit is busy, current flows through the earphone to make an audible click.

[54] Evolution of switching systems

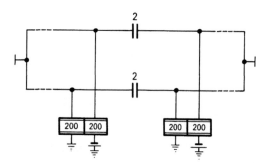

Figure 3.4 Transmission bridge.

The same principle is used in later automatic systems. Switches are multipled together and the trunks have three wires. These are usually known as the +, − and P wires, but are sometimes still called T, R and S. A switch tests the potential of the P wire to find whether a trunk is free before making connection to it. Thus, the P wire performs the function of *guarding*; it prevents any subsequent call from interfering with an established connection.

A line circuit (containing a line and cut-off relay and a calling lamp) and switchboard jacks must be provided for every customer's line. However, the number of calls in progress at any time is always much smaller than the total number of customers. Consequently, the number of cord circuits required is an order of magnitude less than the number of lines. The traffic from the large number of lightly loaded customers's lines is said to be *concentrated* onto the much smaller number of cord circuits. The lamps and keys used by the operators to supervise and control connections are located in the cord circuits instead of in the line circuits. Thus, many fewer are required and a considerable economy is obtained. Automatic exchanges also perform the supervisory function where the traffic has been concentrated onto the minimum number of trunks. Sometimes the supervisory trunk is still called a cord circuit.

The cord circuit contains a *transmission bridge* to feed current to the calling and called line after the connection has been made. A typical transmission bridge is shown in Figure 3.4. The series capacitance and the shunt inductances of the two relays provide a high-pass filter to transmit the AC speech signals, while the relays respond independently to the DC loop/disconnect signals from the calling and called customer respectively.

The supervisory function performed by the operator includes charging fees for calls, in addition to monitoring connections to clear them down when calls end. Two methods are used for making call charges; these are known as *ticketing* and *metering*. In ticketing, the operator, when setting up the connection, writes down on paper (i.e. on a ticket) the numbers of the calling and called customers and the time when the called customer answers. The operator also records the time when the connection is cleared down. Thus, the duration of the call and the charge for it can subsequently be calculated for billing to the caller.

Ticketing was too laborious to use for large numbers of unit-fee local calls. Consequently, some operating administrations made no separate charge for these calls and other administrations adopted metering. A *meter* or *message register*, consisting of a solenoid-operated counter, is connected to the P wire at the line circuit. By operating a key in the cord circuit the operator sends a pulse of current over the P wire, which steps the calling customer's meter.

Automatic switching systems have continued to use these two methods of call charging. In *periodic pulse metering*, pulses are generated at intervals determined by the charging rate for the call; the higher the charging rate, the more frequently they occur. The number of times that the meter is operated is determined by the duration of the call, so the fee charged is proportional to both charging rate and call duration. When ticketing is employed, the 'ticket' is now an electronic record in a data store instead of a piece of paper. The data can be obtained by a central processor as it establishes and clears down a connection. Metering can only provide the customer with a bulk bill for the total of calls made, but ticketing enables the customer to receive an itemized bill listing each call.

The ways in which both outgoing and incoming calls are handled differ for different kinds of customers. These are determined by the *class of service* (COS) of a customer's line. This is in two parts, concerned with originating and terminating calls respectively. Examples of different originating COS are lines with ordinary telephones and lines with payphones. Charges for the former are made by ticketing or metering; those for the latter are collected through the coin box. Some lines may be barred from making long-distance calls. Others may be barred from making any outgoing calls, because the customer's bill is unpaid. Examples of different terminating COS are an ordinary customer having a single line and a customer having a group of lines to a private branch exchange. In the latter case a caller is not told that the customer is busy until every line has been tested and found to be engaged.

In a manual system different classes of service can be indicated to the operator by calling lamps of different colours and by labels attached to the switchboard jacks. In electromechanical systems lines having different classes of service are served by different groups of switches. In a stored-program-controlled system a customer's COS forms part of the data stored for that customer. Thus, any COS can be associated with any line terminal and more classes of service can be provided.

A cord circuit is occupied throughout a call, which is typically of several minutes' duration. However, it only takes a fraction of a minute for an operator to set up a connection and to clear it down, so a single operator can control several cord circuits. Thus, *common control* is a feature of the manual system. Common control is also used in many more recent automatic switching systems; the control functions for a large number of connections are concentrated in a much smaller number of control equipments.

In order to respond to 'call' and 'clear' signals the operator's eyes are continually scanning the lamps on the switchboard, looking for changes in their states. The technique of *scanning* is also used in modern electronic switching systems.[7] A common control scans the speech paths cyclically. It connects to each in turn for a short time, in order to detect any change in its state and perform the appropriate control action.

The operator in a manual exchange provides many different services, using the same basic switchboard apparatus, by performing different sequences of actions which are stored in his or her memory. The same principle is applied in *stored-program control* (SPC), which is employed in modern switching systems. A central computer causes the switching equipment to perform different functions by executing different programs stored in its electronic memory.

So far, only local calls (i.e. calls between customers on the same exchange) have been discussed. However, customers on one exchange also need to make calls to customers on a different one and these need to be processed by an operator at each exchange. The time taken to set up such a connection can be reduced if the operators at the two exchanges can process it concurrently instead of sequentially. This was done by the operators communicating over a circuit separate from the switched junction circuits and linking the operators' positions directly. This was known as *order-wire working*[8] and it was the forerunner of *common-channel signalling*, which is now used between the central processors of exchanges in integrated digital networks, as described in Chapter 8.

3.5 Functions of a switching system

The manual exchange example demonstrates the basic functions that all switching systems must perform. These are as follows:

1. *Attending* The system must be continually monitoring all lines to detect call requests. The 'calling' signal is sometimes known as a 'seize' signal because it obtains a resource from the exchange.
2. *Information receiving* In addition to receiving call and clear signals the system must receive information from the caller as to the called line (or other service) required. This is called the *address signal*.
3. *Information processing* The system must process the information received, in order to determine the actions to be performed and to control these actions. Since both originating and terminating calls are handled differently for different customers, class of service information must be processed in addition to the address information.
4. *Busy testing* Having processed the received information to determine the required outgoing circuit the system must make a busy test to determine whether it is free or already engaged on another call. If a call is to a customer with a group of lines to a PBX or to an outgoing junction route, each line in the group is tested until a free one is found. In an automatic system, busy testing is also required on trunks between switches in the exchange.
5. *Interconnection* For a call between two customers, three connections are made in the following sequence:
    (a) A connection to the calling terminal
    (b) A connection to the called terminal
    (c) A connection between the two terminals.

In the manual system connections (a) and (b) are made at the two ends of the cord circuit and connection (c) merely joins them together in the cord circuit. Many automatic systems also complete connection (c) by joining (a) and (b) at the transmission bridge. However, some modern systems release the initial connections (a) and (b) and establish connection (c) over a separate path through the switching network. This is known as *call-back* or *crank-back*. The calling line is called back and the connection to the called line is cranked back to it.

6. *Alerting* Having made the connection, the system sends a signal to alert the called customer to the call, e.g. by sending ringing current to a customer's telephone.
7. *Supervision* After the called terminal has answered, the system continues to monitor the connection in order to be able to clear it down when the call has ended. When a charge for the call is made by metering, the supervisory circuit sends pulses over the P wire to operate a meter in the line circuit of the calling customer. When automatic ticketing is employed, the system must send the number of the caller to the supervisory circuit when the connection is set up. This process is known as *calling-line identification* (CLI) or *automatic number identification* (ANI).[10] In a stored-program-controlled system the data for call charging can be generated by a central processor as it sets up and clears down connections.
8. *Information sending* If the called customer's line is located on another exchange the additional function of information sending is required. The originating exchange must signal the required address to the terminating exchange (and possibly to intermediate exchanges if the call is to be routed through them).

3.6 The Strowger step-by-step system

The first practical system of automatic telephony was invented by Almon B. Strowger in 1891 and it has been used worldwide. Although the system is now obsolete, in some places it is still in service today.

Strowger invented the *two-motion selector*, [6,9,10] one type of which is shown in Figure 3.5. This connects an incoming trunk to any one of a hundred outgoing trunks. The switch has three banks of contacts, in order to make connections to the +, − and P wires of each outgoing trunk. The bank contacts are arranged in semi-circular arcs, each containing ten contacts, on ten levels. Connection is made to the bank contacts by three wipers mounted on a vertical shaft and joined to the incoming trunk by flexible cords. The selector replaces the action of an operator, in searching over the multiple and inserting a plug in a jack, by raising the wipers to one of the ten levels and rotating them to make contact with one of the ten outlets on that level. These actions are produced by pawl-and-ratchet mechanisms, operated by a vertical magnet and a rotary magnet.

In order to select one of the hundred outgoing trunks, indicated by a two-digit decimal number, the vertical magnet of the selector is pulsed a number of times

[58] *Evolution of switching systems*

Figure 3.5 Strowger two-motion selector.

corresponding to the first digit and the rotary magnet is pulsed a number of times corresponding to the second digit of the number. Relays are mounted above the selector mechanism to receive these pulses from the incoming trunk and direct the first pulse train to the vertical magnet and the second pulse train to the rotary magnet. Another relay responds to a clear signal at the end of the call and causes the wipers to be returned to their unoperated position.

Since Strowger eliminated the operator, the calling customer must provide the pulses to drive the selector. (Thus, as far as the customer is concerned, an automatic exchange is less automatic than a manual exchange!) The means for generating the pulse trains was obtained by adding the dial to the telephone. Loop/disconnect signalling is used. The number of breaks in the loop current corresponds to a dialled digit (except that dialling '0' produces ten interruptions and steps the selector 10 times). The subsequent pause in dialling indicates the end of a digit.

A 100-line exchange can be constructed, as shown in Figure 3.6, by providing a

The Strowger step-by-step system [59]

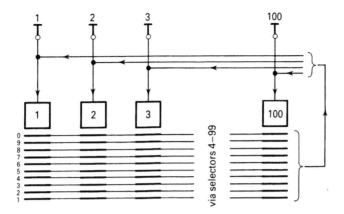

Figure 3.6 Elementary 100-line exchange with two-motion selectors.

selector for each customer's line and multipling together their corresponding outlets. Each customer can select any other by dialling a two-digit number. This arrangement has two obvious defects:

1. It is too expensive because each line has its own Strowger selector. (The concentration obtained in the manual system has been lost.)
2. Its capacity is limited to 100 lines.

To enable a large number of lines to share a smaller number of two-motion selectors, concentration was introduced. Each line is provided with a much cheaper single-motion switch, called a *uniselector*, [6,9] which is shown in Figure 3.7. A large number of customers' uniselectors have their bank contacts multipled to trunks leading to a much smaller number of two-motion selectors.

The uniselector has a self-drive mode of operation. Each time the magnet is energized, the pawl and ratchet wheel move the wipers round by one step and operate interrupter springs to release the magnet, so that it can make another step. The wipers step past busy outlets, which are indicated by an earth potential on the P wire, until a free outlet is found. A relay then stops the uniselector from driving any further.

When the line relay detects a 'calling' signal, the uniselector mechanism is energized and drives its contacts round the bank until a free two-motion selector is found. It is useless for the caller to start dialling before this process is complete, so *dial tone* was introduced. The two-motion selector sends back dial tone as a 'proceed to send' signal when it is seized by the uniselector and it disconnects the dial tone when it begins to receive a dialled digit.

Alternatively, the uniselectors may be associated with the two-motion selectors and customers' lines connected to the multipled bank contacts of the uniselectors. A uniselector then hunts for the calling line instead of for a selector. Fewer uniselectors are needed, but a common control is required to allot a uniselector to each call that originates. A uniselector used in this way is called a *linefinder*. Two-motion selectors have also been used as line finders.[9]

Figure 3.7 Uniselector.

The size of a Strowger exchange can be increased to 10 000 lines by using the multi-stage switching network shown in Figure 3.8. The arrangement of switches and interconnecting trunks in an exchange is called its *trunking* and Figure 3.8 is an example of a *trunking diagram*. It is not necessary to show every switch in the exchange; the connections to them can be represented adequately by showing one selector in each switching stage.

The exchange shown in Figure 3.8 has three ranks of two-motion selectors and a four-digit numbering scheme. The selectors in the first two stages each step vertically in response to a dialled digit. However, during the inter-digit pause they operate in a self-drive mode like a uniselector. Thus, the wipers hunt for a free outlet on the selected level, which leads to a selector in the next switching stage. These selectors are called *group selectors*. The selectors in the last stage respond to two dialled digits and are called *final selectors*. Figure 3.8 shows the path through the exchange for a call set up by dialling the digits '2121'.

Since the connection is set up in stages, in response to the digits dialled, the system is called the *step-by-step system*. This an example of a system with *progressive control*, since each step in setting up the connection is controlled by relays mounted on the selector which operates at that stage. The uniselectors act as a *concentrator*, since

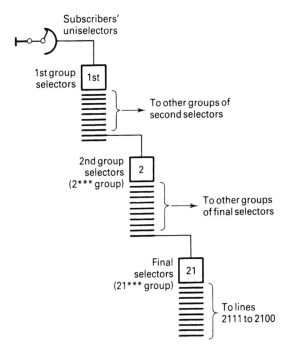

Figure 3.8 Trunking diagram of 10 000 line step-by-step exchange.

there are many fewer trunks between switching stages than there are customers' lines. The final selectors act as an *expander*, to connect the heavily loaded trunks to the much larger number of lightly loaded customers' lines. Operation of the system is controlled by relay circuits mounted on individual selectors. Thus, it uses a form of *distributed control*, in contrast to the common control inherent in the manual system. The controlling relay circuits[11] are an example of *wired logic*. This is characteristic of electromechanical switching systems and so differs from the stored-program control used both in the old manual system and in modern electronic switching systems.

The basic step-by-step system described above has been used extensively, but it has several disadvantages. In practice, its four-digit numbering scheme provides access to fewer than 10 000 customers' lines, because dialling codes are also required for other purposes. In a typical example, codes of the form '1xx' provide access to various services (e.g. code '100' for operator assistance, '154' for fault reports and '192' for directory enquiries), the code '999' is dialled for emergency services (ambulance, fire and police) and '0' is used a a trunk prefix digit (which routes calls to the trunk exchange). Dialling codes are also needed for making junction calls to other local exchanges. For example, codes of the form '7x' may be used to access junctions from levels of 2nd group selectors. In this example, levels 1, 7, 9 and 0 of 1st group selectors cannot be used for obtaining access to customers of the exchange and its capacity is reduced from 10 000 to only 6000 customers' lines.

[62] *Evolution of switching systems*

For junction calls, dialled digits are used to make a connection through each exchange on the route and there is a rigid association between the digits dialled for a call and its routing. Thus, different dialling codes are required to reach the same destination from different originating exchanges. For example, a customer on exchange A may dial '2345' to reach another on this exchange. A customer on exchange B may need to dial '74 2345', and a customer on exchange C may need to dial '68 2345'. If calls from exchange D to exchange A use a tandem connection via C, a customer on exchange D may need to dial '45 68 2345' to reach customer number 2345 on exchange A.

It is a disadvantage to have to dial different numbers to reach the same destination, depending on where a call originates. This disadvantage can be minimized by using a *linked numbering scheme*.[9] The numbering scheme of an area and the trunking of its exchanges are arranged so that a customer can be reached by dialling the same digits from any exchange in the area. For example, a town with a 10 000-line main exchange and several 1000-line satellite exchanges can use a five-digit linked numbering scheme. This requires an extra rank of selectors. Each first-selector level gives access to a maximum of 10 000 lines and each second-selector level gives access to 1000 lines, which may be part of the main exchange or all of a satellite exchange. Six-digit linked numbering schemes have been used for larger areas.

3.7 Register–translator–senders

Although linked numbering schemes have been widely used, a more general solution to the routing problem was required for very large cities. Ideally, the route for a call should be completely divorced from the digits dialled to establish the connection. The dialling code to reach a particular terminating exchange can then be the same from every originating exchange.

The solution obtained to this problem was to install equipment known as a *register–translator–sender*, usually simply called a *register*. This receives the number dialled by a customer and stores it (i.e. the digits are registered). The stored digits are then analyzed to determine the routing. If necessary, part of the number is translated into a different number which is sent out to establish the connection. By using different translations, customers on different originating exchanges can all dial the same digits to reach a terminating exchange, although different routes are used to make the connections.

A register is used only for a short period at the beginning of a call. The time taken to receive a customer's dialled digits, perform the translation and send digits out is only a fraction of a minute. Thus, registers are provided as common controls. A register is connected to a trunk by an auxiliary switch when it receives a calling signal, performs its functions and is then released for use on other calls.

The register function consists of receiving digits and storing them. In electromechanical systems this is done on switches or relays. In modern systems, it is done electronically. Digits may be sent out again as trains of loop/disconnect pulses.

However, modern systems use more-rapid signalling methods, as described in Chapter 8, in order to reduce the *post-dialling delay* which is inherent when registers are employed. (It may be noted that the basic step-by-step system has zero post-dialling delay; a connection is complete as soon as dialling has finished.) Some systems that send out different kinds of signals over junctions from those received from customers' lines segregate the receiving and sending functions in different equipments. Fewer senders than registers are then needed, so some economy is obtained.

The translation function consists essentially of looking up a table of data. The dialled digits stored in the register are used to access an address in a store and the number read out from that address is the translation, i.e. the digits to be used for establishing the required connection through the network. Since the charge for a call depends on its destination, the translation may also contain the charging rate for the call. In electromechanical systems translations are stored by means of wire strappings on a field of terminals[11] and these can be rearranged manually when required. Electronic systems use semiconductor memories and these are alterable electronically. The time required to obtain a translation is much less than that taken to receive and send digits. Some systems therefore use a common translator to serve a large number of registers. In stored-program-controlled systems the register function and translation can both be performed by a central processor.

Registers were added to step-by-step exchanges for use in large cities. A uniform seven-digit numbering scheme is used throughout the area. The first three digits (ABC) are an exchange code (which is translated) and the last four digits are the customer's number on that exchange (which is not translated). The registers added to step-by-step exchanges are called *directors*[9] and the cities using them are called director areas. Registers are an inherent part of more recent systems and some of these systems translate subscribers' numbers in addition to routing digits.

In a director exchange[9] there are three ranks of group selectors, as shown in Figure 3.9. The first, known as the 1st-code selector, routes a local call from its level 1 to a final selector via a 1st-numerical and a 2nd-numerical selector. Junction calls are routed from other levels of the 1st-code selector, either directly (for a heavy-traffic route), or via a 2nd-code selector (for a medium-traffic route), or via 2nd- and 3rd-code selectors (for a light-traffic route). The supervisory function is provided by relays in the circuit of the 1st-code selector and this is connected to a director by a uniselector known as the A-digit-selector hunter.

Some systems used registers even for own-exchange calls, because their selectors were driven by external motors instead of being controlled directly by a customer's dialled digits. Examples of such *power-drive systems*[1,9] are the Ericsson 500-line selector system, the International Standard Electric Rotary System and the Bell Panel System. Instead of pulse trains being sent forward as a connection was established, *revertive pulsing* was used. When a selector was driven it sent a pulse back to the register as it stepped onto each outlet. The register counted these pulses and, when it had received the correct number, it sent a signal to the switch which halted its motion.

In order to enable customers to dial long-distance calls directly it was necessary to introduce national numbering schemes with the same code for a terminating

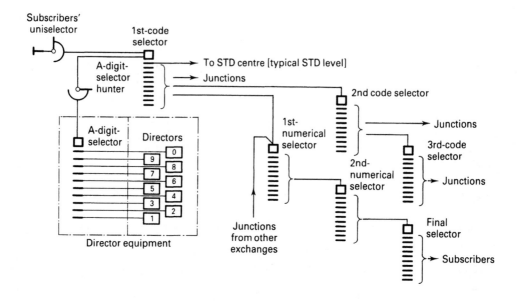

Figure 3.9 Trunking diagram of director exchange.

exchange corresponding to different routes from originating exchanges in different parts of the country. This is called *direct distance dialing* (DDD) in North America[12] and *subscriber trunk dialling* (STD) in the UK.[13] Its implementation required the use of register–translators in trunk exchanges. Finally, the introduction of *international subscriber dialling* (ISD), with a world numbering plan, necessitated the use of registers in international exchanges to translate the codes for different countries.[14]

3.8 Distribution frames

Many changes occur during the life of a telephone exchange. New customers join and old ones leave. Customers move from one part of the exchange area to another and those with PBXs may increase their number of exchange lines. The total number of lines may increase over the years, say from an initial installation of 2000 lines to an ultimate 10 000 lines. Growth of traffic may require additional switches in the exchange and more junctions to other exchanges. Great flexibility is therefore required in the trunking of an exchange. This is obtained by inserting *distribution frames* into the permanent exchange cabling. Typically, these frames contain an array of terminal blocks and the terminals are linked in a less permanent fashion (with soldered, wire-wrapped or crimped connections) by wires called *jumpers*.

The distribution frames in a typical step-by-step exchange are shown in Figure 3.10. The *main distribution frame* (MDF)[6,8] is the place where the cables of the customers' distribution network terminate. The arrangement of terminals on the *line*

Distribution frames [65]

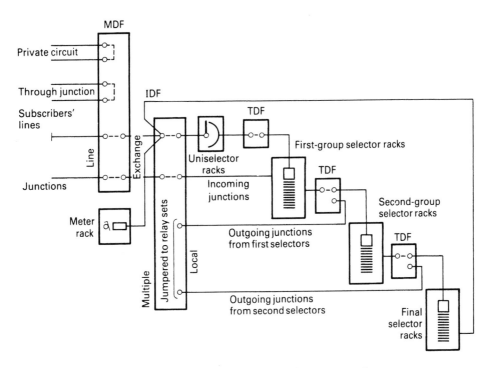

Figure 3.10 Distribution frames in Strowger exchange.

side of the MDF corresponds to the street cabling and so reflects the geography of the area. The terminals on the *exchange side* of the MDF are arranged in directory-number (DN) order. Thus, the number of a line is changed by moving its jumper. Protectors and fuses are mounted on the MDF to guard exchange apparatus against any high-voltage surges on the external lines. The MDF also provides a convenient point of access for testing lines. Circuits which are not switched in the exchange (for example private circuits and through junctions) are strapped together at the MDF, as shown in the figure.

Some customers originate much traffic, but others very little. The *intermediate distribution frame* (IDF)[6,8] is used to distribute incoming traffic evenly over the groups of first selectors. On the *multiple side* of the IDF, lines are arranged in directory-number order. On the *local side*, the order can be arbitrary to obtain the desired result. The terminals on this side of the IDF can be said to correspond to *equipment numbers* (EN) of the lines. Customers' uniselectors are therefore connected to the local side of the IDF. If the exchange is equipped with meters, these are required to be associated with directory numbers; they are therefore cabled to the multiple side of the IDF. Since meter pulses on the P wires arrive at the local side, the IDF provides *equipment-number* to *directory-number* (EN-to-DN) *translation*. Incoming calls for a customer terminate at a final selector on an outlet corresponding to the directory number. The final-selector multiples are therefore cabled to the multiple side of the

[66] *Evolution of switching systems*

IDF. Some subscribers receive much traffic, others very little. However, there is no possibility of redistributing this traffic between final selectors without changing customers' directory numbers. Alteration of a customer's number may be necessitated by a change in class of service, for example a change from a single exchange line to a PBX group of lines. Modern systems provide *directory-number to equipment number* (DN-to-EN) *translation*, in order to enable customers' incoming traffic to be redistributed in addition to their outgoing traffic.

Between the ranks of selectors there are *trunk distribution frames* (TDF). If additional selectors are needed at any switching stage, to cater for growth in traffic, these can be accommodated by rearranging connections in a TDF.

The flexibility arrangements in a modern system may differ from those described above for the Strowger system. For a digital switching system, digital circuits are terminated on a *digital distribution frame* (DDF). In an SPC system, EN-to-DN and DN-to-EN translation may be performed by a central processor reading data from an electronic memory. Use of an IDF is then no longer necessary. However, some means equivalent to those described above must be provided to accommodate the changes occurring during the life of the exchange. Moreover, it must be possible to make these changes, at any time, without taking the system out of service.

3.9 Crossbar systems

Strowger switches require regular maintenance. The banks need cleaning, mechanisms need lubrication and adjustment and wipers and cords wear out. This disadvantage led to the development of several other forms of switch.[1,9] One idea was to replace the manually operated switch of Figure 3.2 by a matrix of telephone relays, with their contacts multipled together horizontally and vertically as shown in Figure 3.11. Since a switch with N inlets and N outlets requires N^2 relays for its crosspoints, this was uneconomic except for small private exchanges requiring small switches.

A more economic solution was provided by the invention of the *crossbar switch* by G. A. Betulander[1] in 1917. This is shown in Figure 3.12. The crossbar switch retains a set of contacts at each crosspoint, but these are operated through horizontal and vertical bars by magnets at the sides of the switch. Thus, a switch with N inlets and N outlets only needs $2N$ operating magnets and armatures, instead of N^2. The magnets which operate the horizontal bars are called *select magnets* and those operating the vertical bars are called *hold magnets* or *bridge magnets*.

The mechanism of a crossbar switch is shown in more detail in Figure 3.13. Operation of a select magnet tilts one of the horizontal bars up or down. This causes flexible fingers to engage with the contact assemblies of one row of crosspoints and provides the link which was missing from their operating mechanisms. One of the bridge magnets is then operated and this closes the contacts of the crosspoint at the coordinates corresponding to the horizontal and vertical magnets. The select magnet is then released, but the finger remains trapped and the crosspoint contacts remain closed for as long as the bridge magnet is energized. Current flows in this magnet for as long as

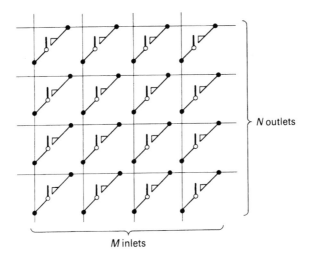

Figure 3.11 Matrix of crosspoints.

Figure 3.12 General view of crossbar switch.

the P wire is at earth potential. This persists until a 'clear' signal causes the earth to be removed at the end of a call.

Strowger selectors perform counting and searching. However, the crossbar switch has no 'intelligence'. Something external to the switch must decide which magnets to operate. This is called a *marker*. Since it takes less than a second to operate the switch, a marker can control many switches and serve many registers, as shown in

[68] *Evolution of switching systems*

Figure 3.14. Thus, even a large exchange needs few markers. This is a further stage of common control, which we shall call *centralized control*.

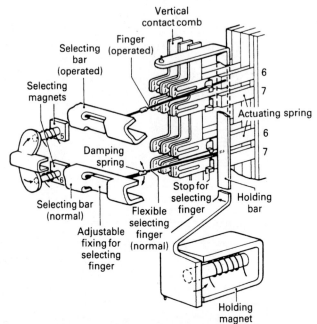

Figure 3.13 Mechanism of crossbar switch.

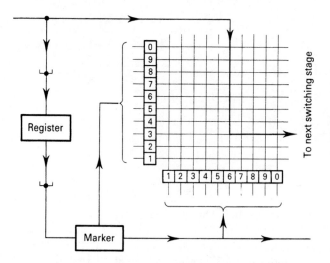

Figure 3.14 Marker control of crossbar switch.

Crossbar systems

Unlike the two-motion selector, a crossbar switch can make more than one connection at a time. It can make as many connections as it has vertical bars. Thus, it can be used as if it were a group of uniselectors instead of a single two-motion selector. For example, a switch of size 10 × 10 can make up to ten simultaneous connections between ten incoming trunks and ten outgoing trunks.

In order to produce larger switches, a two-stage link system of primary and secondary switches is used, as shown in Figure 3.15. This is called a *link frame*. The figure shows twenty switches of size 10 × 10 used to connect 100 incoming trunks to 100 outgoing trunks. There is one link fom each primary switch to each secondary switch and these links are arranged systematically. The number of an outlet on a primary switch corresponds to the number of the secondary switch to which its link goes and the number of an inlet on a secondary switch corresponds to the number of the primary switch from which its link comes. For example, link 23 connects outlet 3 of primary switch 2 to inlet 2 of secondary switch 3.

When a marker is instructed to set up a connection from a given incoming trunk to a given outgoing trunk, this also defines the link to be used and the select and bridge magnets to be operated to make the connection. The marker does not make the

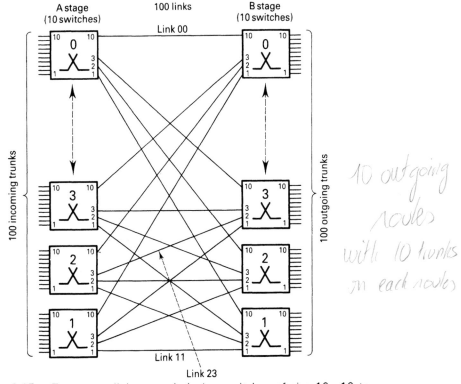

Figure 3.15 Two-stage link network (using switches of size 10 × 10 to interconnect 100 incoming and 100 outgoing trunks).

connection until it has interrogated the busy/free condition of the outgoing trunk and of the relevant link. Only if both are found to be free does it operate the switches. This is called *conditional selection*.

A concentrator can be constructed by multipling together the horizontals of a number of primary switches, as shown in Figure 3.16. This shows a network with 500 incoming trunks and 100 outgoing trunks using switches of size 10 × 10. An expander would, instead, have a number of secondary switches multipled together.

In order to provide a larger network, four stages can be used. This is implemented as a two-stage network in which each switch is itself a two-stage network. Figure 3.17 shows a four-stage network, constructed from 400 switches of size 10 × 10, to serve 1000 incoming and outgoing trunks.

Although these networks provide access from every incoming trunk to every outgoing trunk, it may not always be possible to make a connection even when the required outgoing trunk is free. In the two-stage network of Figure 3.15 there is only one link from a primary switch to a secondary switch. When a connection is required from an incoming trunk on the primary switch to an outgoing trunk on the secondary switch, the link may be busy because it is already in use for a connection from another incoming trunk on that primary switch to another outgoing trunk on that secondary switch. The call attempt fails, although the outgoing trunk is free. This situation is called *blocking*. It is also sometimes called a *mismatch*; there are free outgoing trunks and free links but they cannot be used together. The traffic capacity of link networks is therefore affected by internal blocking in addition to congestion of the external trunks. The traffic performance of link networks is analyzed in Chapter 5.

Marker control enables conditional selection to be made over networks having several stages. Thus, if a free path through the network exists, it can always be found. In contrast, in a step-by-step exchange, a call may fail due to all trunks being busy at a late stage when it would have succeeded if a different choice had been made at the first

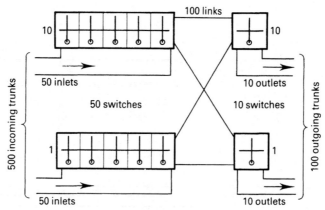

Figure 3.16 Concentrator connecting 500 incoming trunks to 100 outgoing trunks (switches 2 to 9 not shown).

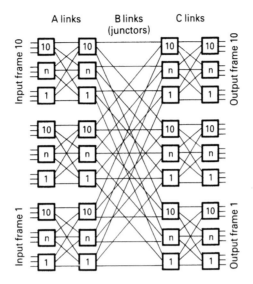

Figure 3.17 Four-stage switching network for 1000 incoming trunks and 1000 outgoing trunks using 10 × 10 switches.

group selector. A further advantage of marker control is that, since the marker has access to both ends of a connection that it sets up through a network, it can test the connection for continuity before it releases. If the connection is found to be faulty, the marker can output a fault report and attempt to set up the connection over a different path through the network.

A large crossbar exchange needs several markers in order to handle its traffic and this introduces a complication. It is essential to prevent two or more markers from attempting to set up connections in the same link frame at the same time. For example, if one marker attempts to connect inlet i to outlet j of a switch by operating crosspoint (i,j) and, at the same time, another marker attempts to to connect inlet m to outlet k by operating crosspoint (m,k), the result would be to operate crosspoints (i,j), (i,k), (m,j) and (m,k). Thus, two double-connections would result. Consequently, the markers in a crossbar system incorporate quite complex lock-out cicuits[11] to prevent this happening.

The trunking of a typical crossbar system is shown in Figure 3.18. This is the Plessey 5005 system[6,15] (designated the TXK1 system by British Telecom). Each customer's line is terminated on a two-stage network (the distributor) which serves both as concentrator and expander (i.e. it is equivalent to the uniselectors and the final selectors of a Strowger exchange). This is possible because the switches do not respond to dialled digits but are set by a marker and this provides DN-to-EN translation.

When a customer calls, the distributor connects the line to a supervisory trunk and this obtains connection to a register. In this system, the supervisory trunk is called

[72] *Evolution of switching systems*

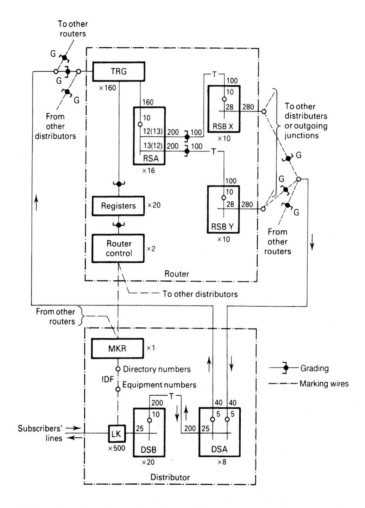

Figure 3.18 Trunking diagram of 5005 crossbar system. TRG = transmission relay group, RSA = A route switch, RSB = B route switch, LK = line and cut-off relays of subscriber's line circuit, MKR = marker, DSA = A distributor switch, DSB = B distributor switch.

the transmission relay group (TRG), since it contains the transmission bridge. The TRG has access to outgoing junctions via a two-stage network called the router (which corresponds to the group selectors in a Strowger exchange). Connection is made to a called customer's line via the router and the distributor. When the register has received sufficient digits to determine the call destination, it accesses a marker (MKR) which operates the switches needed to connect the TRG to the required outgoing line. A wire from the marker to the customer's terminal on the distributor is included in the jumper across the IDF, in order to obtain DN-to-EN translation.

Crossbar systems [73]

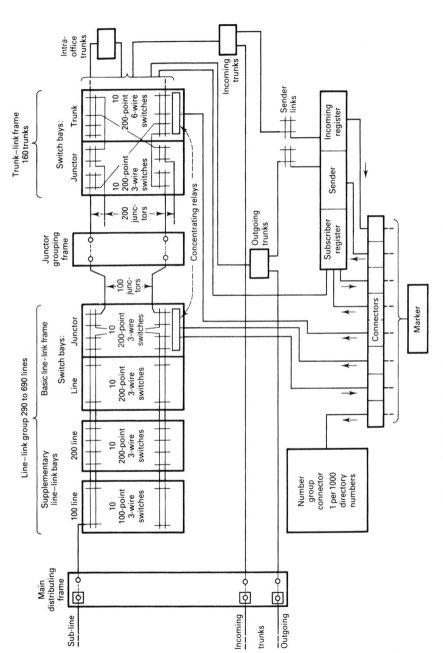

Figure 3.19 Trunking diagram of Bell No.5 crossbar system.

[74] *Evolution of switching systems*

The Bell No. 5 crossbar system,[10,16] shown in Figure 3.19, is more complex. Two-stage switching networks are used, called the line link frame and the trunk link frame, corresponding to the distributor and router of the TXK1 system. However, there are separate registers and senders. Very large multi-contact relays are used as connectors to transfer information between the switching frames and the markers, between registers and markers and between registers and senders.

This system uses the crank-back principle. An initial connection is made from a calling customer's line to a subscribers' register via the line link frame and trunk link frame. Thus, there are no auxiliary switches between trunks and these registers. After the register has received dialled digits from the calling customer, this connection is released and a marker sets up another connection between the calling line and the required outgoing circuit.

If another customer's line is to be called, the subsequent connection is made through the line link frame of the caller and the trunk link frame to one side of an intra-office trunk (which contains a supervisory circuit). The other end of this trunk is connected to the called line via the trunk link frame and a line link frame. This increases the efficiency of the trunking, since a larger number of paths are available through the complete switching network than between a supervisory trunk and a called line through only part of the network.

For an outgoing junction call the subsequent connection is made through the line link frame and the trunk link frame to an outgoing trunk. A sender is connected to the outgoing trunk through an auxiliary crossbar switch, the subscribers' register transfers the necessary address information to the sender and this is sent out over the junction.

For an incoming junction call the incoming trunk is connected to an incoming register by an auxiliary crossbar switch and receives the number of the called customer. A connection is then made from this trunk to the called line via the trunk link frame and the line link frame. Thus, crank-back is not used for these connections.

3.10 A general trunking

A general trunking diagram for a telephone exchange is shown in Figure 3.20. It contains three kinds of switching network: concentrators, expanders and a central route switch. A customer's call is connected to a supervisory trunk by a concentrator. For a local call, the route switch connects the supervisory trunk to the particular expander on which is terminated the line of the called customer. For a junction call, it connects the supervisory trunk to an outgoing junction on the route to the required destination. The route switch also connects incoming junction calls to the expanders of called customers. A register, which is connected to the supervisory trunk by an auxiliary concentrator, receives address information and processes it to determine the destination of a call. The central processor interrogates all trunks which can be used for a required connection to determine which are free. It then selects a suitable set of these trunks and marks them to make the connection. Any particular system can be represented by the elements shown in Figure 3.20, or by a subset of them.

A general trunking [75]

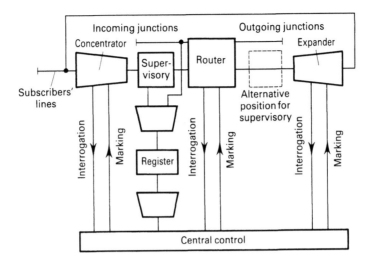

Figure 3.20 General trunking diagram for a switching system.

In the manual system, the concentrators consist of groups of answering jacks and plugs in front of the operators. The supervisory trunks are the cord circuits. The operator performs the functions of a register and the central control. The expanders are provided by the jacks in the multiple. The manual system is unique, because a single switch can provide access to all lines on even the largest exchange. Thus, there is no need for the route switch shown in Figure 3.20.

In the basic Strowger step-by-step system each concentrator is a group of customers' uniselectors or line finders and each expander is a group of final selectors. The group selectors form the route switch. Since distributed control is used, there are no registers and no central control. Because there is no register to interface with a supervisory circuit, the supervisory function can, in principle, be located at either of the two positions shown in Figure 3.20. In practice, this function is normally performed at the outgoing side of the route switch. For a local call, it is provided by relays in the final-selector circuit. For a junction call, it is provided by an outgoing junction relay set.

In a director exchange the directors perform the register function, but there is no central control. The route switch contains three ranks of selectors: the 1st-code selectors, together with the 2nd- and 3rd-code selectors and the 1st- and 2nd-numerical selectors. Since registers are used, the supervisory circuit is located in the incoming trunk of the route switch, i.e. in the 1st-code selector. This is connected to a director by the A-digit-selector hunter.

In a crossbar system the markers comprise the central control shown in Figure 3.20. Each of the switching networks consists of a two-stage link network. Since DN-to-EN translation is provided, the concentrator can also act as an expander. It serves two groups of trunks. One group connects it to supervisory units and the other is folded back to it from the outgoing side of the route switch, as shown in Figure 3.18.

[76] *Evolution of switching systems*

There are variants of this trunking. The Ericsson ARF crossbar system[1] uses progressive control, with separate markers controlling the route switch (called the group-selector stage) and the concentrator (called the subscriber stage). The Bell No.5 system has separate registers and senders and it uses crank-back.

In a stored-program-controlled system all call-processing functions may be performed by a central processor. Thus, there are no separate supervisory units and no registers between the trunks and the central control.

A tandem exchange or a trunk exchange has no subscribers, so it contains no concentrators or expanders. Concentration is not required, because incoming junctions are more heavily loaded than customers' lines. The originating local exchange has already concentrated the outgoing traffic from its customers onto these junctions.

3.11 Electronic switching

Research on electronic switching[2] started soon after the Second World War, but commercial fully electronic exchanges began to emerge only about 30 years later. However, electronic techniques proved economic for common controls much earlier. In electromechanical exchanges, common controls mainly use switches and relays which were originally designed for use in switching networks. In common controls, they are operated much more frequently and wear out earlier. In contrast, the life of an electronic device is almost independent of its frequency of operation. This gave an incentive for developing electronic common controls and resulted in electronic replacements for registers, markers, etc., having much greater reliability than their electromechanical predecessors.

Advances made in computer technology were incorporated[2] and led to the development of stored-program control (SPC). This enables a digital computer to be used as a central control and perform different functions with the same hardware by executing different programs. As a result, SPC exchanges can offer a wider range of facilities than earlier systems. In addition, the facilities provided to an individual customer can be readily altered by changing the customer's class-of-service data stored in a central electronic memory. Moreover, since the processor's stored data can be altered electronically, some of these facilities can be controlled by customers. Examples include:

1. *Call barring (outgoing or incoming)* The customer can prevent unauthorized calls being made and can prevent incoming calls when wishing to be left in peace.
2. *Repeat last call* If a called line is engaged, the caller can try again later without having to redial the full number.
3. *Reminder calls* The exchange can be instructed to call the customer at a pre-arranged time (e.g. for a wake-up call).
4. *Call diversion* The exchange can be instructed to connect calls to a different number when the customer goes away.

5. *Three-way calls* The customer can instruct the exchange to connect a third party to a call that is already in progress.
6. *Charge advice* As a result of the caller sending the appropriate instruction when starting a call, the exchange calls back at the end of the call to indicate the call duration and charge.

In order to develop a fully electronic exchange it was necessary to replace electromechanical switches in the speech path, in addition to using electronic common controls. One approach is to replace the relay contacts of the switch shown in Figure 3.11 by electronic devices multipled together. A diode could be used as a crosspoint, as shown in Figure 3.21. If point A is positive the diode is reverse biased and the crosspoint is open. If point A is negative the diode conducts and the crosspoint is closed. However, an external circuit is needed to keep point A negative for the duration of a connection. Thus, in order to implement a crosspoint, a one-bit memory is required in addition to the switching element. For example, a crossbar switch provides this memory by means of the finger which is trapped in each crosspoint when it is operated. Research was carried out into devices which could combine both functions in a single component. Cold-cathode gas tubes and PNPN semiconductor devices have been used.[2]

Another approach is to use a multiplex system instead of multipled elements. A frequency-division multiplex (FDM) system could be used as a switch by bringing the two ends of its transmission path together. If the modems at one end of the path operate at fixed frequencies, but those at the other end can operate at the frequency of any channel, then any trunk at one side of the switch can be connected to any trunk at the other side. On one side of the switch it must be possible to tune the filters to any channel and to supply any carrier frequency to any modem. This proved too expensive to put into practice.

Nevertheless, the multiplex principle was not abandoned. A multipled-element switch with N incoming and outgoing trunks needs N^2 elements. A multiplex switch of the same size needs N devices on each side, i.e. a total of only $2N$. Thus, if N is sufficiently large the devices used can be much more costly than those of a multipled-element switch. A time-division multiplex (TDM) system, as shown in Figure 2.8, can be used. If any of the N receiving gates is operated by a train of pulses coincident with those applied to one of the N sending gates, then a transmission path is provided from the incoming trunk to the outgoing trunk via the common highway. For a transmission system, fixed pulse timings are used. If the pulse timings can be altered, then any incoming trunk can be connected to any outgoing trunk, i.e. an $N \times N$ switch

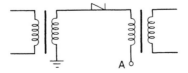

Figure 3.21 Diode crosspoint.

is obtained. Moreover, a memory which stores the appropriate pulse timings can be implemented relatively cheaply. Research on TDM switching proved successful and systems using it are now in service world wide. They are described in Chapter 6.

Consequently, switching systems may be classified as:

1. *Space-division (SD) systems*   Each connection is made over a different path in space which exists for the duration of the connection.
2. *Time-division (TD) systems*   Each connection is made over the same path in space, but at different instants in time.

All electromechanical exchanges are, of course, SD systems. Both SD and TD electronic switching systems have been developed, but the latter now predominate.

The distinction between SD and TD applies to the control arrangements of an exchange as well as to its switching network. For example, if an exchange uses individual registers, this is space division. If the register function is performed by a central processor, handling each call in turn, this is time division.

Some fully electronic systems were given field trials over 30 years ago.[2] They worked, but they were too expensive to enter commercial production. The chief reason for this was that the switches (whether SD or TD) were suitable for speech signals, but they could not handle the relatively high currents and voltages required for line feeding and ringing. These had to be provided from every customer's line circuit, instead of from a much smaller number of supervisory trunks as in electromechanical systems. It is also more difficult to provide adequate protection from high-voltage surges on external lines for electronic components than for electromechanical apparatus. Testing of the insulation resistance of a line is not possible without disconnecting it from an electronic line circuit. If time-division switching is used, the speech path is a four-wire circuit, so this requires a hybrid transformer in the customer's line circuit. It took about another ten years for these problems to be solved in an economic fashion.

3.12 Reed–electronic systems

The solution to the problem of high-cost line circuits was to exploit the advantages of electronic control, but to revert to electromechanical switches, with a much-improved crosspoint, in the form of the reed relay shown in Figure 3.22. A single operating coil, as shown in Figure 3.23, can contain three such reed inserts to switch +, − and P wires, together with a fourth which acts as the one-bit memory to hold the relay operated for the duration of a connection. A matrix of reed relays is equivalent to a crossbar switch and similar trunking principles can be used.

An alternative design of crosspoint, the ferreed,[17] uses a remanent ferrite material in its magnetic circuit instead of the latching contact. It is therefore released by a pulse of current in the opposite direction from its operating current. An improved form of ferreed, known as the remreed, is used in the Bell No.1 ESS system.[18,19] This system uses a *map in memory*. The busy/free conditions of all the trunks are stored in a central memory. To find a free path for establishing a connection, the central processor

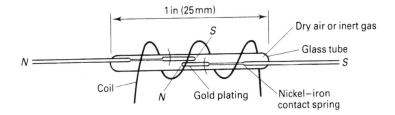

Figure 3.22 Principle of reed relay.

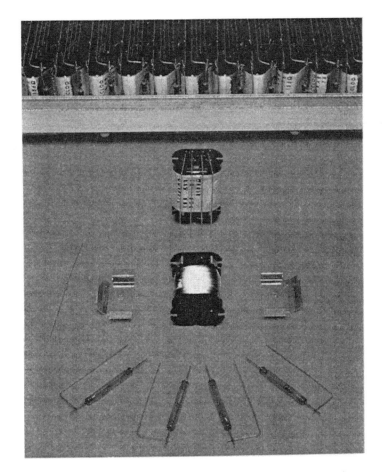

Figure 3.23 Construction of reed-relay crosspoint and switch.

[80] *Evolution of switching systems*

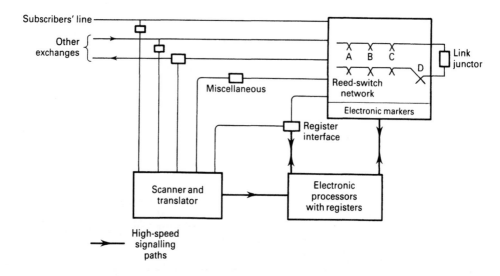

Figure 3.24 Trunking of TXE4 reed–electronic exchange.

interrogates this store, so no P wires are required. Consequently, the ferreed contains only two reed inserts, used for switching the + and − wires of a connection.

The reed relay is not only more reliable than the crossbar switch (because of its sealed precious metal contacts) but it is much faster in operation (less than 1 ms). As a result, even the largest exchange only needs to set up one connection at a time. Only one marker is required, instead of the plurality needed in a large crossbar exchange. This opened the way for stored-program control[20,21] since a single central processor can handle all the switches in the exchange.

Consequently, the first generation of so-called electronic exchanges were really semi-electronic exchanges. They use reed-relay switches but electronic central control. Some examples[18] are the Bell No.1 ESS system, the British Telecom TXE2 and TXE4 systems, the Ericsson AXE system and the ITT Metaconta system. A trunking diagram of the TXE4 system[6,18] is shown in Figure 3.24.

3.13 Digital switching systems

While reed–electronic exchanges were being developed, TDM transmission was being introduced for trunk and junction circuits, in the form of pulse-code modulation (PCM). If time-division transmission is used with space-division tandem switching, as shown in Figure 3.25(a), it is necessary to provide demultiplexing equipment to demodulate every channel to audio before switching and multiplexing equipment to retransmit it after switching. If time-division switching is used, as shown in Figure

Digital switching systems [81]

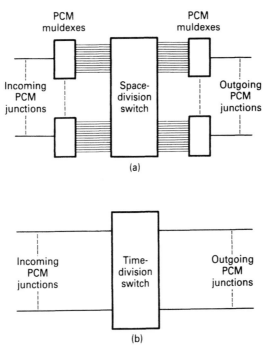

Figure 3.25 Tandem exchange with PCM junctions. (a) Space-division switching. (b) Time-division switching.

3.25(b), no multiplexing and demultiplexing equipment is needed. A considerable economy is thereby obtained.

If, as shown in Figure 3.26(a), a tandem exchange has a mixture of PCM junctions and analog audio junctions, PCM terminal equipment is needed instead for the analog junctions. However, these are a decreasing proportion of the total, so large cost savings can be made by using time-division switching for trunk and tandem exchanges.

In addition, tandem exchanges have no customers' lines. Thus, the disadvantage of high-cost customers' line circuits, which made fully electronic switching uneconomic for local exchanges, did not apply. Consequently, digital exchanges, such as the Bell No.4 ESS system[22] and the French E 10 system,[23] were first introduced for trunk and junction switching. This has led to the conversion of trunk networks into integrated digital networks (IDN), in which all transmission and switching are digital.

From Figure 3.20 it can be seen that a local exchange can be formed by adding concentrators to the periphery of a tandem exchange. Systems were developed[22] using reed-relay concentrators, as shown in Figure 3.26(b). This enabled cheap line circuits to be retained and a large number of subscribers to share a PCM equipment to access the time-division routing switch.

[82] *Evolution of switching systems*

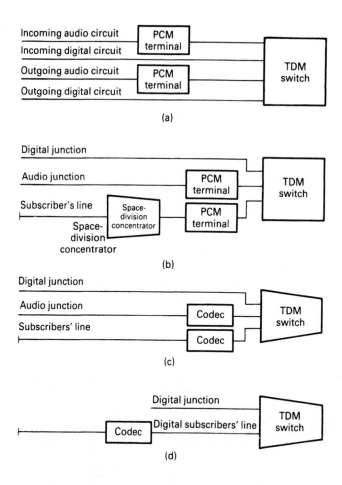

Figure 3.26 Evolution of digital switching systems. (a) Trunk or tandem exchange. (b) Local exchange with space-division concentrators. (c) Local exchange with codecs in customers' line circuits. (d) Local exchange with digital customers' lines.

Before such systems were installed very extensively, enormous developments had taken place in semiconductor technology. In particular, large-scale integrated circuits appeared. This enabled a PCM coder/decoder (codec) to be manufactured on a single 'chip' and made it practicable to have one for each customer's line. As a result, it became possible to implement all the necessary functions economically on a subscriber' line-interface circuit (SLIC), as shown in Figure 3.27. These functions can be summarized by the acronym BORSCHT, as follows:

Digital switching systems [83]

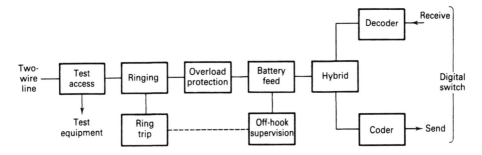

Figure 3.27 Block diagram of subscriber's line interface circuit for a digital exchange.

Battery feed
Over-voltage protection
Ringing
Supervisory signalling
Coding
Hybrid
Testing

These developments enabled electromechanical concentrators to be eliminated, as shown in Figure 3.26(c), resulting in a fully digital local exchange.[21,24,25] Examples of such systems[22] include the AXE-10 system (developed in Sweden), the DMS-10 system (Canada), the E-12 system (France), the No.5 ESS system (USA), the EWS-D system (Germany), the NEAX system (Japan) and System X (UK). At last, fully electronic exchanges became economic for the full range of applications: trunk and tandem exchanges, local exchanges and private branch exchanges (PBXs).

Since a digital concentrator is connected to the route switch by a PCM link, it can be remote from it. It can be controlled by the central processor of the exchange by control signals sent over the PCM link in addition to the speech channels. As a result, the size of a local-exchange area is no longer restricted by the DC resistance and attenuation limits of customers' lines. This enables a local exchange to serve many more customers (e.g. 50 000 instead of 10 000). Many small local exchanges can thus be replaced by remote concentrators, parented on large main exchanges.

The next evolutionary step after that of Figure 3.26(c) is to move the PCM codec from the exchange end of the customer's line to the customer's end, as shown in Figure 3.26(d). This provides digital transmission over the customer's line, which can have a number of advantages. Consider data transmission. If there is an analog customer's line, a modem must be added and data can only be transmitted at relatively slow speeds. If the line is digital, data can be transmitted by removing the codec (instead of adding a modem). Moreover, data can be transmitted at 64 kbit/s instead of at, say, 2.4 kbit/s. Indeed, any form of digital signal can be transmitted whose rate does not exceed

64 kbit/s. This can include high-speed fax and slow-scan TV, in addition to speech and data.

This concept has led to the integrated services digital network (ISDN), in which the customer's terminal equipment and the local digital exchange can be used to provide many different services, all using 64 kbit/s digit streams.[26–28]

Access to an ISDN is provided in two forms:

1. *Basic-rate access* The customer's line carries two 64 kbit/s 'B' channels plus a 16 kbit/s 'D' channel (for a common signalling channel) in each direction.
2. *Primary-rate access* Two lines are provided to carry a complete PCM frame at 1.5 Mbit/s or 2 Mbit/s in each direction. This gives the customer 23 or 30 circuits at 64 kbit/s plus a common signalling channel, also at 64 kbit/s.

Primary-rate access provides a very economical means of connecting a digital PBX to the main exchange. Private circuits between digital PBXs can also be provided by PCM links. Thus, integrated private digital networks exist within the integrated digital networks of PTOs. At present, most organizations still handle their data traffic on LANs and their voice traffic on PBXs. However, since business customers adopt new services more rapidly than domestic customers, their networks can be expected to develop into private ISDNs linking integrated-services private branch exchanges (ISPBX). Basic-rate access is expected to meet the needs of small businesses and residential customers. Its two B channels can provide two independent voice connections or simultaneous speech and data.

The ISDN can be seen as the culmination of developments in digital transmission, digital switching, stored-program control and common-channel signalling. It enables customers to obtain access, through their local exchanges, to a wide range of teleservices and bearer services (including circuit-switched telephony and packet-switched data communication) provided by a national telecommunications network. At present, these services are limited to those which can be provided by 64 kbit/s bearers. The networks are now called narrowband ISDN (N-ISDN), because plans are already being made for broadband ISDN networks (B-ISDN) to cater for services, such as high-definition television and colour graphics, which need much greater bandwidth. Switching technology for B-ISDN is discussed in Section 9.5.3.

Problems

1. Discuss the advantages obtained by the use of a system with registers, compared with the use of a non-register step-by-step system, in a large area with a high telephone density (such as New York or London).

2. Furnace End is a small exchange on the fringe of the Birmingham area in the UK. The unit-fee range for Furnace End spans the areas of Nuneaton and Coventry (each of which had 1st-, 2nd- and 3rd-group selectors) and the Birmingham director area.

When Furnace End had a Strowger exchange, its subscribers were provided with a list of dialling codes, some of which are given below.

Draw a trunking diagram showing the routing of calls from Furnace End customers to those destinations. Show the selector levels used in each exchange through which the calls passed.

Bedworth	731
Berkswell	735943
Claverdon	79784
Coventry	735
Kenilworth	73592
Leamington Spa	797
Meriden	73594
Nuneaton	7
Southam	735924

3. Explain briefly the meanings of the following terms:
Concentration, expansion, multiple, supervision, class of service.

4. Using switch units of size 10 × 10, devise an arrangement for connecting 100 incoming trunks to 50 outgoing trunks by means of a two-stage switching network. Provide 50 links and arrange these so that access from the primary switches to the secondary switches is equalized.

5. (a) Design a two-stage switching network, using 4 × 4 switches, to connect 16 incoming trunks to 16 outgoing trunks, using a consistent numbering arrangement for associating the switches with the links.
(b) Draw an arrangement using two input grids and two output grids of the above type for connecting 32 incoming trunks to 32 outgoing trunks, with a possibility of 32 simultaneous connections. Each incoming trunk shall have access to every outgoing trunk.
(c) What is the minimum number of through connections in each of the above arrangements which can block a connection from an incoming trunk to a particular free outgoing trunk?

6. Explain briefly the meanings of the following terms: Progressive control, common control, stored-program control, register, marker, link system, conditional selection, crank-back.

7. A four-stage switching network is shown in Figure 3.17.
(a) How many crosspoints does it contain?
(b) How many alternative paths are there for a connection between an incoming trunk and an outgoing trunk?

8. A link network is required to connect 64 incoming trunks to 64 outgoing trunks. All its switches are to be of equal size and 64 links are to be provided at each stage.
Suggest a suitable size of switch to use in (i) a two-stage network, (ii) a four-stage network. Sketch these networks. How many crosspoints does each contain?

9. For local calls, a telephone exchange operates the calling customer's meter once, when the called customer answers. Thus, there is a fee of one unit (U) for all local calls. The average duration, h, of a local call is 3 minutes. It can be assumed that the probability of the duration of a call, T, exceeding time t is given by:

$$P(T \geq t) = e^{-t/h}$$

(a) The operating company decides to introduce periodic-pulse metering for local calls. The meter is to be operated when the called customer answers and at subsequent intervals of 3 minutes during the call. By how much will this increase the company's revenue from local calls?
(b) If automatic ticketing is introduced the charge for a call should, ideally, be exactly proportional to its duration. What should be the fee per minute in order to obtain the same local-call revenue as with periodic-pulse metering?
(c) The operating company decides instead to charge a unit fee when the called customer answers and for each subsequent complete minute. What should be the fee per minute in order to obtain the same local-call revenue as with periodic-pulse metering?

10. The approximate cost of a telephone exchange can be expressed in the form:
$C = K + N(A + BE)$
where C = total cost
N = number of customers' lines
E = traffic per line (erlangs)
A, B and K are constants depending on the design of the system.
Justify this expression by considering Figure

3.20 and discuss its validity. For a practical system, the cost curve departs from this linear law. Explain why, for any two systems you may choose. (The erlang is the unit of traffic and is defined in Chapter 4. For the purpose of this problem, the number of trunks on a route may be considered proportional to its traffic in erlangs.)

References

[1] Chapuis, R.J. (1982), *100 Years of Telephone Switching*, North-Holland, Amsterdam.
[2] Chapuis, R.J. and Joel, A.E. (1990), *Electronics, Computers and Telephone Switching: a book of technological history*, North-Holland, Amsterdam.
[3] Daniels, E.B. (1975), 'Telegraph and Telex services', Chapter 9 in Flood, J.E. (ed.), *Telecommunication Networks*, Peter Peregrinus, Stevenage.
[4] Rubin, M. and Haller, C.E. (1966), *Communication Switching Systems*, Reinhold, New York.
[5] Lockwood, T.D. (1884), *Practical Information for Telephonists*, W.T. Johnson, New York.
[6] Smith, S.F. (1978), *Telephony and Telegraphy*, 3rd edn, Oxford University Press, Oxford.
[7] Flowers, T.H (1976), *Introduction to Exchange Systems*, Wiley, Chichester.
[8] Atkinson, J. (1948), *Telephony*, Vol. 1, Pitman, London.
[9] Atkinson, J. (1950), *Telephony*, Vol. 2, Pitman, London.
[10] Talley, D. (1979), *Basic Telephone Switching Systems*, 2nd edn, Hayden, Rochelle Park, NJ.
[11] Keister, W., Ritchie, A.E. and Washburn, S.H. (1952), *The Design of Switching Circuits*, Van Nostrand, New York.
[12] Clarke, A.B. and Osborne, H.S. (1952), 'Automatic switching for nation-wide telephone service', *Bell Syst. Tech. Jour.*, **31**, 823–31.
[13] Barron, D.A. (1959), 'Subscriber trunk dialling', *Proc. IEE.*, **106B**, 341–54.
[14] Munday, S. (1967), 'New international switching and transmission plan recommended by the CCITT for public telephony', *Proc. IEE*, **114**, 619–27.
[15] Corner, A.C. (1966), 'The 5005 crossbar telephone-exchange switching system', *PO Electr. Engrs. Jour.*, **59**, 170–7.
[16] Davis, R.C. (1949), 'No.5, the post-war crossbar', *Bell Labs. Record*, **27**, 85–8.
[17] Feiner, A. (1964), 'The ferreed', *Bell Syst. Tech. Jour.*, **43**, 1–14.
[18] Joel, A.E. (ed.) (1976), *Electronic Switching: central offices of the world*, IEEE Press, New York.
[19] Talley, D. (1982), *Basic Electronic Switching for Telephone Systems*, Hayden, Rochelle Park, NJ.
[20] Hills, M.T. (1979), *Telecommunications Switching Principles*, Allen and Unwin, London.
[21] Redmill, F.J. and Valdar, A.R. (1990), *SPC Digital Telephone Exchanges*, Peter Peregrinus, Stevenage.
[22] Joel, A.E. (ed.) (1982), *Electronic Switching: digital central office systems of the world*, IEEE Press, New York.
[23] GRINSEC (le groupe des ingenieurs du secteur commutation) (1981), *La commutation electronique*, Eyrolle, Paris.
[24] McDonald, J.C. (ed.) (1983), *Fundamentals of Digital Switching*, Plenum Press, New York.
[25] Ronayne, J. (1986), *Introduction to Digital Communication Switching*, Pitman, London.
[26] Griffiths, J.M. (1992), *ISDN explained*, 2nd edn, Wiley, Chichester.
[27] Helgert, H.J. (1991), *Integrated Services Digital Networks*, Addison-Wesley, Reading, MA.
[28] Stallings, W. (1992), *ISDN and Broadband ISDN*, 2nd edn, Macmillan, New York.

CHAPTER 4

Telecommunications traffic

4.1 Introduction

When any industrial plant is to be designed, an initial decision must be made as to its size, in order to obtain the desired throughput. For example, for an oil refinery, this is the number of barrels per day; for a machine shop, it is the number of piece parts per day. In the case of a telecommunications system, it is the traffic to be handled. This determines the number of trunks to be provided.

In teletraffic engineering[1–3] the term *trunk* is used to describe any entity that will carry one call. It may be an international circuit with a length of thousands of kilometres or a few metres of wire between switches in the same telephone exchange. The arrangement of trunks and switches within a telephone exchange is called its *trunking*.

If a record is made over a few minutes of the number of calls in progress on a large telecommunications system, such as a telephone exchange or a transmission route, it appears as shown in Figure 4.1. The number of calls varies in a random manner, as individual calls begin and end.

If this random variation is smoothed out by taking a running average, the

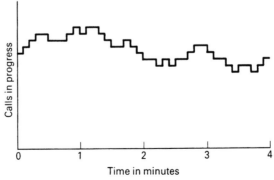

Figure 4.1 Short-term traffic variation.

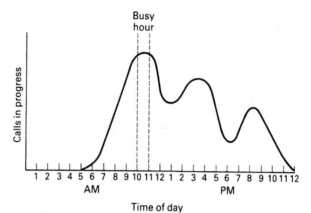

Figure 4.2 Traffic variation during a day.

number of calls in progress is found to vary during the day, for example as shown in Figure 4.2. There are very few calls during the night. The number of calls rises as people go to work and reaches a maximum by the middle of the morning. It falls at mid-day, as people go to lunch, and it rises again in the afternoon. It decreases as people go home from work and it has a further peak in the evening as people make social calls.

Figure 4.2 is typical of a telephone exchange serving the whole of a town. For an exchange serving the centre of a city, where few people live, the evening peak may be entirely absent. For an exchange serving a suburban residential area, the evening peak may be the largest. In addition to varying during the day, the number of calls carried may vary over the week. For example, a city-centre exchange may have very few calls during the weekend. The number of calls may also vary between different seasons of the year. For example, an exchange in a holiday resort will be busy during the summer and lightly loaded during the winter.

The number of trunks to be provided obviously depends on the traffic to be carried. Moreover, the number must be sufficient for the busiest time of the day. A period of one hour is chosen which corresponds to the peak traffic load and this is called the *busy hour*. In the example of Figure 4.1 the busy hour is from 10 a.m. to 11 a.m.

Since the amount of equipment provided must be sufficient to cope with the busy hour, much of it is idle during most of the day. It is for this reason that telecommunications operating companies offer their customers cheaper calls at off-peak periods. It costs them almost nothing to carry such calls. Moreover, if this induces some customers to make calls at off-peak periods that would otherwise be made in the busy hour, less equipment is needed and capital expenditure is reduced.

4.2 The unit of traffic

The *traffic intensity*, more often called simply the traffic, is defined as the average number of calls in progress. Although this is a dimensionless quantity, a name has been

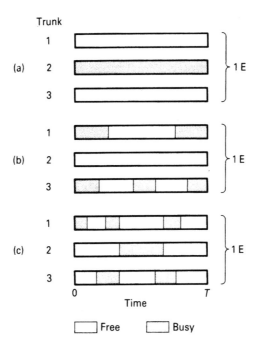

Figure 4.3 Examples of 1 erlang of traffic carried on three trunks.

given to the unit of traffic. This is the *erlang* (abbreviation E), which is named after A. K. Erlang, the Danish pioneer of traffic theory[4], whose results are given in Sections 4.6.1 and 4.7.1.

On a group of trunks, the average number of calls in progress depends on both the number of calls which arrive and their duration. The duration of a call is often called its *holding time*, because it holds a trunk for that time. The example in Figure 4.3 shows how one erlang of traffic can result from one trunk being busy all of the time, from each of two trunks being busy for half of the time, or from each of three trunks being busy for one third of the time.

In North America, traffic is sometimes expressed in terms of *hundreds of call seconds per hour* (CCS). Since an hour contains 3600 seconds, 1 erlang = 36 CCS.

It follows from the definition of the erlang that the traffic carried by a group of trunks is given by:

$$A = \frac{Ch}{T} \tag{4.1}$$

where A = traffic in erlangs
C = average number of call arrivals during time T
h = average call holding time.

From equation (4.1), if $T = h$, then $A = C$. Thus, the traffic in erlangs is equal to the mean number of calls arriving during a period equal to the mean duration of the calls.

Since a single trunk cannot carry more than one call, $A \leq 1$. The traffic is a fraction of an erlang equal to the average proportion of time for which the trunk is busy. This is called the *occupancy* of the trunk. The probability of finding the trunk busy is equal to the proportion of time for which the trunk is busy. Thus, this probability equals the occupancy (A) of the trunk.

Example 4.1

On average, during the busy hour, a company makes 120 outgoing calls of average duration 2 minutes. It receives 200 incoming calls of average duration 3 minutes. Find (1) the outgoing traffic, (2) the incoming traffic, (3) the total traffic.

1. The outgoing traffic is $120 \times 2/60 = 4$ E.
2. The incoming traffic is $200 \times 3/60 = 10$ E.
3. The total traffic is $4 + 10 = 14$ E.

Note: on average, four outgoing calls are made during their average holding time of 2 minutes and ten incoming calls are received during their average holding time of 3 minutes (i.e. if $T = h$, then $C = A$).

Example 4.2

During the busy hour, on average, a customer with a single telephone line makes three calls and receives three calls. The average call duration is 2 minutes. What is the probability that a caller will find the line engaged?

Occupancy of line $= (3 + 3) \times 2/60 = 0.1$ E $=$ probability of finding the line engaged.

4.3 Congestion

It is uneconomic to provide sufficient equipment to carry all the traffic that could possibly be offered to a telecommunications system. In a telephone exchange it is theoretically possible for every subscriber to make a call simultaneously. The cost of meeting this demand would be prohibitive, but the probability of it happening is negligible.

The situation can therefore arise that all the trunks in a group of trunks are busy, and so it can accept no further calls. This state is known as *congestion*. In a message-switched system, calls that arrive during congestion wait in a queue until an outgoing trunk becomes free. Thus, they are delayed but not lost. Such systems are therefore called *queuing systems* or *delay systems*. In a circuit-switched system, such as a telephone exchange, all attempts to make calls over a congested group of trunks are unsuccessful. Such systems are therefore called *lost-call systems*. In a lost-call system the result of congestion is that the traffic actually carried is less than the traffic offered to the system. We may therefore write:

$$\text{Traffic carried} = \text{traffic offered} - \text{traffic lost}.$$

The proportion of calls that is lost or delayed due to congestion is a measure of the service provided. It is called the *grade of service*. For a lost-call system, the grade of service, B, may be defined as:

$$B = \frac{\text{Number of calls lost}}{\text{Number of calls offered}}$$

Hence, also:

$$B = \frac{\text{Traffic lost}}{\text{Traffic offered}}$$

= proportion of the time for which congestion exists[1]
= probabilty of congestion[1]
= probability that a call will be lost due to congestion

Thus, if traffic A erlangs is offered to a group of trunks having a grade of service B, the traffic lost is AB and the traffic carried is $A(1-B)$ erlangs.

The larger the grade of service, the worse is the service given. (Perhaps it should have been called the grade of disservice!) The grade of service is normally specified for the traffic at the busy hour. At other times, it is much better. Specifying a grade of service is a matter of judgement. If it is too large, users make many unsucessful calls and are dissatisfied. If it is too small, unnecessary expenditure is incurred on equipment which is rarely used. In practice, busy-hour grades of service can vary from, say, 1 in 1000 for cheap trunks inside an exchange to 1 in 100 for inter-exchange connections and 1 in 10 for expensive international routes.

The basic problem of determining the size of a telecommunications system, known as the *dimensioning problem*, is: given the offered traffic, A, and the specified grade of service, B, find the number of trunks, N, that is required. This problem is dealt with in Section 4.6.4.

Example 4.3

During the busy hour, 1200 calls were offered to a group of trunks and six calls were lost. The average call duration was 3 minutes. Find:

1. The traffic offered
2. The traffic carried
3. The traffic lost
4. The grade of service
5. The total duration of the periods of congestion.

[92] *Telecommunications traffic*

1. $A = Ch/T = 1200 \times 3/60 = 60$ E
2. $1194 \times 3/60 = 59.7$ E
3. $6 \times 3/60 = 0.3$ E
4. $B = 6/1200 = 0.005$
5. $0.005 \times 3600 = 18$ seconds

4.4 Traffic measurement

It is important for an operating company to know how much busy-hour traffic its systems are handling. In particular, it needs to know when a system is becoming overloaded and additional equipment should be installed. Thus, the traffic should be measured regularly and records kept. Since equipment must be manufactured, installed and commissioned before it enters service, it is always specified to carry the traffic which is forecast for a future date. In order for the forecast to be as accurate as possible, it should be derived from figures for the present traffic which are as accurate as possible.

By definition, measuring the traffic carried amounts to counting the calls in progress at regular intervals during the busy hour and averaging the results. In the past, engineers counted the plugs inserted in a manual switchboard or the number of selectors off-normal in an automatic exchange. The results were known as a *peg count* or a *switch count*. Since this method was laborious, automatic traffic recorders (operating on the same principle) were developed and installed in automatic exchanges.[5] In modern stored-program-controlled systems, the central processors generate records of the calls that they set up.[6]

Example 4.4

Observations were made of the number of busy lines in a group of junctions at intervals of 5 minutes during the busy hour. The results obtained were:

$$11, 13, 8, 10, 14, 12, 7, 9, 15, 17, 16, 12$$

It is therefore estimated that the traffic carried, in erlangs, was:

$$\frac{11 + 13 + 8 + 10 + 14 + 12 + 7 + 9 + 15 + 17 + 16 + 12}{12} = 12\,E$$

4.5 A mathematical model

In order to obtain analytical solutions to teletraffic problems it is necessary to have a mathematical model of the traffic offered to telecommunications systems. A simple model is based on the following assumptions:

- Pure-chance traffic
- Statistical equilibrium.

The assumption of *pure-chance traffic* means that call arrivals and call terminations are independent random events. Calls made by an individual user, of course, are not made at random. However, the total traffic generated by a large number of users is observed to behave as if calls were generated at random. If call arrivals are independent random events, their occurrence is not affected by previous calls. The traffic is therefore sometimes called *memoryless traffic*. It also implies that the number of sources generating calls is very large. If the number of sources is small and several are already busy, then the rate at which new calls can be generated is less than it would be if all the sources were free.

This assumption of random call arrivals and terminations leads to the following results:

1. The number of call arrivals in a given time has a Poisson distribution (a proof is given in Appendix 1), i.e.:

$$P(x) = \frac{\mu^x}{x!} e^{-\mu}$$

   where x is the number of call arrivals in time T and μ is the mean number of call arrivals in time T. For this reason, pure-chance traffic is also called *Poissonian traffic*.

2. The intervals, T, between call arrivals are the intervals between independent random events and it is shown in Appendix 1 that these intervals have a negative exponential distribution, i.e.:

$$P(T \geq t) = e^{-t/\bar{T}}$$

   where $\bar{T}$ is the mean interval between call arrivals.

3. Since the arrival of each call and its termination are independent random events, call durations, T, are also the intervals between two random events and have a negative exponential distribution, i.e.:

$$P(T \geq t) = e^{-t/h}$$

   where h is the mean call duration (holding time).

The assumption that call terminations are random may seem odd, because it implies that a call is as likely to end if it has just started as if it has been going on for a long time. However, in practice, some calls are short and others are long, with the result that the distribution of holding times is observed to fit the negative exponential distribution.

The assumption of *statistical equilibrium* means that the generation of traffic is a stationary random process, i.e. probabilities do not change during the period being considered. Consequently, the mean number of calls in progress remains constant. Figure 4.2 shows that this condition is satisfied during the busy hour and it is, of course, the busy-hour grade of service that one wishes to determine. (Statistical equilibrium is not obtained immediately before the busy hour, when the calling rate is increasing, nor at the end of the busy hour, when the calling rate is falling.)

Example 4.5

On average, one call arrives every 5 seconds. During a period of 10 seconds, what is the probability that:

1. No call arrives?
2. One call arrives?
3. Two calls arrive?
4. More than two calls arrive?

$$P(x) = \frac{\mu^x}{x!} e^{-\mu}, \text{ where } \mu = 2$$

1. $P(0) = \frac{2^0}{0!} e^{-2} = e^{-2} = 0.135$

2. $P(1) = \frac{2}{1!} e^{-2} = 0.270$

3. $P(2) = \frac{2^2}{2!} e^{-2} = 0.270$

4. $P(>2) = 1 - P(0) - P(1) - P(2)$
   $= 1 - 0.135 - 0.270 - 0.270$
   $= 0.325$

Example 4.6

In a telephone system the average call duration is 2 minutes. A call has already lasted 4 minutes. What is the probability that:

1. The call will last at least another 4 minutes?
2. The call will end within the next 4 minutes?

These probabilities can be assumed to be independent of the time which has already elapsed.

1. $P(T \geq t) = e^{-t/h} = e^{-2} = 0.135$
2. $P(T \leq t) = 1 - P(T \geq t) = 1 - 0.135 = 0.865$

For a group of N trunks the number of calls in progress varies randomly, as shown in Figure 4.1. This is an example of a *birth and death process* or a *renewal process*. The number of calls in progress is always between 0 and N. It thus has $N+1$ states and its behaviour depends on the probability of change from each state to the one above and to the one below it. Such a process is called a *simple Markov chain*.[7] It is represented in Figure 4.4, where $P(j)$ is the probability of state j and $P(k)$ is the probability of the next higher state k. $P_{j,k}$ is the probability of a state increase to k, given that the present state is j. $P_{k,j}$ is the probability of a decrease to j, given that the present state is k. The probabilities $P(0), P(1),...,P(N)$ are called the *state probabilities* and the conditional probabilities $P_{j,k}, P_{k,j}$ are called the *transition probabilities* of the Markov chain. If there

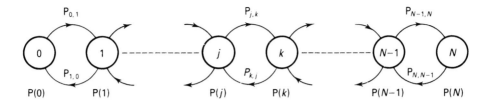

Figure 4.4 State transition diagram for N trunks.

is statistical equilibrium, these probabilities do not change and the process is said to be a *regular Markov chain*.

Consider a very small interval of time δt, starting at time t. Since δt is very small, the probability of something happening during it is small. The probability of two or more events during δt is therefore negligible. The events which can happen during δt are thus as follows:

- One call arriving, with probability P(a)
- One call ending, with probability P(e)
- No change, with probability $1 - P(a) - P(e)$.

Equation (4.1) shows that the mean number of calls arriving during the average holding time, h, is $C = A$. Thus, the mean number arriving during time δt is $A\delta t/h$. Since δt is very small, $A\delta t/h \ll 1$ and represents the probability, P(a), of a call arriving during δt.

$$\therefore \quad P_{j,k} = P(a) = A\ \delta t/h \tag{4.2}$$

If the mean holding time is h and the number of calls in progress is k, one expects an average of k calls to end during a period h. The average number of calls ending during δt is therefore $k\delta t/h$. Since δt is very small, $k\delta t/h \ll 1$ and represents the probability, P(e), of a call ending during δt.

$$\therefore \quad P_{k,j} = P(e) = k\delta t/h \tag{4.3}$$

If the probability of j calls in progress at time t is P(j), then the probability of a transition from j to k busy trunks during δt is:

$$P(j \rightarrow k) = P(j)\ P(a) = P(j)\ A\delta t/h \tag{4.4}$$

If the probability of k calls at time t is P(k), then the probability of a transition from k to j busy trunks during δt is:

$$P(k \rightarrow j) = P(k)\ P(e) = P(k)\ k\delta t/h \tag{4.5}$$

The assumption of statistical equilibrium requires that $P(j \rightarrow k) = P(k \rightarrow j)$. Otherwise, the number of calls in progress would steadily increase or decrease. Thus, from equations (4.4) and (4.5):

[96] *Telecommunications traffic*

$$k\ P(k)\ \delta t/h = A\ P(j)\delta t/h$$

$$\therefore \quad P(k) = \frac{A}{k} P(j) \tag{4.6}$$

Hence:
$$P(1) = \frac{A}{1} P(0)$$

$$P(2) = \frac{A}{2} P(1) = \frac{A^2}{2.1} P(0)$$

$$P(3) = \frac{A}{3} P(2) = \frac{A^3}{3 \times 2 \times 1} P(0)$$

and, in general:

$$P(x) = \frac{A^x}{x!} P(0) \tag{4.7}$$

The assumption of pure-chance traffic implies a very large number of sources. Thus, x can have any value between zero and infinity and the sum of their probabilities must be unity. Then:

$$1 = \sum_{x=0}^{\infty} P(x) = \sum_{x=0}^{\infty} \frac{A^x}{x!} P(0) = e^A P(0)$$

$\therefore$ and
$$P(0) = e^{-A}$$
$$P(x) = \frac{A^x}{x!} e^{-A}$$

Thus, if call arrivals have a Poisson distribution, so does the number of calls in progress. This requires an infinite number of trunks to carry the calls. If the number of trunks available is finite, then some calls can be lost or delayed and the distribution is no longer Poissonian. The distributions which then occur are derived in Sections 4.6 and 4.7.

4.6 Lost-call systems

4.6.1 Theory

Erlang determined the grade of service (i.e. the loss probability) of a lost-call system having N trunks, when offered traffic A, as shown in Figure 4.5. His solution depends on the following assumptions:

- Pure-chance traffic
- Statistical equilibrium
- Full availability
- Calls which encounter congestion are lost.

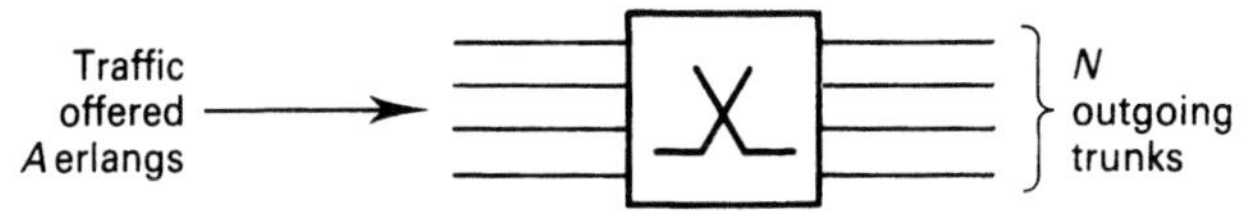

Lost-call systems

```
Traffic                          N
offered     →    ===[X]===  }   outgoing
A erlangs                        trunks
```

Figure 4.5 Lost-call system.

The assumption of pure-chance traffic implies that call arrivals and call terminations are independent random events and statistical equilibrium implies that probabilities do not change, as discussed in Section 4.5.

Full availability means that every call that arrives can be connected to any outgoing trunk which is free. If the incoming calls are connected to the outgoing trunks by switches, each switch must have sufficient outlets to provide access to every outgoing trunk. (In many practical cases, this condition is not satisfied; the switches have insufficient outlets and so can provide only limited availability. This problem is considered in Chapter 5.)

The lost-call assumption implies that any attempted call which encounters congestion is immediately cleared from the system. When this happens, the user is likely to make another attempt shortly afterwards. Thus, the traffic offered during the busy hour is slighly greater than it would have been if there were no congestion. The subject of repeat attempts has engaged the attention of teletraffic specialists.[1] Here, it will simply be assumed that the traffic offered is the total arising from all successful and unsuccessful calls.[2]

If there are x calls in progress, then equation (4.7) gives:

$$P(x) = \frac{A^x}{x!} P(0)$$

However, there cannot be a negative number of calls and there cannot be more than N. Thus, we know with certainty that $0 \leq x \leq N$.

$$\therefore \quad \sum_{x=0}^{N} P(x) = 1 = \sum_{x=0}^{N} \frac{A^x}{x!} P(0)$$

Hence,
$$P(0) = 1 \bigg/ \sum_{x=0}^{N} \frac{A^x}{x!}$$

Substituting in equation (4.7) gives:

$$P(x) = \frac{A^x/x!}{\sum_{k=0}^{N} A^k/k!} \tag{4.8}$$

[98] *Telecommunications traffic*

This is the first Erlang distribution.[3] Of particular importance is $P(N)$, since this is the probability of congestion, i.e. the probabilty of a lost call, which is the grade of service, B. This is given the symbol $E_{1,N}(A)$, which denotes the loss probability for a full-availability group of N trunks offered traffic A erlangs.

$$\therefore \quad B = E_{1,N}(A) = \frac{A^N/N!}{\sum_{k=0}^{N} A^k/k!} \qquad (4.9)$$

Many engineers remember equation (4.9) as 'Erlang's lost-call formula' and are unaware of equation (4.8). It is preferable to remember equation (4.8); then equation (4.9) is just a special case.

The grade of service of a loss system with N full-availability trunks, offered A erlangs of traffic, is given by $E_{1,N}(A)$. It may be computed directly or by the iterative application of a simple recurrence relation obtained as follows:

$$E_{1,N-1} = \frac{A^{N-1}}{(N-1)!} \bigg/ \sum_{k=0}^{N-1} \frac{A^k}{k!} \qquad \text{(from equation (4.9))}$$

$$\therefore \quad \sum_{k=0}^{N} \frac{A^k}{k!} = \frac{A^{N-1}/(N-1)}{E_{1,N-1}(A)} + \frac{A^N}{N!}$$

Substituting in equation (4.9):

$$E_{1,N}(A) = \frac{A\, E_{1,N-1}(A)}{N + A\, E_{1,N-1}(A)}$$

Since $E_{1,0} = 1$, this iterative formula enables $E_{1,N}(A)$ to be computed for all values of N. Tables of $E_{1,N}(A)$ have also been published.[3]

Example 4.7

A group of five trunks is offered 2 E of traffic. Find:

1. The grade of service
2. The probability that only one trunk is busy
3. The probability that only one trunk is free
4. The probability that at least one trunk is free.

1. From equation (4.9):

$$B = E_{1,N}(A) = \frac{32/120}{1 + \frac{2}{1} + \frac{4}{2} + \frac{8}{6} + \frac{16}{24} + \frac{32}{120}}$$

$$= \frac{0.2667}{7.2667} = 0.037$$

2. From equation (4.8): $P(1) = 2/7.2667 = 0.275$
3. $P(4) = \dfrac{16/24}{7.2667} = 0.0917$
4. $P(x<5) = 1 - P(5) = 1 - B = 1 - 0.037 = 0.963$

Example 4.8

A group of 20 trunks provides a grade of service of 0.01 when offered 12 E of traffic.

1. How much is the grade of service improved if one extra trunk is added to the group?
2. How much does the grade of service deteriorate if one trunk is out of service?

1.
$$E_{1,21}(12) = \dfrac{12\, E_{1,20}(12)}{21 + 12\, E_{1,20}(12)}$$
$$= \dfrac{12 \times 0.01}{21 + 12 \times 0.01}$$
$$= 0.0057$$

2.
$$E_{1,20}(12) = 0.01 = \dfrac{12\, E_{1,19}(12)}{20 + 12\, E_{1,19}(12)}$$
$$0.2 + 0.12\, E_{1,19}(12) = 12\, E_{1,19}(12)$$
$$\therefore \quad E_{1,19}(12) = 0.017$$

4.6.2 Traffic performance

If the offered traffic, A, increases, the number of trunks, N, must obviously be increased to provide a given grade of service. However, for the same trunk occupancy the probability of finding all trunks busy is less for a large group of trunks than for a small group. Thus, for a given grade of service a large group of trunks can have a higher occupancy than a small one, i.e. the large group is more efficient. This is shown by Figure 4.6 for a grade of service of 0.002 (i.e. one lost call in 500). For example, 2 E of traffic requires seven trunks and their occupancy is 0.27 E. However, 20 E requires 32 trunks and their occupancy is 0.61 E.

Since large groups of trunks are more efficient than small groups, it is better to concentrate traffic onto a single large group of trunks than to handle it by several small groups of trunks. The principle of concentration is applied widely. For example, in a local exchange, the traffic from a large number of low-occupancy subscribers' lines is concentrated onto a smaller number of high-occupancy trunks. Tandem exchanges enable a single route of high-occupancy junctions to be used for traffic to several destinations. This requires fewer junctions than for separate routes, each of which would carry less traffic with a lower occupancy.

The penalty paid for the high efficiency of large groups of trunks is that the grade of service (GOS) deteriorates more with traffic overloads than for small groups of trunks. Figure 4.7 shows how the grade of service varies with offered traffic for different

[100] *Telecommunications traffic*

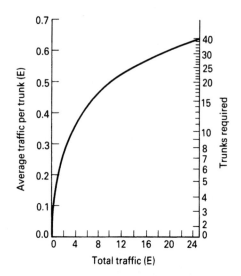

Figure 4.6 Trunk occupancies for full-availability groups of various sizes. (Grade of service = 0.002.)

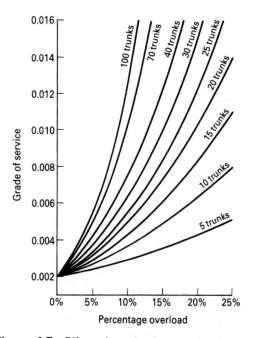

Figure 4.7 Effect of overload on grade of service.

sizes of group, which were all dimensioned to provide a GOS of 0.002 at their normal traffic load. For a group of five trunks, a 10% overload increases the GOS by 40%. However, for a group of 100 trunks, it causes the GOS to increase by 550%.

For this reason, most telecommunications operating companies adopt a dual criteria. Two GOS are specified: one for the normal traffic load and another, larger, GOS for a given percentage overload: for example, a GOS of B at the normal load and $5B$ for a 20% overload. The number of trunks to be provided is determined by which criterion requires the greater number. For small groups, the number is determined by the normal-load criterion; for large groups, it is determined by the overload criterion.

Example 4.9

Groups of trunks are to be dimensioned to provide a grade of service of 0.002 under normal load and 0.01 under 20% overload. For what range of nominal traffic loads does the normal-load criterion apply and when does the overload criterion apply?

From Figure 4.7 the grade of service deteriorates to 0.01 for 20% overload when there are 19 trunks. From Figure 4.6, 19 trunks can handle 9.5 E with a grade of service of 0.002.

∴ Normal-load criterion applies for $A \leq 9.5$ E
Overload criterion applies for $A \geq 9.5$ E

In many switching systems, trunks in a group are selected by means of a sequential search. A call is not connected to trunk no. 2 unless no. 1 is busy. It is not connected to no. 3 unless both no. 1 and no. 2 are busy, and so on. Calls finding the last-choice trunk busy are lost. As a result, the first trunk has a very high occupancy and the traffic carried by subsequent trunks is less. The last-choice trunk is very lightly loaded indeed. This behaviour is illustrated in Figure 4.8 for a group of 20 trunks.

The performance of such an arrangement can be analyzed as follows. Let traffic A erlangs be offered to the group of trunks. From equation (4.9), the GOS of a single-trunk group is

$$E_{1,1}(A) = A/(1+A).$$

Traffic overflowing from the first trunk to the second is

$$A\, E_{1,1}(A) = A^2/(1+A)$$

∴ traffic carried by the first trunk is

Traffic offered − traffic lost = $A - A^2/(1+A) = A/(1+A)$

In general:

Traffic carried by the kth trunk = traffic lost from group of first k-1 trunks
− traffic lost from group of first k trunks
= $A\, [E_1,k-1\,(A) - E_1,k(A)]$

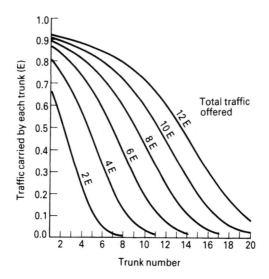

Figure 4.8 Distribution of traffic over trunks of a group with sequential search.

It may be thought simpler to consider each trunk as a single-trunk group offered the overflow traffic from the previous trunk. However, this would be incorrect. The traffic offered to the first trunk is Poissonian. However, the traffic overflowing to the second trunk is more peaky, because the traffic carried by the first trunk has been removed.[4] Traffic offered to subsequent trunks is still more peaky. Calculating the traffic carried on the kth trunk by considering the traffic offered to a group of the first k trunks ensures that the conditions for the Erlang lost-call formula are satisfied.

Example 4.10

If sequential selection is used for the group of trunks in Example 4.7, how much traffic is carried by:

1. The first-choice trunk?
2. The last-choice trunk?

1. Traffic carried by the first-choice trunk is:

$$A/(1 + A) = 2/(1 + 2) = 0.67 \text{ E}$$

2. $E_{1,5}(2) = 0.037$

$$E_{1,4}(2) = \frac{16/24}{1 + \frac{2}{1} + \frac{4}{2} + \frac{8}{6} + \frac{16}{24}} = 0.095$$

Traffic carried by the last-choice trunk $= 2 (0.095 - 0.037) = 0.12$ E

4.6.3 Loss systems in tandem

The user is interested in the grade of service of a complete connection, which may have several links in tandem, rather than the grade of service of a single link in the connection. If a connection consisting of two links, having grades of service B_1, B_2, is offered traffic A erlangs, then:

Traffic offered to second link $= A(1 - B_1)$
∴ Traffic reaching destination $= A(1 - B_1)(1 - B_2)$
$= A(1 + B_1 B_2 - B_1 - B_2)$

and the overall grade of service is $B = B_1 + B_2 - B_1 B_2$.

If $B_1, B_2 \ll 1$, as they should be, then $B_1 B_2$ is negligible and the overall grade of service is simply $B = B_1 + B_2$.

In general, for an n-link connection, we may write:

$$B = \sum_{k=1}^{n} B_k$$

In practice, this equation is an over-simplification, for two reasons. First, grades of service are specified for the busy hour and the busy hours of the different links may not coincide. Thus, at the busy hour of one link the loss probabilities of the other links may be much smaller than their specified grades of service, so the overall loss is little more than that of the busy link.

Second, when traffic on a system is growing, it is not economic to instal equipment incrementally to match the traffic growth exactly. Instead, it is installed in sufficient quantity to meet the traffic which is forecast for the end of a provision period of several years. Consequently, the grade of service of any link will be much better than the specified value at the beginning of this period and will increase to the specified value at the end of the period. However, if the traffic forecast is wrong and the traffic grows faster than expected, the grade of service will exceed the specified value before the end of the provision period. This has often happened in the history of telecommunications!

4.6.4 Use of traffic tables

Equation (4.9) for $E_{1,N}(A)$ is suitable for solving the problem: given A and N, find B. However, the dimensioning problem is: given A and B, find N. Equation (4.9) is most unsuitable for this! Fortunately, our predecessors have calculated B for a wide range of values of A and N. Rearrangement of the data enabled tables to be published which show the traffic that can be handled for different values of B and N. An example is shown in Table 4.1. To use the table, one selects the specified value of B and looks down the column to find the first value of A exceeding the specified traffic load. For example, Table 4.1 shows that, if the specified GOS is 0.01, then 10 E requires 18 trunks and 50 E requires 64 trunks.

If an operating company adopts a dual criterion, it can use modified tables, in which the number of trunks for each entry corresponds to whether the normal-traffic or

Telecommunications traffic

Table 4.1 Traffic-capacity table for full-availability groups

Number of trunks	1 lost call in				Number of trunks	1 lost call in			
	50 (0.02)	100 (0.01)	200 (0.005)	1000 (0.001)		50 (0.02)	100 (0.01)	200 (0.005)	1000 (0.001)
	E	E	E	E		E	E	E	E
1	0.020	0.010	0.005	0.001	51	41.2	38.8	36.8	33.4
2	0.22	0.15	0.105	0.046	52	42.1	39.7	37.6	34.2
3	0.60	0.45	0.35	0.19	53	43.1	40.6	38.5	35.0
4	1.1	0.9	0.7	0.44	54	44.0	41.5	39.4	35.8
5	1.7	1.4	1.1	0.8	55	45.0	42.4	40.3	36.7
6	2.3	1.9	1.6	1.1	56	45.9	43.3	41.2	37.5
7	2.9	2.5	2.2	1.6	57	46.9	44.2	42.1	38.3
8	3.6	3.2	2.7	2.1	58	47.8	45.1	43.0	39.1
9	4.3	3.8	3.3	2.6	59	48.7	46.0	43.9	40.0
10	5.1	4.5	4.0	3.1	60	49.7	46.9	44.7	40.8
11	5.8	5.2	4.6	3.6	61	50.6	47.9	45.6	41.6
12	6.6	5.9	5.3	4.2	62	51.6	48.8	46.5	42.5
13	7.4	6.6	6.0	4.8	63	52.5	49.7	47.4	43.4
14	8.2	7.4	6.6	5.4	64	53.4	50.6	48.3	44.1
15	9.0	8.1	7.4	6.1	65	54.4	51.5	49.2	45.0
16	9.8	8.9	8.1	6.7	66	55.3	52.4	50.1	45.8
17	10.7	9.6	8.8	7.4	67	56.3	53.3	51.0	46.6
18	11.5	10.4	9.6	8.0	68	57.2	54.2	51.9	47.5
19	12.3	11.2	10.3	8.7	69	58.2	55.1	52.8	48.3
20	13.2	12.0	11.1	9.4	70	59.1	56.0	53.7	49.2
21	14.0	12.8	11.9	10.1	71	60.1	57.0	54.6	50.1
22	14.9	13.7	12.6	10.8	72	61.0	58.0	55.5	50.9
23	15.7	14.5	13.4	11.5	73	62.0	58.9	56.4	51.8
24	16.6	15.3	14.2	12.2	74	62.9	59.8	57.3	52.6
25	17.5	16.1	15.0	13.0	75	63.9	60.7	58.2	53.5
26	18.4	16.9	15.8	13.7	76	64.8	61.7	59.1	54.3
27	19.3	17.7	16.6	14.4	77	65.8	62.6	60.0	55.2
28	20.2	18.6	17.4	15.2	78	66.7	63.6	60.9	56.1
29	21.1	19.5	18.2	15.9	79	67.7	64.5	61.8	56.9
30	22.0	20.4	19.0	16.7	80	68.6	65.4	62.7	58.7
31	22.9	21.2	19.8	17.4	81	69.6	66.3	63.6	58.7
32	23.8	22.1	20.6	18.2	82	70.5	67.2	64.5	59.5
33	24.7	23.0	21.4	18.9	83	71.5	68.1	65.4	60.4
34	25.6	23.8	22.3	19.7	84	72.4	69.1	66.3	61.3
35	26.5	24.6	23.1	20.5	85	73.4	70.1	67.2	62.1
36	27.4	25.5	23.9	21.3	86	74.4	71.0	68.1	63.0
37	28.3	26.4	24.8	22.1	87	75.4	71.9	69.0	63.9
38	29.3	27.3	25.6	22.9	88	76.3	72.8	69.9	64.8
39	30.1	28.2	26.5	23.7	89	77.2	73.7	70.8	65.6
40	31.0	29.0	27.3	24.5	90	78.2	74.7	71.8	66.6
41	32.0	29.9	28.2	25.3	91	79.2	75.6	72.7	67.4
42	32.9	30.8	29.0	26.1	92	80.1	76.6	73.6	68.3
43	33.8	31.7	29.9	26.9	93	81.0	77.5	74.3	69.1
44	34.7	32.6	30.8	27.7	94	81.9	78.4	75.4	70.0
45	35.6	33.4	31.6	28.5	95	82.9	79.3	76.3	70.9
46	36.6	34.3	32.5	29.3	96	83.8	80.3	77.2	71.8
47	37.5	35.2	33.3	30.1	97	84.8	81.2	78.2	72.6
48	38.4	36.1	34.2	30.9	98	85.7	82.2	79.1	73.5
49	39.4	37.0	35.1	31.7	99	86.7	83.2	80.0	74.4
50	40.3	37.9	35.9	32.5	100	87.6	84.0	80.9	75.3

overload criterion is more onerous.[5] Tables have also been published for traffic which is not pure-chance (for example, for 'smooth' traffic[5]) and this is discussed in Appendix 2.

Example 4.11

The company in Example 4.1 wishes to obtain a grade of service of 0.01 for both incoming and outgoing calls. How many exchange lines should it rent if:

1. Incoming and outgoing calls are handled on separate groups of lines?
2. A common group of lines is used for both incoming and outgoing calls?

    1. From Table 4.1:
       4 E of outgoing traffic needs 10 lines
       10 E of incoming traffic needs 18 lines
       Total number of lines required is 28
    2. The total traffic of 14 E requires only 23 lines.

4.7 Queueing systems

4.7.1 The second Erlang distribution

Erlang determined the probability of encountering delay when traffic A is offered to a queuing system with N trunks, as shown in Figure 4.9. In queuing systems, the trunks are often called *servers*. This is because the theory has been applied in many fields other than telecommunications.[8,9] For example, it is applicable to queues of people waiting to be served in a post office or at a supermarket checkout. Erlang's solution depends on the following assumptions:

1. Pure-chance traffic
2. Statistical equilibrium
3. Full availability
4. Calls which encounter congestion enter a queue and are stored there until a server becomes free.

Such a system is sometimes known as a $M/M/N$ system.[5]

Assumptions (1) to (3) are common to the theory of lost-call systems. However, assumption (2) implies that $A \leq N$. If $A \geq N$, calls are entering the system at a greater

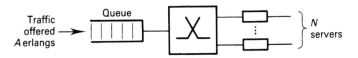

Figure 4.9 Queuing system.

rate than they leave. As a result, the length of the queue must continually increase towards infinity. This is not statistical equilibrium!

Let x be the total number of calls in the system. Thus, when $x < N$, then x calls are being served and there is no delay. When $x > N$, all the servers are busy and incoming calls encounter delay; there are N calls being served and $x - N$ calls in the queue.

If $x \leq N$:

There is no queue and the behaviour of the system is the same as that of a lost-call system in the absence of congestion. Thus, from equation (4.7):

$$P(x) = \frac{A^x}{x!} P(0) \qquad \text{for } 0 \leq x \leq N \tag{4.7}$$

If $x \geq N$:

The probability of a call arrival in a very short period of time, δt, from equation (4.2) is given by:

$$P(a) = A \, \delta t / h$$

where h is the mean service time.

Thus, the probability of a transition from $x - 1$ to x calls in the system during δt, from equation (4.4), is given by:

$$P(x - 1 \to x) = P(x - 1) \, A \delta t / h$$

Since all servers are busy, only the N calls being served can terminate (instead of x calls in a lost-call system). Therefore, equation (4.3) is modified to

$$P(e) = N \, \delta t / h$$

and the probability of a transition from x to $x - 1$ calls is given by:

$$P(x \to x - 1) = P(x) \, P(e) = P(x) \, N \, \delta t / h$$

For statistical equilibrium, $P(x - 1 \to x) = P(x \to x - 1)$,

$$\therefore \qquad P(x) \, N \delta t / h = P(x - 1) \, A \delta t / h$$

and

$$P(x) = \frac{A}{N} P(x - 1) \tag{4.10}$$

But

$$P(N) = \frac{A^N}{N!} P(0) \qquad \text{(from equation (4.7))}$$

$$\therefore \qquad P(N + 1) = \frac{A}{N} P(N) = \frac{A^{N+1}}{N \cdot N!} P(0)$$

$$P(N + 2) = \frac{A}{N} P(N + 1) = \frac{A^{N+2}}{N^2 \cdot N!} P(0)$$

and so on.

In general, for $x \geq N$:

$$P(x) = \frac{A^x}{N^{x-N} \cdot N!} P(0) = \frac{N^N}{N!} \left(\frac{A}{N}\right)^x P(0) \qquad (4.11)$$

If there is no limit to the possible length of queue, then x can have any value between zero and infinity.

$$\therefore \quad \sum_{x=0}^{\infty} P(x) = 1$$

Thus, from equations (4.7) and (4.11):

$$\frac{1}{P(0)} = \sum_{x=0}^{N-1} \frac{A^x}{x!} + \frac{N^N}{N!} \left(\frac{A}{N}\right)^N \sum_{k=0}^{\infty} \left(\frac{A}{N}\right)^k \qquad (4.12)$$

where $k = x - N$. Since $\frac{A}{N} \leq 1$, then

$$\sum_{k=0}^{\infty} \left(\frac{A}{N}\right)^k = \left[1 - \frac{A}{N}\right]^{-1}$$

$$\therefore \quad \frac{1}{P(0)} = \sum_{x=0}^{N-1} \frac{A^x}{x!} + \frac{A^N}{N!}\left[1 - \frac{A}{N}\right]^{-1}$$

i.e
$$P(0) = \left[\frac{N A^N}{N!(N-A)} + \sum_{x=0}^{N-1} \frac{A^x}{x!}\right]^{-1} \qquad (4.13)$$

Thus, $P(x)$ is given by equations (4.7) or (4.11), depending on whether $x \leq N$ or $x \geq N$, where $P(0)$ is given by equation (4.13). This is the second Erlang distribution.[6]

4.7.2 Probability of delay

Delay occurs if all servers are busy, i.e. $x \geq N$. Now, from equation (4.11), the probability that there are at least z calls in the system (where $z \geq N$) is given by:

$$P(x \geq z) = \sum_{x=z}^{\infty} P(x)$$

$$= \frac{N^N}{N!} P(0) \sum_{x=z}^{\infty} \left(\frac{A}{N}\right)^x$$

$$= \frac{N^N}{N!} P(0) \left(\frac{A}{N}\right)^z \sum_{k=0}^{\infty} \left(\frac{A}{N}\right)^k$$

where $k = x - N$.

$$\therefore \quad P(x \geq z) = \frac{N^N}{N!} \left(\frac{A}{N}\right)^z P(0) \left[1 - \frac{A}{N}\right]^{-1}$$

[108] *Telecommunications traffic*

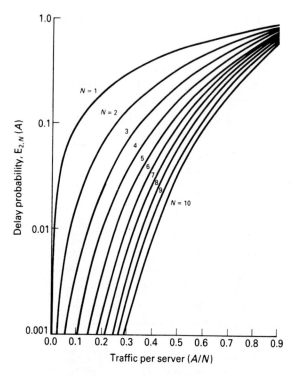

Figure 4.10 Delay probabilities for queuing systems (A = traffic in erlangs, N = number of servers).

$$\therefore \quad P(x \geq z) = \frac{N^N}{N!} \left(\frac{A}{N}\right)^z \frac{N}{N-A} P(0) \quad (4.14)$$

The probability of delay is $P_D = P(x \geq N)$.

$$\therefore \quad P_D = \frac{A^N}{N!} \frac{N}{N-A} P(0) \quad (4.15)$$
$$= E_{2,N}(A)$$

The probability of delay, for a system with N servers offered traffic A erlangs, is thus given by equation (4.15), where P(0) is given by equation (4.13). This formula for $E_{2,N}(A)$ is known as the *Erlang delay formula*.

Some results are plotted in Figure 4.10. The probability of delay increases towards 1.0 as A increases towards N. When $A > N$, the length of the queue grows indefinitely.

4.7.3 Finite queue capacity

A practical system cannot contain an infinite queue. Thus, when the queue has become full, calls that arrive subsequently are lost. If the queue can only hold up to Q calls, then $x \leq Q + N$ and equation (4.12) becomes:

$$\frac{1}{P(0)} = \sum_{x=0}^{N-1} \frac{A^x}{x!} + \frac{N}{N!}\left(\frac{A}{N}\right)^N \sum_{k=0}^{Q} \left(\frac{A}{N}\right)^k$$

$$\therefore \quad \frac{1}{P(0)} = \sum_{x=0}^{N-1} \frac{A^x}{x!} + \frac{A^N}{N!} \frac{1 - (A/N)^{Q+1}}{1 - A/N} \tag{4.16}$$

However, if the loss probability is small, there is negligible error in using equation (4.13).

The loss probability can be estimated by first assuming that the queue capacity is infinite and then calculating $P(x \geq Q + N)$.

Now from equation (4.14):

$$P(x \geq Q + N) = \frac{N^N}{N!}\left(\frac{A}{N}\right)^{Q+N} \frac{N}{N-A} P(0)$$

$$= \left(\frac{A}{N}\right)^Q P_D \tag{4.17}$$

Hence, the queue capacity, Q, needed to obtain an adequately low loss probability can be found.

4.7.4 Some other useful results

Equations (4.11) to (4.15) lead to further results, [1] as follows:

1. Mean number of calls in the system:

   (i) When there is delay, the mean number of calls is

   $$\bar{x}' = \frac{A}{N-A} + N \tag{4.18a}$$

   (ii) Averaged over all time, the mean number of calls is

   $$\bar{x} = \frac{A}{N-A} E_{2,N}(A) + A \tag{4.18b}$$

2. Mean length of queue:
   (i) When there is delay, the mean queue length is

   $$\overline{q'} = \overline{x'} - N = \frac{A}{N-A} \qquad \frac{p}{1-p} \tag{4.19a}$$

   (ii) Mean length of queue averaged over all time is

   $$\bar{q} = \overline{q'}\, P_D = \frac{A}{N-A} E_{2,N}(A) \tag{4.19b}$$

3. Mean delay time when the queue discipline is 'first in first out' (FIFO):
   (i) When there is delay, the mean delay, $\overline{T'}$, is

   $$\overline{T'} = h/(N-A) \tag{4.20a}$$

   where h is the mean holding time.
   (ii) Averaged over all time, the mean delay, $\overline{T}$, is

   $$\overline{T} = E_{2,N}(A)\,\overline{T'}$$
   $$= E_{2,N}(A)\, h/(N-A) \tag{4.20b}$$

   The variation of mean delay, $\overline{T}$, with traffic is shown in Figure 4.11.

4. Distribution of delays (FIFO queue discipline):
   Since the holding times have a negative exponential probability distribution, so do the delays, T_D. Hence:
   (i) When there is delay, $P(T_D \geq t) = e^{-t/\overline{T'}}$ \hfill (4.21a)

   (ii) Averaged over all time,

   $$P(T_D \geq t) = E_{2,N}(A) e^{-t/\overline{T'}} \tag{4.21b}$$

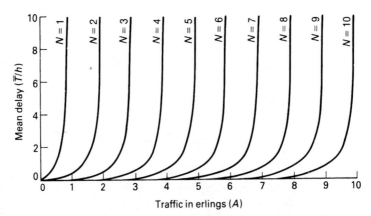

Figure 4.11 Mean delays for queuing systems with FIFO queue discipline (T = mean delay, h = mean service time, N = number of servers.)

Queueing systems [111]

From equation (4.21) one can therefore calculate the probability of exceeding any given delay.
The formulae for $\overline{x'}$, $\overline{q'}$ and $\overline{T'}$ are needed because, when $E_{2,N}(A)$ is small, delays which occur are often much greater than $\overline{T'}$.

Formulae have also been derived for more-complex queuing situations.[1,11,12] Examples are: constant holding time (instead of a negative-exponential distribution), random service (instead of a FIFO queue discipline) and queues with priorities.

4.7.5 System with a single server

When there is only a single server, the probability of it being busy is simply its occupancy, A, and this is the probability of delay, i.e. $E_{2,1}(A) = A$. As a result, the expressions in Sections 4.7.2 and 4.7.4 are simplified, as follows:

$$P_D = A \quad \overline{x'} = 1/(1-A) \quad \bar{x} = A/(1-A)$$
$$P(0) = 1 - A \quad \overline{q'} = A/(1-A) \quad \bar{q} = A^2/(1-A)$$
$$P(x) = A^x(1-A) \quad \overline{T'} = h/(1-A) \quad \overline{T} = Ah/(1-A)$$
$$P(x \geq z) = A^z$$

Example 4.11

A PBX has three operators on duty and receives 400 calls during the busy hour. Incoming calls enter a queue and are dealt with in order of arrival. The average time taken by an operator to handle a call is 18 seconds. Call arrivals are Poissonian and operator service times have a negative exponential distribution.

1. What percentage of calls have to wait for an operator to answer them?
2. What is the average delay, for all calls and for those which encounter delay?
3. What percentage of calls are delayed for more than 30 seconds?

1. $A = 400 \times 18/3600 = 2.0$ E
   From equation (4.13):

   $$\frac{1}{P_0} = \frac{3 \times 8}{3 \times 2 \times 1 \times 1} + 1 + \frac{2}{1} + \frac{4}{2} = 4 + 1 + 2 + 2 = 9$$

   $\therefore \quad P_0 = 1/9$

   From equation (4.15):

   $$P_D = \frac{8}{3 \times 2 \times 1} \times \frac{3}{1} \times \frac{1}{9} = \frac{4}{9}$$

   i.e. 44 % of calls have delay on answer.
2. When there is delay, the mean delay is:

   $$\overline{T'} = h/(N - A) = 18/(3 - 2) = 18 \text{ seconds}$$

The delay averaged over all calls is:
$$\bar{T} = \overline{T'} P_D = 18 \times 4/9 = 8 \text{ seconds}$$
3. When there is delay:
$$P(T_D \geq t) = e^{-t/\overline{T'}} = e^{-30/18} = 18.9\%$$
Averaged over all calls:
$$P(T_D \geq 30) = 18.9 \times 0.44 = 8.3\%.$$

Example 4.13

A message-switching centre sends messages on an outgoing circuit at the rate of 480 characters per second. The average number of characters per message is 24 and the message lengths have a negative exponential distribution. The input of messages is a Poisson process and they are served in order of arrival.

How many messages can be handled per second if the mean delay (averaged over all messages) is not to exceed 0.5 second?

For a single server the mean delay is

$$\bar{T} = Ah/(1 - A)$$
∴ $$A = \bar{T} / (h + \bar{T})$$
Now $$h = 24//480 = 0.05 \text{ second}$$
∴ $$A = 0.5 / (0.05 + 0.5) = 0.909 = Ch$$
where C = number of messages per second
∴ $$C = 0.909/0.05 = 18.2$$

4.7.6 Queues in tandem

When queuing systems are connected in tandem the delays are cumulative. If the first stage has a Poissonian input and a negative-exponential distribution of holding times, the inputs to the second and subsequent stages are also Poissonian. Thus, the queues can be considered as independent for calculating their delays.

The delay probability and the mean delay for the complete system are the sums of these for the individual stages. However, the probability distribution of the sum of several random variables is obtained by convolution of the separate distributions.[7] This computation is difficult, so it is usual to specify for each stage the probability of delay exceeding a given value and add these probabilities to obtain a measure of the overall grade of service. This will be a pessimistic estimate, because the probability of a long delay at more than one stage should be small.

4.7.7 Delay tables

Tables have been published[3,13] of $E_{2,N}(A)$. However, it is possible to calculate $E_{2,N}(A)$ from $E_{1,N}(A)$, as follows:

$$\sum_{k=0}^{N} \frac{A^k}{k!} = \frac{A^N}{N! E_{1,N}(A)} \qquad \text{(from equation (4.9))}$$

$$\therefore \sum_{k=0}^{N-1} \frac{A^k}{k!} = \frac{A^N}{N! E_{1,N}(A)} - \frac{A^N}{N!}$$

Substituting in equation (4.13):

$$\frac{1}{P(0)} = \frac{N A^N}{N!(N-A)} + \frac{A^N}{N! E_{1,N}(A)} - \frac{A^N}{N!}$$

$$= \frac{A^N}{N!} \frac{AE_{1,N}(A) + (N-A)}{(N-A)E_{1,N}(A)}$$

Substituting in equation (4.15):

$$E_{2,N}(A) = \frac{A^N}{N!} \frac{N}{N-A} \frac{N!}{A^N} \frac{(N-A)E_{1,N}(A)}{AE_{1,N}(A) + (N-A)}$$

$$= \frac{N E_{1,N}(A)}{N - A + AE_{1,N}(A)}$$

$$\doteq \frac{N}{N-A} E_{1,N}(A) \quad \text{if } E_{1,N}(A) \text{ is small}$$

Hence, values of $E_{2,N}(A)$ can be readily calculated from tables of $E_{1,N}(A)$.

Tables have also been published of other useful data for queuing systems, [13–15], such as the probabilities of delay exceeding given values.

4.7.8 Applications of delay formulae

A message switch, or a packet switch, is obviously a queuing system. If outgoing trunks are busy, messages or packets queue until an outgoing trunk becomes free. A system should be dimensioned to meet a specified delay probability or a specified mean delay. Since delays can be tolerated, the occupancy of trunks can approach unity. Thus, for given traffic, fewer trunks are required than for a circuit-switched system.

A telephone exchange is a circuit-switched system and its switching network is a lost-call system. However, common controls in an exchange form queuing systems. In an exchange with registers, when all registers are busy, incoming calls are not lost: callers merely wait for the dialling tone. In an exchange with markers, when two registers need to use the same marker, one waits until the other has been served. In a stored-program-controlled system, a central processor performs various different tasks. These wait in a queue until the processor has completed previous tasks. Thus, common controls are dimensioned to meet delay criteria.

4.8 Simulation

Rigorous analyses can only be made of simple systems, such as the full-availability systems considered in Sections 4.6 and 4.7 and extensions of them. In more difficult

[114] *Telecommunications traffic*

cases it is sometimes possible to obtain approximate solutions, as in Chapter 5. However, systems and networks are often so complex that the approximations required to arrive at an analytical solution are such as to make the results very inaccurate. It is then necessary to resort to computer simulation.[1,16–18]

A computer program is written to represent the behaviour of the system and the call arrivals and terminations are modelled by generating pseudo-random numbers. General-purpose programming languages have been used. However, special simulation languages have been developed to free the user from having to program low-level details of the simulation.[19] Software packages are now available commercially for simulation studies on various types of network.

Since the proportion of lost or delayed calls is normally required to be small, a very large number of calls must be simulated to determine this small proportion with sufficient accuracy. This necessitates very long computer runs and can be expensive!

Notes

1. The proportion of time for which congestion exists is termed the *time congestion* and the probability that a call will encounter congestion is termed the *call congestion*. Usually, these are equal, as is assumed here. However, there can be circumstances in which they are unequal. This is explained in Appendix 2.
2. In North America, Molina's formula[1] has been used, which is based on the 'lost-calls-held' assumption. It is assumed that a call which finds all trunks busy continues to demand service for a period equal to the holding time it would have had if successful. Thus, the grade of service (B) equals the probability of more than N calls in an infinite group of trunks, i.e.:

$$B = e^{-A} \sum_{x=N}^{\infty} \frac{A^x}{x!}$$

   This is no more than a convenient fiction, but it has been widely used in the USA.
3. It is also called the Erlang B distribution. (The author does not know of an Erlang A distribution.)
4. Conversely, the traffic carried by the first-choice traffic is less peaky than Poissonian traffic. This is an example of *smooth traffic*. The properties of smooth traffic and peaky traffic are considered in Appendix 2.
5. This is the notation introduced by Kendall.[10] It describes a queueing system as $X/Y/N$, where X is the input process, Y is the service time distribution and N is the number of servers. The following symbols are used:

   M: Markov process (i.e. random arrivals and terminations)
   D: Constant holding time
   G: General distribution

   Thus, the assumptions made above correspond to an $M/M/N$ system.
6. It is also called the Erlang C distribution.

Problems

1. During the busy hour a group of trunks is offered 100 calls having an average duration of 3 minutes. One of the calls fails to find a disengaged trunk. Find the traffic offered to the group and the traffic carried by the group.

2. A group of 20 trunks was found to have ten trunks engaged at 10 a.m., 15 at 10.10 a.m., 16 at 10.20 a.m. and 11 at 10.30 a.m. Calculate the average traffic intensity during this period.

3. Four junctions are arrranged in a full-availability group. If the traffic offered to the group in the busy hour is 0.8 E, what grade of service is given?

4. During the busy hour, on average, 30 E is offered to a group of trunks. On average, the total period during which all trunks are busy is 12 seconds and two calls are lost.
Find the average number of calls carried by the group and the average call duration. Show that the average number of calls offered to the group during a period equal to the average call duration is 30.

5. 10 E of traffic is offered to switches that hunt sequentially over a group of trunks. Estimate the traffic carried by each of the first three trunks.

6. Use Table 4.1 to find the number of trunks required to give a grade of service of 0.01 for the following loads of offered traffic:

1 E, 2 E, 4 E, 10 E, 40 E, 80 E

In each case, determine the occupancy and the number of trunks required per erlang.

7. A telecommunications company dimensions its routes by the following criteria:

(i) Grade of service for normal load: 0.005.
(ii) Grade of service for 10% overload: 0.02.

From Table 4.1, find the numbers of trunks required for:

10 E, 40 E, 50 E, 60 E, 79 E

In each case, state which is the determining criterion.

8. A crossbar exchange is required to set up 5000 calls in the busy hour. The marker holding times may be assumed to be exponentially distributed, with an average holding time of 0.5 second.

(a) Find the number of markers required, if the probability of a register having to wait for a marker is to be less than 0.2.
(b) What will happen if a local crisis causes 25 000 calls to originate in the busy hour?

9. In an automatic telegraph switching system incoming messages are stored in a queue until the retransmitting equipment of an outgoing trunk can send them. Messages arrive at the rate of 120 per hour. The times taken to retransmit messages may be assumed to have an exponential distribution with a mean time of 10 seconds.
If the probability of losing a message because the store is full must be less than 10^{-6}, how many messages must the equipment be able to store?

10. A common-control device in a telephone exchange is required to commence operation within an average period of 10 milliseconds after receiving a calling signal.

(a) If the device is held, on average for 50 milliseconds per call, how many calls can it handle per hour?
(b) If the device is required to handle 18 000 calls per hour, what is the maximum permissible average holding time?

References

[1] Bear, D. (1988), *Principles of Telecommunication Traffic Engineering*, 3rd edn, Peter Peregrinus, Stevenage.
[2] Syski, R. (1986), *Introduction to Congestion Theory in Telephone Systems*, 2nd edn, Oliver Boyd, Edinburgh.
[3] Beckmann, P. (1968), *Introduction to Queuing Theory and Telephone Traffic*, Golen Press, New York.
[4] Brockmeyer, B., Halstron, H. L. and Jensen, A. (1950), *The Life and Works of A. K. Erlang*, Copenhagen Telephone Company.
[5] Atkinson, J. (1950), *Telephony*, vol. 2, Pitman, London.
[6] Redmill, F. J. and Valdar, A. R. (1990), *SPC Digital Telephone Exchanges*, Peter Peregrinus, Stevenage.
[7] Beckmann, P. (1967), *Probability in Communication Engineering*, Harcourt, Brace and World, New York.
[8] Lee, A. M. (1966), *Applied Queuing Theory*, Macmillan, London.
[9] Cooper, R.B. (1972), *Introduction to Queueing Theory*', Macmillan, London.
[10] Kendall, D.G. (1951), 'Some problems in the theory of queues', *Jour. Royal Stat. Soc.*, **13**, Pt B, 151–73.
[11] Khintchine, A.Y. (1969), *Mathematical Methods in the Theory of Queueing*, 2nd edn, Griffin, New York.
[12] Kleinrock, L. (1975), *Queueing Systems*, Wiley, New York.
[13] Sarkowski, H. (1966), *Teletraffic Engineering Manual*, Standard Elektrik Lorenz, Stuttgart.
[14] Freeman, R.L. (1985), *Reference Manual for Telecommunications Engineering*, Wiley, New York.
[15] Hunter, J.M., Lawrie, N. and Peterson, M. (1988), *Telecommunications Tariffs, Traffic and Performance*, Com Ed Publishing, London.
[16] Shannon, R.E. (1975), *Systems Simulation*, Prentice Hall, Englewood Cliffs, NJ.
[17] Fishman, G.S. (1978), *Principles of Discrete Event Simulation*', Wiley, Chichester.
[18] Bulgren, W.C. (1982), *Discrete System Simulation*, Prentice Hall, Englewood Cliffs, NJ.
[19] Saver, C.H. and MacNair, E.A. (1983), *Simulation of Computer Communications Systems*, Prentice Hall, Englewood Cliffs, NJ.

CHAPTER 5

Switching networks

5.1 Introduction

A basic requirement for constructing switching systems, such as telephone exchanges, is to be able to design switching networks having a greater number of outlets than the switches from which they are built. This can be done by connecting a number of switching stages in tandem. For example, the Strowger exchange shown in Figure 3.8 gives access to up to 10 000 line terminations by using three ranks of 100-outlet selectors (two ranks of group selectors and a rank of final selectors). Figure 3.15 shows a two-stage network of ten-outlet crossbar switches giving access to 100 outgoing trunks. Figures 3.18 and 3.19 illustrate supervisory trunks connected to called customers' lines via four stages of crossbar switches. Figure 3.24 shows supervisory trunks connected to calling lines via three stages and to called lines via four stages of reed-relay switches.

In his monumental book on traffic theory[1] Syski wrote: 'At the present stage of development, the theoretical analysis of the telephone exchange as a whole has not yet been attempted.' In his book on switching networks[2] Benes said: 'The general theory of switching systems now consists of some apparently unrelated theorems, hundreds of models and formulas for simple parts of systems, and much practical lore associated with specific systems.' This chapter will attempt to develop such models and formulas for some of the basic networks used in switching systems. It will concentrate on space-division networks and the results will be extended to time-division networks in Chapter 6.

5.2 Single-stage networks

Figure 3.11 shows a single-stage network having M inlets and N outlets, consisting of a matrix of crosspoints. These may, for example, be separate relays or electronic devices or the contacts of a crossbar switch. The network could also be constructed by multiplying the banks of M uniselectors or one level of a group of M two-motion

Switching networks

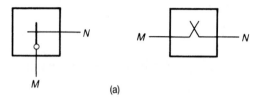

(a)

Figure 5.1 Switch symbols.

selectors, each having N outlets. Future systems may employ *photonic* switches,[3] in which opto-electronic devices are used as crosspoints to make connections between optical-fiber trunks. The network of Figure 3.11 may be represented, in simplified form, by the symbols shown in Figure 5.1. When Figure 5.1(a) is used to represent electromechanical switches, the circle indicates the side of the switch associated with the control mechanism (e.g. the wipers of a Strowger switch or the bridge magnet of a crossbar switch).

The switch shown in Figure 3.11 gives full availability; no calls are lost unless all outgoing trunks are congested. The number of simultaneous connections that can be made is either M (if $M < N$) or N (if $N < M$). The switch contains MN crosspoints. If $M = N$, the number of crosspoints is:

$$C_1 = N^2 \tag{5.1}$$

Thus, cost (as indicated by the number of crosspoints) increases as the square of the size of the switch. However, efficiency (as indicated by the proportion of the crosspoints which can be used at any time, i.e. $N/N^2 = 1/N$) decreases inversely with N. It is therefore uneconomic to use a single-stage network for large numbers of inlets and outlets. For example, a switch with 100 inlets and outlets requires 10 000 crosspoints and only 1% of these can be in use at any time. Switches for making connections between large numbers of trunks are therefore constructed as networks containing several stages of switches.

If the switch shown if Figure 3.11 is used to make connections between N similar circuits, then each circuit is connected to both an inlet and an outlet. Operation of the crosspoint at coordinates (j,k) to connect inlet j to outlet k thus performs the same function as operating crosspoint (k,j) to connect inlet k to outlet j. Consequently, half

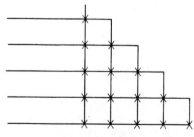

Figure 5.2 Triangular crosspoint matrix for connecting both-way trunks.

the crosspoints are redundant and can be eliminated. This results in the triangular crosspoint matrix shown in Figure 5.2. The number of crosspoints required is

$$C_1 = \tfrac{1}{2}N(N-1) \tag{5.2}$$

Triangular switches are not usually found in telephone switching systems because both-way trunks are not used. The trunks are operated on a one-way basis to facilitate supervision. For example, ringing tone and ringing current are sent over separate one-way trunks depending on whether a customer's line is calling or being called.

5.3 Gradings

5.3.1 Principle

For a route switch or a concentrator it is not necessary for each incoming trunk to have access to every outgoing trunk. It is adequate if each incoming trunk has access to a sufficient number of trunks on each route to give the required grade of service. This is known as *limited availability*. The number of outgoing trunks to which an incoming trunk can obtain connection is called the *availability* and corresponds to the outlet capacity of the switches used.

Figure 5.3(a) shows 20 trunks on an outgoing route to which incoming trunks have access by means of switches giving an availability of only ten (e.g. 20 circuits on an outgoing junction route from a selector level in a step-by-step exchange having 100-outlet two-motion group selectors). In Figure 5.3(a), the outlets of the switches are multipled together in two separate groups and ten outgoing trunks are allocated to

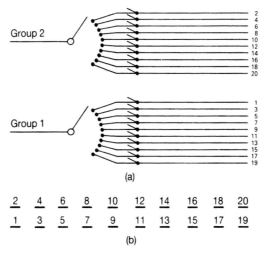

Figure 5.3 Twenty trunks connected in two separate groups to switches of availability 10. (a) Full diagram. (b) Grading diagram.

[120] *Switching networks*

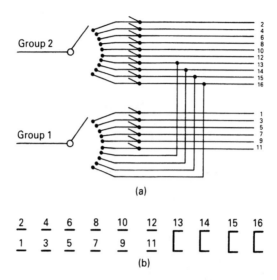

Figure 5.4 Sixteen trunks interconnected to two groups of switches of availability 10. (a) Full diagram. (b) Grading diagram.

each group. If the total traffic offered by the incoming trunks is, say, 8 E, each group of outgoing trunks is offered 4 E and will provide a grade of service (GOS) of better than 0.01. (Table 4.1 shows that a full-availability group of 10 trunks will cater for 4.5 E with a GOS of 0.01). The arrangement shown in Figure 5.3(a) is clearly less efficient than a single full availability group. (Table 4.1 shows that only 15 trunks are required to provide the same GOS for 8 E of traffic.)

If the traffic offered to the two groups of incoming trunks is random, peak loads will seldom occur simultaneously in the two groups. Efficiency can therefore be improved through mixing the traffic by interconnecting the multiples of the two groups so that some of the outgoing trunks are available to both groups of switches. If the switches search sequentially for free outlets, the later-choice outlets carry the least traffic (as shown in Figure 4.8). It is therefore desirable to connect the later-choice trunks to both groups of selectors, as shown in Figure 5.4(a). In this arrangement, the first six outlets are in two separate full-availability groups; the last four outlets are common to both groups and carry the traffic that overflows when the first six outlets of either group are busy. It is shown in Section 5.3.4 that this arrangement will still give a GOS of about 0.01, although it requires only 16 trunks instead of 20. The arrangement shown in Figure 5.4(a) requires only one more outgoing trunk than a full-availability group having a similar grade of service.

The technique described above of interconnecting the multiples of switches is called *grading*. An interconnection of trunks based on this principle is called grading. The conventional diagramatic representation of a grading is shown in Figure 5.4(b). A grading enables a single switching stage to provide access to a number of trunks greater than the availability (i.e. the outlet capacity) of the switches, but not exceeding it by an order of magnitude. A grading provides a poorer grade of service than a full-availability group with the same number of trunks. Some lost calls occur, even when there are free outgoing trunks, when these trunks are in part of the grading which is not accessible to the group of selectors containing the incoming trunk requiring connection.

Gradings of the form shown in Figure 5.4 were extensively studied by G. F. O'Dell [4] in the 1920s and are therefore called *O'Dell gradings*. They are the most widely used form of grading in a class of gradings known as *progressive gradings* because the switches hunt over the outlets sequentially from a fixed home position.

5.3.2 Design of progressive gradings

In order to form a grading, the switches having access to the outgoing route are multipled into a number of separate groups, known as *graded groups*. On early choices, each group has access to individual trunks and on late choices trunks are common, as shown in Figure 5.4. This diagram shows a small grading for only two groups of switches. For larger numbers of outgoing trunks, gradings may contain four or more groups. For example, Figure 5.5 shows four-group gradings. Since the traffic decreases with later choices of outlet, the number of groups connected together increases from individual connections on the early choices through partial commons (doubles) to full commons on the late choices.

In designing a grading to provide access to N outgoing trunks from switches having availability k, the first step is to decide on the number of graded groups g. If all the choices were individual trunks, we would have $N = gk$. If all the choices were full commons, $N = k$. Since the grading contains a mixture of individuals, partial commons and full commons, then $k < N < gk$. A reasonable choice for N is $N = \frac{1}{2}gk$ and traffic simulations have shown that the efficiency of such gradings is near the optimum. The number of groups is thus chosen to be:

$$g = \frac{2N}{k} \tag{5.3}$$

Since the grading should be symmetrical, g must be an even number, so the value of g given by equation (5.3) is rounded up to the next even integer.

It is now necessary to decide how the gk trunks entering the grading are to be interconnected to N outgoing trunks. For a two-group grading there is only one solution. If the number of columns of 'singles' is s and the number of commons is c, then:

$$\text{Availability} = k = s + c$$
$$\text{No. of trunks} = N = 2s + c$$
$$\therefore \quad s = N - k \text{ and } c = 2k - N$$

If the grading has more than two groups, there is no unique solution. It is necessary to choose from the possible solutions the best one, i.e. the grading with the greatest traffic

[122] *Switching networks*

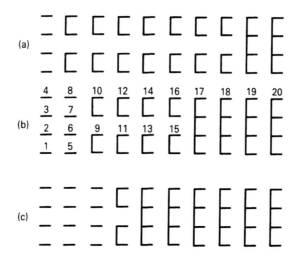

Figure 5.5 Four-group gradings for 20 trunks (availability 10).

capacity. The traffic offered to adjacent outlets will not differ greatly, so they should not be connected to very different sizes of common. There should thus be a smooth progression on the choices from individuals to partial commons, from smaller partial commons to larger ones, and from partial commons to full commons. The numbers of choices of each type in a group should therefore be as nearly equal as possible. This is achieved by minimizing the sum of the successive differences between the number of choices of one type and those of the type following it.

Let g have q factors: $f_1 < f_2 < \ldots < f_i \ldots < f_q$, where

$$f_1 = 1 \text{ and } f_q = g$$

Let r_i be the number of choices having their incoming trunks connected as f_i tuples.

$$\therefore \sum_{i=1}^{q} r_i = k \tag{5.4}$$

Now each f_i tuple contains g/f_i outgoing trunks.

$$\therefore \sum_{i=1}^{q} r_i g/f_i = N \tag{5.5}$$

Since there are only two equations and more than two unknowns (if $q > 2$), there are a number of different solutions for $(r_1, \ldots, r_q)$. These are found and, for each, the sum of the successive differences, D, is given by:

$$D = |r_{-1} - r_2| + |r_2 - r_3| + \ldots + |r_{q-1} - r_{1q}| \tag{5.6}$$

The best grading is that having the smallest value of D.

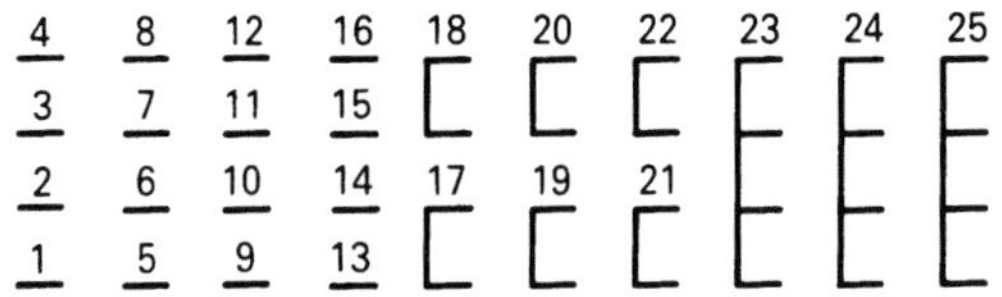

```
  4    8   12   16  18  20  22  23  24  25
  —    —   —    —   ⌐   ⌐   ⌐   ⌐   ⌐   ⌐
  3    7   11   15  |   |   |   |   |   |
  —    —   —    —   |   |   |   |   |   |
  2    6   10   14  17  19  21  |   |   |
  —    —   —    —   ⌐   ⌐   ⌐   |   |   |
  1    5    9   13  |   |   |   |   |   |
  —    —   —    —   ⌐   ⌐   ⌐   ⌐   ⌐   ⌐
```

Figure 5.6 Grading of Figure 5.5(b) modified to accommodate 25 trunks.

Example 5.1

Design a grading for connecting 20 trunks to switches having ten outlets.

The number of graded groups, given by equation (5.3) is $g = 40/10 = 4$, and the factors of g are 1, 2 and 4. Let

> the number of choices having singles $= s$
> the number of choices having doubles $= d$
> the number of choices having quadruples $= q$

Substituting in equations (5.4) and (5.5):

$$s + d + q = 10$$
$$4s + 2d + q = 20$$
$$\therefore \quad 3s + d = 10$$

If
$s = 1 : d = 7$ and $q = 10 - 8 = 2$
$s = 2 : d = 4$ and $q = 10 - 6 = 4$
$s = 3 : d = 1$ and $q = 10 - 4 = 6$
$s \geqslant 4 : d < 0$, so this is not possible.

There are thus three possible gradings, which are shown in Figure 5.5. The sums of the successive differences for these gradings are respectively given by:

$$D_1 = 6 + 5 = 11$$
$$D_2 = 2 + 0 = 2$$
$$D_3 = 2 + 5 = 7$$

The second grading (shown in Figure 5.5(b)) is therefore the best.

If a growth in traffic makes it necessary to increase the number of trunks connected to a grading, this can be done by reducing the number of commons and partial commons and increasing the number of individuals. Figure 5.6 shows the grading of Figure 5.5(b) rearranged to provide access to 25 trunks.

5.3.3 Other forms of grading

In an O'Dell grading, the partial commons are arranged as separate groups, so each is available to only some of the incoming trunks. For example, in Figure 5.5(b) the upper row of pairs serves only the first two groups. However, the principle of grading is based on the sharing of outgoing trunks between different sets of incoming trunks. Efficiency can be improved if this principle can be applied to the whole of a grading instead of only

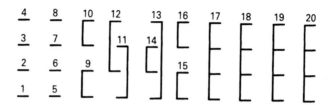

Figure 5.7 Skipped grading.

to parts of it. This can be done by connecting non-adjacent groups, in addition to adjacent groups, as shown in Figure 5.7. This is known as *skipping*.[5]

As an example of the improvement provided, consider the O'Dell grading of Figure 5.5(b) at a time when the upper two groups are carrying heavy traffic but the lower two groups are lightly loaded. Outgoing trunks 11 and 13 may be free but cannot be used. The skipped grading of Figure 5.7 enables these trunks to be used at such a time.

Progressive gradings are intended to be used with switches that hunt sequentially from a fixed home position. However, if switches do not hunt from a fixed home position or they select outlets at random, there is no advantage in connecting some outlets to singles and others to partial or full commons. The grading should then be designed to share each trunk between an equal number of groups, as shown in Figure 5.8. Such gradings are known as *homogeneous gradings*. Sequential hunting is more efficient than random selection because late-choice trunks are left free as long as possible to cater for traffic peaks. However, in practice, the difference between the traffic capacities of sequential and homogeneous gradings is often quite small.[6]

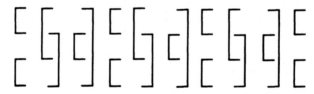

Figure 5.8 Homogeneous grading.

5.3.4 Traffic capacity of gradings

In an ideal grading, the interconnections would ensure that each outgoing trunk carried an identical traffic load.[1] Thus, if total traffic A is carried by N trunks, the occupancy of each trunk is A/N. It is assumed that each trunk being busy is an independent random event. Each call has access k to trunks (where k is the availability), and the probability of all k trunks being busy is thus:

$$B = (A/N)^k$$

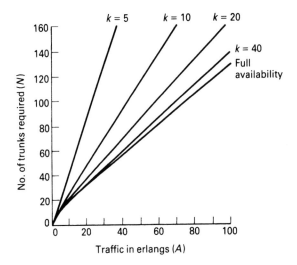

Figure 5.9 Traffic capacity of gradings (from modified Erlang formula, $B = 0.002$).

The number of trunks required to carry A erlangs with a GOS of B is therefore given by:

$$N = AB^{-1/k} \qquad (5.7)$$

This is *Erlang's ideal grading formula* and gives a linear relationship between the traffic and the number of trunks required.

Practical gradings do not satisfy the conditions for Erlang's ideal grading. However, it has been found[6] that they do have a linear relationship between traffic capacity A and number of trunks for a given grade of service B. An approximate curve of A against N can therefore be derived from Erlang's full-availability theory for $N \leq k$ and extended as a straight line for $N \geq k$. From equation (5.7) this line is given by:

$$A = A_k + (N - k)B^{1/k} \qquad (5.8)$$

where A_k is the traffic carried by a full-availability group of k trunks (with GOS $= B$).

Figure 5.9 shows a family of curves plotted from the above modified Erlang formula. This method is, of course, an approximation. More accurate methods have been developed. These include the Palm–Jacobeus formula[6] and the *equivalent random method* of Wilkinson, which is described in Appendix 2. Traffic tables for gradings have also been published.[7]

Example 5.2

Find the traffic capacity of the two-group grading shown in Figure 5.4 if the required grade of service is 0.01.

[126] *Switching networks*

$k = 10$ and, from Table 4.1, $A_k = 4.5$ E.
From equation (5.8):
$$\begin{aligned} A &= A_k + (N - k)\, B^{1/k} \\ &= 4.5 + (16 - 10) \times 0.01^{0.1} \\ &= 4.5 + 6 \times 0.631 \\ &= 8.3 \text{ E} \end{aligned}$$

(Note: Table 4.1 shows that a full-availability group of 16 trunks can handle 8.9 E with 0.01 grade of service.)

5.3.5 Applications of gradings

Gradings have been widely employed in step-by-step systems. In Figure 3.10, the trunk distribution frames (TDF) between the ranks of selectors provide cross-connections in the form of gradings.

The use of gradings is not confined to the Strowger system. In the Plessey 5005 crossbar system[8] (the TXK1 system of British Telecom), which is shown in Figure 3.18, there is a grading on each side of the router switch.

Another example of the use of a grading in a link system is in the Bell No.1 ESS system.[9] The subscribers' concentrator of this system is shown in Figure 5.10(a). The number of crosspoints required in the primary switches is reduced by omitting them in a systematic manner. Each primary switch is equivalent to four groups of four-outlet selectors having access to eight trunks through the homogeneous grading shown in Figure 5.10(b). Full availability between the primary-switch inlets and the links would have required eight-outlet selectors, thus doubling the number of crosspoints needed.

In the Ericsson AXE digital switching system[10] a subscribers' concentrator serving 2048 lines contains 16 modules, each with 128 lines. Each module has access to two PCM highways. One is individual to it and the other is common to all the modules of the concentrator. Traffic overflows to the latter when all PCM channels on the former are busy.

Automatic alternative routing (AAR) presents a situation similar to the use of grading. Traffic to a given destination is offered first to a group of direct circuits. If these are all busy, the traffic overflows to a common indirect route, where it is mixed with traffic to other destinations. In a grading, when all the 'singles' belonging to a group are busy, traffic overflows to common trunks where it mixes with traffic from other groups. Thus, the same theory is applicable to grading and AAR and similar dimensioning methods can be used. The equivalent random method is described in Appendix 2. It was developed by Wilkinson for AAR, but it is equally applicable to gradings.

5.4 Link systems

5.4.1 General

Examples of two-stage link systems are shown in Figures 3.2, 3.15 and 3.16 and a four-stage link system is shown in Figure 3.17. In general, a link system may have any

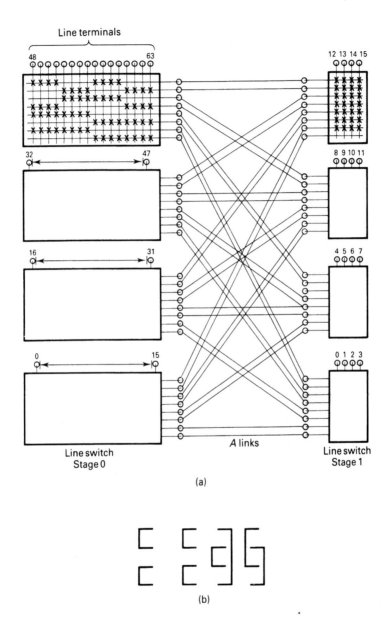

Figure 5.10 Two-stage concentrator used in Bell No.1 ESS system. (a) Arrangement of trunks. (b) Four-group homogeneous grading incorporated at A stage. (Copyright © 1964 AT&T. All rights reserved. Reprinted with permission.)

number of stages and the number of possible patterns of links between stages is very large. Only a few commonly used configurations will be considered here.

In the two-stage network of Figure 3.15 there is only one link between each primary switch and each secondary switch. Thus, it may be impossible to make a connection from a given incoming trunk to a selected outgoing trunk because the link is already being used for another connection between that primary switch and that secondary switch. This situation is called *blocking*. It is also known as a *mismatch*, because free links exist but none of them can be used for the required connection. If connection must be made to one particular outgoing trunk (e.g. an individual customer's line), the probability of blocking is unacceptably high. For this application, it is therefore necessary to use a network with more stages (e.g. the four-stage network of Figure 3.17), in order to have a choice of paths through the network.

The two-stage network of Figure 3.15 can be used as a route switch. If it serves ten outgoing routes with ten trunks on each route, then trunk no. 1 of each route is connected to secondary switch no. 1, trunk no. 2 is connected to switch no.2, and so on. Thus, an incoming trunk can obtain connection to the selected outgoing route via any of the links outgoing from its primary switch. The call is only lost if all the paths to free outgoing trunks are blocked. The probability of this occurring simultaneously for all links is obviously much smaller than the probability of a single link being busy. Similarly, if the incoming trunks are from several different routes, one trunk from each route is normally terminated on each primary switch.

Step-by-step selection is unsuitable. If the link is chosen before the outgoing trunk, then a free link could be seized that leads to a secondary switch whose trunk on the required outgoing route is already busy. Instead, *conditional selection* is used. The marker does not set up a connection until it has interrogated the busy/free conditions of all the relevant outgoing trunks and internal links. Only when it has found a match between a free outgoing trunk and a free internal link does it operate the switches. With this method of selection, if a free path through the network exists, it can be used. In a multistage step-by-step network, it is possible to encounter congestion at a late stage which would have been avoided if a different choice had been made at an earlier stage of switching, i.e. a free path exists but it has been omitted from the limited search made by the selectors.

A further advantage of conditional selection is that the marker has access at the same time to both ends of the connection through the network. Having set up the connection, it can test it for continuity. If the connection is found to be faulty, the marker can produce a fault record and make a second attempt to set up the connection by choosing a different path through the network.

If any free trunk may be used, as when the network is acting as a concentrator, then an incoming trunk can use any free link from its primary switch. If there are as many links as outgoing trunks, a connection can always be made if there is a free outgoing trunk. Except in this case, a link system will normally give a poorer grade of service than a single full-availability switching stage. This is because calls are lost by internal blocking in addition to congestion of the external trunks.

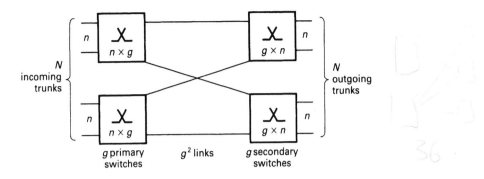

Figure 5.11 Two-stage switching network.

It has been seen that the grade of service of a link system depends on the way it is used. We may classify these uses as follows:

Mode 1: Connection is required to one particular free outgoing trunk. (Since conditional selection is used, an attempt will not be made to set up this connection unless the trunk *is* free.)

Mode 2: Connection is required to a particular outgoing route, but any free trunk on that route may be used.

Mode 3: Connection may be made to any free outgoing trunk.

It will be seen from Figure 3.20 that a concentrator operates in mode 3, a route switch operates in mode 2 and an expander operates in mode 1. The grades of service obtained are considered in Section 5.5.

5.4.2 Two-stage networks

If the two-stage network shown in Figure 5.11 has N incoming and N outgoing trunks and contains primary switches having n inlets and secondary switches having n outlets, then no. of primary switches (g) = no. of secondary switches = no. of outlets per primary switch = no. of inlets per secondary switch, where

$$g = N/n$$

The no. of crosspoints per primary switch = no. of crosspoints per secondary switch = $gn = N$. The total no. of crosspoints (C_2) in the network = (no. of switches) × (crosspoints per switch) i.e.

$$C_2 = 2g\, N = 2N^2/n \tag{5.9}$$

Since there is one link from each primary switch to each secondary switch, the number of links is equal to no. of primary switches × no. of secondary switches, i.e.

$$\text{No. of links} = g^2 = (N/n)^2 \tag{5.10}$$

[130] *Switching networks*

The number of crosspoints thus varies as $1/n$, but the numbers of link varies as $1/n^2$. If n is made very large to reduce the number of crosspoints, there will be too few links to carry the traffic. Let the number of links be equal to the number of incoming and outgoing trunks, a reasonable choice, since each set of trunks carries the same total traffic.

Then $\quad g^2 = N$

Substituting in equation (5.10) gives

$$n = \sqrt{N} \tag{5.11}$$

Then the total number of crosspoints (from equation (5.9)) is

$$C_2 = 2N^{3/2} \tag{5.12}$$

Equation (5.11) can be only a guide; one should select the nearest integer to n that is a factor of N. Also, in practice, designers are often constrained to use switch units of fixed sizes. For example, crossbar switches may be of sizes 10×10 or 10×20. The Bell No.1 ESS system[9] uses switches constructed from modules of size 8×8 and the British Telecom TXE2 system[11] uses modules of size 5×5.

The number of crosspoints per incoming trunk (from equation (5.12)) is $2\,N^{1/2}$. The cost per trunk therefore increases fairly slowly with the number of trunks. For large networks, however, it becomes more economic to use networks with more than two stages.

Example 5.3

Design a two-stage switching network for connecting 200 incoming trunks to 200 outgoing trunks.

Now, $\sqrt{200} = 14.14$. However, n must be a factor of 200, so the nearest practicable values are $n = 10$ and $n = 20$. Two possible networks are shown in Figure 5.12. Each contains 6000 crosspoints. The network of Figure 5.12(a) is suitable for 20 outgoing routes, each having 10 trunks, and that of Figure 5.12(b) is suitable for 10 outgoing routes, each having 20 trunks.

The network in Figure 5.11 has the same number of outgoing trunks as incoming trunks. However, a concentrator has more incoming than outgoing trunks and an expander has more outgoing than incoming trunks.

Consider a concentrator with M incoming trunks and N outgoing trunks ($M > N$). Let each primary switch have m inlets and each secondary switch have n outlets. Then

>No. of primary switches $= M/m$
>No. of secondary switches $= N/n$
>No. of crosspoints per primary switch $= mN/n$
>No. of crosspoints per secondary switch $= nM/m$

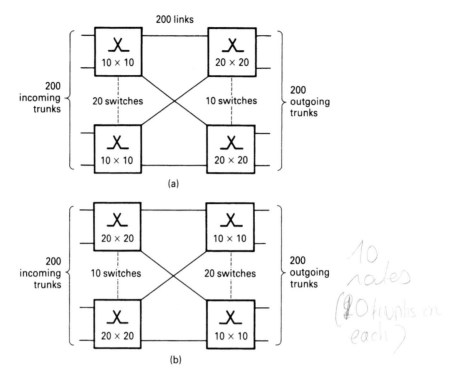

Figure 5.12 Examples of two-stage networks. (a) For 20 outgoing routes (10 trunks on each). (b) For 10 outgoing routes (20 trunks on each).

The total number of crosspoints is:

$$C_2 = \frac{M}{m}\frac{mN}{n} + \frac{N}{n}\frac{nM}{m}$$

$$= MN\left[\frac{1}{n} + \frac{1}{m}\right] \tag{5.13}$$

The number of links = no. of primary switches × no. of secondary switches

$$= \frac{MN}{mn}$$

Since the traffic capacity is limited by the number of outgoing trunks, there is little point in providing more than this number of links, so let the number of links be N.

$$\therefore \quad \frac{MN}{mn} = N$$

and

$$n = M/m \tag{5.14}$$

Substituting in equation (5.13) from equation (5.14):

$$C_2 = MN\left[\frac{m}{M} + \frac{1}{m}\right]$$

In order to minimize C_2, treat m as if it were a continuous variable and differentiate with respect to it:

$$\frac{dC_2}{dm} = MN\left[\frac{1}{M} - \frac{1}{m^2}\right]$$

$$= 0 \text{ when } m = \sqrt{M}$$

Hence, from equation (5.14):

$$m = n = \sqrt{M} \tag{5.15}$$

Thus, the number of crosspoints is a minimum when the number of inlets per primary switch equals the number of outlets per secondary switch.

Substituting in equation (5.13):

$$C_2 = MN\left[\frac{1}{\sqrt{M}} + \frac{1}{\sqrt{M}}\right]$$

$$= 2M^{\frac{1}{2}}N \tag{5.16}$$

Again, equation (5.15) is no more than a guide; m and n must be integers and factors of M and N, respectively. Moreover, the designer may also be constrained to use switch units of standard sizes. Since $M > N$, equation (5.15) gives larger, and thus fewer, secondary switches than if $n = \sqrt{N}$ were chosen. Consequently, a poorer grade of service is obtained when the network is operated in mode 2. Thus, practical networks sometimes use $n = \sqrt{N}$, $m = M/\sqrt{N}$. For example, this is the case for the crossbar concentrator shown in Figure 3.16. To obtain an expander, M is exchanged with N and m with n.

5.4.3 Three-stage-networks

Figure 5.13 shows a three-stage switching network. There is one link from each primary switch to each secondary switch and one link from each secondary switch to each tertiary switch. A connection from a given inlet on a primary switch to a selected outlet on a tertiary switch may thus be made via any secondary switch, unles its link to the primary switch or its link to the secondary switch is busy. The call can be set up unless this condition applies simultaneously to every secondary switch. The probability of being unable to set up a connection because of blocking is thus much less than for a two-stage network. This network is therefore suitable for operation in mode 1.

If the three-stage network has N incoming trunks and N outgoing trunks and has primary switches with n inlets and tertiary switches with n outlets, then:

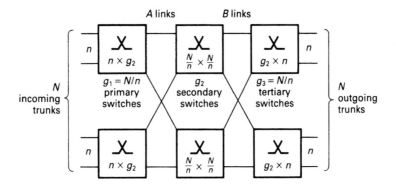

Figure 5.13 Fully interconnected three-stage switching network.

No. of primary switches (g_1) = no. of tertiary switches (g_3) = N/n.
∴ The secondary switches have N/n inlets and outlets.

If the number of primary–secondary links (A links) and secondary–tertiary links (B links) are each N, then the number of secondary switches is

$$g_2 = N \div (N/n) = n$$

= no. of outlets per primary switch = no. of inlets per tertiary switch.

No. of crosspoints in primary stage = $n^2(N/n) = nN$
No. of crosspoints in secondary stage = $n(N/n)^2 = N^2/n$
No. of crosspoints in tertiary stage = $n^2(N/n) = nN$

and the total number of crosspoints is

$$C_3 = N(2n + N/n) \tag{5.17}$$

By differentiating equation (5.17) with respect to n and equating to zero, it can be shown that the number of crosspoints is a minimum when

$$n = \sqrt{N/2} \tag{5.18}$$

and then

$$\begin{aligned} C_3 &= 2\sqrt{2}\, N^{3/2} \\ &= \sqrt{2}\, C_2 \\ &= 2^{3/2} N^{-1/2} C_1 \end{aligned} \tag{5.19}$$

Switching networks

If a three-stage concentrator has M incoming trunks and N outgoing trunks ($M > N$), its primary switches each have m inlets and its tertiary switches each have n outlets, then:

$$\text{No. of primary switches} = M/m$$
$$\text{No. of tertiary switches} = N/n$$

If there are g_2 secondary switches, then

$$\text{Crosspoints per primary switch} = m\,g_2$$
$$\text{Crosspoints per secondary switch} = \frac{M}{m}\frac{N}{n}$$
$$\text{Crosspoints per tertiary switch} = g_2\,n$$

The total number of crosspoints is:

$$C_3 = \frac{M}{m} \times m g_2 + g_2 \times \frac{M}{m}\frac{N}{n} + \frac{N}{n} \times g_2\,n$$
$$= g_2\left[M + N + \frac{MN}{mn}\right] \quad (5.20)$$

Since $M > N$, let no. of A links = no. of B links = N.

$$\therefore \quad N = g_2\frac{M}{m} = g_2\frac{N}{n}$$

Hence, $g_2 = n$ and $m = n\,M/N$.

Substituting in equation (5.20):

$$C_3 = (M + N)\,n + N^2/n$$

Differentiating with respect to n to find a minimum gives:

$$m = \frac{M}{\sqrt{M+N}}, \quad n = \frac{N}{\sqrt{M+N}} \quad (5.21)$$

$$C_3 = 2N\sqrt{N + M} \quad (5.22)$$

To obtain an expander, M is exchanged with N and m with n.

Example 5.4

Design a three-stage network for connecting 100 incoming trunks to 100 outgoing trunks:

$\sqrt{100/2} = 7.07$ ∴ use $n = 5$ or $n = 10$

1. If $n = 5$, there are:
    20 primary switches of size 5 × 5

Link systems [135]

        5 secondary switches of size 20 × 20
        20 tertiary switches of size 5 × 5.
2.   If $n = 10$, there are 10 primary switches, 10 secondary switches and 10 tertiary switches, each of size 10 × 10.

Both networks contain 3000 crosspoints. However, the second has more secondary switches. It therefore provides a greater number of paths between an incoming and an outgoing trunk and will exhibit less blocking.

Example 5.5

Design a three-stage network for 100 incoming trunks and 400 outgoing trunks.

$$100/\sqrt{100 + 400} = 4.47;\ 400/\sqrt{100 + 400} = 17.89$$

∴ $m = 4$ or 5; $n = 16$ or 20

1.   If $m = 5$, $n = 20$, there are:
        20 primary switches of size 5 × 5
        5 secondary switches of size 20 × 20
        20 tertiary switches of size 5 × 20.
2   If $m = 4$, $n = 16$, there are:
        25 primary switches of size 4 × 4
        4 secondary switches of size 25 × 25
        25 tertiary switches of size 4 × 16.

Both networks contain 4500 crosspoints. However, the first contains more secondary switches and will therefore cause less blocking.

In a three-stage network, the number of the selected outgoing trunk is given by the outlet numbers used on the secondary and tertiary switches. It is not related to the outlet used in the primary switch, since any secondary switch may be used for connection to a given outgoing trunk. For each connection, two sets of links must be interrogated for the busy/free condition and matched to choose a pair connected to the same secondary switch. The control of a three-stage network is thus more complex than that of a two-stage one. For this reason, electromechanical systems usually use trunkings containing a number of separate two-stage networks in tandem. However, systems having electronic central control often employ three-stage switching networks.

A fully interconnected three-stage network (as shown in Figure 5.13) requires a large number of crosspoints when N is large. A reduction can be made in the number of crosspoints (at the expense of an increase in blocking) if the secondary switches have links to only some of the primary and tertiary switches, as shown in Figure 5.14. The secondary and tertiary switches are arranged in separate groups (frames) and are fully interconnected only within their groups. Each primary switch has one link to each of these secondary–tertiary groups. (Alternatively, the primary and secondary switches may be arranged in separate groups, to produce the mirror image of Figure 5.14.)

The number of switches is $3n^2$ and each has n^2 crosspoints, so the total number of crosspoints is:

[136] *Switching networks*

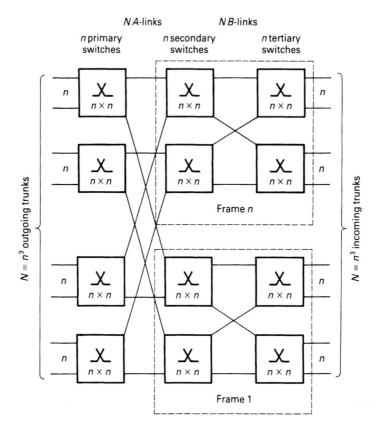

Figure 5.14 Partially interconnected three-stage network.

$$C_3 = 3n^4 = 3N^{4/3} \tag{5.23}$$

and the number of crosspoints per incoming trunk is $3N^{1/3}$.

Since there is only one link from each primary switch to each secondary–tertiary frame, there is only one path from each incoming trunk to each outgoing trunk. Thus, like the two-stage network of Figure 5.11, the network is unsuitable for use in mode 1 (i.e. making a connection from a given incoming trunk to a particular selected outgoing trunk). If the outlets of the tertiary switches serve a number of different routes, having several trunks connected to different secondary–tertiary frames, a sufficient choice of paths is available to give an acceptably low blocking probability in mode 2. A three-stage network of this type is used in some reed-electronic systems. Examples occur in the TXE2 system[11] and TXE4 system[12] of British Telecom.

In Figure 5.13, the third stage added to the two-stage network does not increase the number of outgoing trunks; it increases the mixture of paths available to reach them in order to reduce blocking. The additional stage may therefore be called a *mixing stage.* In Figure 5.14 the added third stage does not increase the number of paths to an outgoing trunk; it increases the number of outgoing trunks over which the incoming traffic can be distributed. It is therefore called a *distribution stage*. In Figure 5.13 a primary switch has a link to every secondary switch, so any secondary switch can be used for a connection to a given outgoing trunk. In Figure 5.14, however, there are many more primary switches and each has a link to only one secondary switch of each two-stage frame.

5.4.4 Four-stage networks

A four-stage network can be constructed by considering a complete two-stage network as a single switch and then forming a larger two-stage array from such switches. Figure 3.17 shows a four-stage network for 1000 incoming and 1000 outgoing trunks constructed from two-stage networks (frames) of 100 inlets and 100 outlets using 10×10 switches. It is necessary that one trunk (B link) be connected from each secondary switch of an incoming frame to a primary switch of an outgoing frame. These trunks are connected to switches of corresponding numbers on the two frames, thus facilitating marking of the network. Four-stage networks of this type are used in crossbar systems.

If a four-stage network with N incoming and N outgoing trunks is constructed with switches of size $n \times n$, then $N = n^3$ and the total number of switches is $4n^2$. Thus, the total number of crosspoints is:

$$C_4 = 4n^2 \cdot n^2$$
$$= 4 N^{4/3} \tag{5.24}$$

The number of crosspoints per incoming trunk is $4 N^{1/3}$.

It should be noted that the partially interconnected three-stage network of Figure 5.14 corresponds to the four-stage network of Figure 3.17 truncated at the A links. Adding the fourth stage has not increased the number of trunks, although it has increased the number of crosspoints by one third. However, it has increased the number of paths between each incoming trunk and outgoing trunk from one to ten: i.e. a mixing stage has been added.

5.4.5 Discussion

If a network has N incoming trunks and N outgoing trunks, then the number of crosspoints per incoming trunk for a single stage is proportional to N, for a two-stage network it is proportional to $N^{1/2}$ and for a three-stage partially interconnected network it is proportional to $N^{1/3}$. Thus, for networks having many trunks, it is economic to use more stages than for networks with fewer trunks.

These networks have only distribution stages; there is only one path between an incoming trunk and an outgoing trunk. If more paths are needed to reduce blocking, mixing stages should be added, for example by changing a two-stage network to a fully interconnected three-stage network or changing a partially interconnected three-stage network to a four-stage network. Consequently, the design of a large switching network[13] involves providing distribution stages to obtain crosspoint economy and mixing stages to reduce blocking.

It is rarely possible to use switches of exactly the optimum sizes given by equations (5.11), (5.15), (5.18) and (5.21). Also, the cost of switches is not exactly proportional to the number of crosspoints. For example, doubling the inlet and outlet capacity of a crossbar switch requires four times as many crosspoints but only twice the number of magnets. In addition, the complexity and cost of the associated control equipment increases with the number of stages. Consequently, many switching networks use fewer stages and larger switches than would be indicated by the above equations.

5.5 Grades of service of link systems

5.5.1 General

A simple theory for calculating the probability of loss in link systems, due to C. Y. Lee,[14] will be given here. The method assumes that trunks and links being busy constitute independent random events. If two random events are independent, the probability of both happening at the same time is given by the product of their separate probabilities of occurrence at that time. If two links are to be connected in tandem, and the probability of one being busy is a and of the other being busy is b, then the probabilities of each being free are $1-a$ and $1-b$, respectively, so the probability of both being free is $(1-a)(1-b)$. Therefore, the probability of the path being blocked is $1 - (1-a)(1-b)$.

The occupancy at each stage is the total traffic carried divided by the number of links at that stage. However, if the loss is small (as it should be), little error is introduced by using the traffic offered instead of the traffic carried.

In a practical system the assumption of independence may not be valid, because there is usually some degree of dependence between links. This reduces the probability of blocking, because traffic peaks at different stages coincide more often than would happen if they were independent random events. This overlapping of peaks tends to reduce the total time during which blocking occurs. Consequently, Lee's method overestimates the loss probability. Nevertheless, the method gives reasonably accurate results in most cases. It also has the merit of simplicity. For these reasons, it is widely used. A more accurate, but still not exact, method was published by Jacobaeus.[6,15] This is described in Appendix 3.

An analytical treatment becomes complex for a network having a large number of stages and handling different kinds of traffic. Under these circumstances, it is necessary to resort to a computer simulation in order to determine the GOS with sufficient accuracy. An approximate theoretical calculation may be adequate to enable the designer to choose between alternative trunking schemes, but it will not be sufficiently accurate for determining the amounts of equipment to be provided in exchanges. A

Grades of service of link systems [139]

small percentage saving in cost on a large network will more than offset the cost of computation!

5.5.2 Two-stage networks

For a two-stage network, as shown in Figure 5.11, let the occupancy of the links be a and the occupancy of the outgoing trunks be b. (If the numbers of links and trunks are equal, then $a = b$.)

For mode 1 (i.e. connection to a particular outgoing trunk) only one link can be used. The probability of this being busy is a and this is the probability of loss. For example, to provide a GOS of $B_1 = 0.01$, each link and outgoing trunk could only carry 0.01 E. This is useless!

For mode 2 (i.e. connection to an outgoing route with one trunk on each secondary switch) any free link can be used. The probability of loss using a particular link is

1 − probability that both link and trunk are free
= 1 − (1 − a)(1 − b)

But there are g paths available. Assuming that each being blocked is an independent random event, the probability of simultaneous blocking for all g paths is:

$$B_2 = [1 - (1-a)(1-b)]^g$$
$$= [a + (1-a)b]^g \qquad (5.25)$$

where g is the number of secondary switches.

If connection may be made to any outgoing trunk that is free (i.e. mode 3) then it is possible to make the connection unless all the outgoing trunks are busy. Thus, if the numbers of incoming trunks, links and outgoing trunks are equal, no calls can be lost. However, this mode of operation is normally used with a concentrator. The number of incoming trunks is then much larger than the number of outgoing trunks, so the grade of service is given by:

$$B_3 = E_{1.N}(A)$$

where A is the total traffic offered to the network.

Example 5.6

1. Find the grade of service when a total of 30 E is offered to the two-stage switching network of Figure 3.15 and the traffic is evenly distributed over the 10 outgoing routes.
   The link and trunk occupancies are $a = b = 30/100 = 0.3$ E.

   $\frac{10}{15} = 0.667$

   B = [1 − (1 − 0.3)(1 − 0.3)]¹⁰ = 0.51¹⁰
   = 0.0012

   $B = 0.172$

[140] *Switching networks*

2. Find the traffic capacity of this network if the grade of service is not to exceed 0.01.

$$B \le 0.01 = [1 - (1-a)^2]^{10}$$
$$1 - (1-a)^2 \le 0.01^{0.1} = 0.631$$
$$\therefore \quad a \le 0.39 \quad \text{and} \quad A \le 39 \text{ E}$$

5.5.3 Three-stage networks

For a fully interconnected three-stage network (as shown in Figure 5.13) let:

Occupancy of A links be a
Occupancy of B links be b
Occupancy of outgoing trunks be c.

For mode 1 (i.e. connection to a particular outgoing trunk), the choice of a secondary switch determines the A and B links.

Probability that both links are free $= (1-a)(1-b)$
$\therefore$ Probability of blocking $= 1 - (1-a)(1-b)$

However, there are g_2 secondary switches.

$\therefore$ Probability that all g_2 independent paths are simultaneously blocked is
$$B_1 = [1 - (1-a)(1-b)]^{g_2} \qquad (5.26)$$
$$= [a + (1-a)b]^{g_2}$$

Thus, for similar occupancies, the three-stage network provides the same GOS for connections to individual trunks as the two-stage network does for connections to a group of trunks (cf. equation (5.25)).

For mode 2 (i.e. a connection to any free trunk in a route having one trunk connected to each tertiary switch):

Probability of blocking for a particular trunk

$$= 1 - (1 - B_1)(1 - c)$$
$$= B_1 + (1 - B_1)c$$

$\therefore$ Probability of simultaneous blocking for all g_3 independent paths is

$$B_2 = [B_1 + c(1 - B_1)]^{g_3} \qquad (5.27)$$

where g_3 is the number of tertiary switches.

Example 5.7

1. Compare the grades of service provided by the two networks of Example 5.4 when each operates in mode 1 and is offered 30 E of traffic:

$$a = b = 30/100 = 0.3 \text{ E}$$

For network (a):
$$B = [1 - (1-0.3)(1-0.3)]^5 = 0.51^5 = 0.035$$

For network (b): $B = 0.51^{10} = 0.0012$

2. What is the traffic capacity of each network if the required grade of service is 0.01?

For network (a):
$$[1 - (1 - a)^2]^5 = 0.01$$
$$1 - (1 - a)^2 = 0.01^{1/5} = 0.398$$
$$\therefore a = 0.224$$
Total traffic capacity $= 100 \times 0.224 = 22.4$ E

For network (b):
$$[1 - (1 - a)^2]^{10} = 0.01$$
$$1 - (1 - a)^2 = 0.01^{1/10} = 0.631$$
$$\therefore a = 0.393$$
Total traffic capacity $= 100 \times 0.393 = 39.3$ E

For a partially interconnected three-stage network, as shown in Figure 5.14, there is only one path between an incoming trunk and an outgoing trunk. The probability that this is free is $(1 - a)(1 - b)$ and the probability of blocking is $1 - (1 - a)(1 - b)$.

For a connection to a trunk on an outgoing route with n trunks, each connected to a different frame, the probability of loss using a particular trunk is

$$1 - (1 - a)(1 - b)(1 - c)$$

But there are n such trunks available. Assuming that each being busy is an independent random event, the probability of simultaneous blocking for all paths is

$$B_2 = [1 - (1 - a)(1 - b)(1 - c)]^n \qquad (5.28)$$

5.5.4 Four-stage networks

For a four-stage network, as shown in Figure 3.17 let:

Occupancy of A links be a
Occupancy of B links be b
Occupancy of C links be c
Occupancy of outgoing trunks be d

For a connection from a given inlet on an input frame to a particular outlet on an output frame (i.e. mode 1), the call may use any primary switch in the output frame. This switch is connected by a B link to only one secondary switch in the particular input frame. From this switch there is only one A link to the primary switch of the given incoming trunk.

Probability of this path being free is

$$(1 - a)(1 - b)(1 - c)$$

$\therefore$ Probability of this path being blocked is

$$1 - (1 - a)(1 - b)(1 - c)$$

Probability that all g_2 independent paths are simultaneously blocked is

[142] *Switching networks*

$$B_1 = [1 - (1 - a)(1 - b)(1 - c)]^{g_2} \tag{5.29}$$

where g_2 is number of secondary switches in input frame = number of primary switches in output frame.

For a route of n outgoing trunks:

Probability of loss for a particular trunk

$$= 1 - (1 - B_1)(1 - d)$$
$$= B_1 + (1 - B_1)d$$

∴ Probability of simultaneous blocking for all n independent paths is

$$B_2 = [B_1 + d(1 - B_1)]^n \tag{5.30}$$

5.6 Application of graph theory to link systems

Switching networks, like other forms of network may be studied by means of the branch of mathematics known as graph theory.[16] Lee[14] used graphs to represent switching networks in 1955 and the method has subsequently been developed by Takaki[17] and others.[13]

A graph is a collection of points, known as *nodes* or *vertices*, connected by lines, known as *edges* or *arcs*. For example, the simple two-stage network shown in the trunking diagram of Figure 5.15(a) (whose links are connected as shown in Figure 3.15) may be represented by the graph shown in Figure 5.15(b). The latter shows the links without details of the switches, whereas Figure 5.15(a) focuses attention on the switches without showing the arrangement of the links.

Most switching networks are symmetrical, so the arrangement of links connecting any particular inlet to any particular outlet is topologically the same as for any other inlet–outlet pair. The representation of a switching network by means of a graph can therefore be simplified by drawing only the paths which can be used for making connections between one particular inlet–outlet pair. This graph is called the *channel graph* of the network. Some examples are shown in Figure 5.16. Expressions for blocking probability (e.g. equations (5.26) and (5.29)) can be obtained directly by inspecting the channel graphs.

An important property of the network which is displayed by the channel graph is its *connectivity*. This may be defined as the minimum number of disjoint paths joining the non-adjacent vertices. Thus, in Figure 5.16(a) and (c) the connectivity is unity, whereas in Figure 5.16(b) and (d) its value is 10. Clearly, the larger the value of connectivity, the lower is the probability of blocking. This is shown by equations (5.26) and (5.29) which are of the form

$$B = (1 - x)^k$$

where $0 < x < 1$ and k is the connectivity of the channel graph.

Takagi[17] has developed a method of using channel graphs to design switching

Application of graph theory to link systems [143]

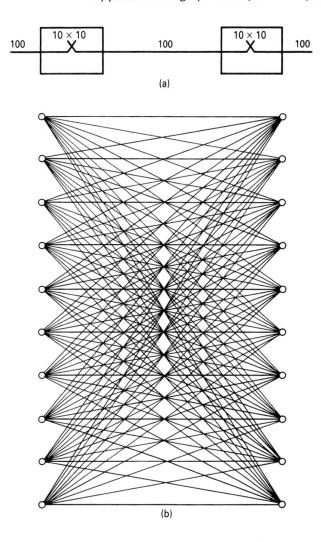

Figure 5.15 Two-stage switching network. (a) Conventional representation. (b) Network graph.

networks. He provides a sufficient number of distribution stages to give access to the required number of outgoing trunks and then adds mixing stages to give sufficient connectivity for the required grade of service. The well-known networks shown in Figures 5.13 and 3.17 can be obtained by Takagi's method. The channel graphs in Figure 5.16(b) and (d) show how adequate connectivity is provided by adding a third stage to the two-stage network and a fourth stage to the partially interconnected three-stage network respectively. Takagi has applied his method to networks containing as many as eight stages.[17]

[144] *Switching networks*

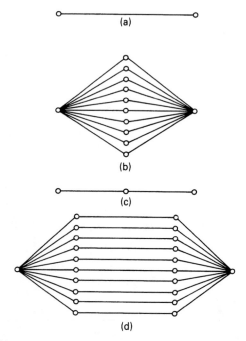

Figure 5.16 Channel graphs of switching networks. (a) Two-stage network of Figure 5.16. (b) Fully interconnected three-stage network of Figure 5.13. (c) Partially interconnected three-stage network of Figure 5.14. (d) Four-stage network of Figure 3.17.

5.7 Use of expansion

It has been shown in Section 5.6 that the blocking loss of a network depends strongly on its connectivity. The loss probability can therefore be reduced by increasing the connectivity by providing more links between switching stages than there are outgoing trunks. This is known as *expansion*.

Expansion is not used in local exchanges. Indeed, because customers' line are lightly loaded, concentration is used, as shown in Figure 5.17(a). Tandem exchanges, as shown in Figure 5.17(b), do not use concentration because junctions are usually heavily loaded. Neither is expansion used, because an adequately low grade of service is obtained without it. Long-distance circuits, particularly international ones, are expensive; however, trunks within an exchange (and their associated switches) are relatively cheap. It is undesirable for expensive circuits to be idle (and losing revenue) because of blocking within an exchange. Expansion is therefore used in trunk-transit and international exchanges, as shown in Figure 5.17(c), to ensure that loss due to blocking is much less than loss due to congestion of outgoing routes.

By using a moderate amount of expansion, the blocking loss of a switching

Use of expansion [145]

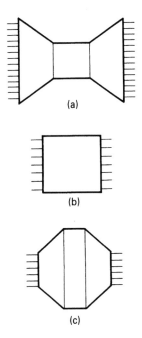

Figure 5.17 Use of concentration and expansion. (a) Local exchange. (b) Tandem exchange. (c) Trunk transit or international exchange.

network can be made negligible compared with the loss due to congestion of the external trunks. Such a network is called a *quasi-nonblocking network*. It will be shown in Section 5.10 that it is actually possible by means of expansion to reduce the blocking to zero.

Example 5.8

A fully interconnected three-stage network has 100 incoming trunks, 100 A links, 100 B links and 100 outgoing trunks. At each stage, it uses ten switches of size 10×10. As shown in Example 5.7, the grade of service in mode 1 is 0.01 when the link occupancy is 0.39 E.

For the same total offered traffic, what grade of service is obtained if the numbers of links and secondary switches are increased by (1) 20%, (2) 50%?

1. The link occupancy of the modified network is

    $b = c = 0.39/1.2 = 0.325$ E

    From equation (5.26), the grade of service is:

    $B = [1 - (1 - 0.325)^2]^{12} = 0.544^{12} = 6.7 \times 10^{-4}$

2. The link occupancy of the modified network is

$$b = c = 0.39/1.5 = 0.26 \text{ E}$$

From equation (5.26), the grade of service is:

$$B = [1 - (1 - 0.26)^2]^{15} = 0.452^{15} = 6.7 \times 10^{-6}$$

5.8 Call packing

In practice, less blocking is obtained when links are not allocated at random than when they are. The chance of a call being blocked is minimized if the number of possible paths available to it can be maximized. Thus, if each call to be set up is routed through the most heavily loaded part of the network which can still take it, subsequent calls will have a greater choice of paths than if the call were routed through a less heavily loaded part of the network. This is known as *call packing*. Simple as the principle is, there is no general proof that it gives better results than any other alternative. However, this can be demonstrated in a few elementary examples[2] and has been verified by computer simulation for some more complex networks.[18] Increases in traffic capacity of up to 10% have been obtained.[6]

A simple call-packing rule that can be applied to the three-stage network of Figure 5.13 is to select the lowest-numbered secondary switch which has free links to the required primary and tertiary switches. Thus, secondary switch 2 is only used if the *A* or *B* link to switch 1 is busy, and so on.

When call packing is used, calls are offered to switches in a predetermined order. A faulty first-choice switch can therefore cause serious degradation of service during periods of low traffic. (This is a well-known defect of the Strowger system.) To obviate this, common-control systems using crosspoint switches usually provide a second-attempt feature. When an attempt to set up a connection fails, the marker makes a second attempt. In order to avoid selecting the same faulty switch, it is necessary to search for suitable free links from a different starting point from that used in the first attempt.

5.9 Rearrangeable networks

Call packing reduces blocking by selecting the most heavily loaded part of a network for each new connection set up. A further reduction in blocking could be obtained if it were possible to ensure that connections which exist through lightly loaded parts of the network were cleared down before those through heavily loaded parts. Clearly, this is not possible, because call terminations are caused by the actions of customers. However, the equivalent result would be obtained if, every time a call ended, all the remaining connections were cleared down and set up again using a call-packing rule. If some of these connections were already made through a heavily loaded part of the network it would not be necessary to move them; it would probably be sufficient to alter only a few connections. Moreover, this need not be done each time a connection clears down; it can be done as part of the selection process when the next connection is set up. A network that is operated in this mode is called a *rearrangeable network*.

Benes has shown[2,19] that by using rearrangement it is possible to obtain networks which completely eliminate blocking. Such networks are said to be *nonblocking in the wide sense*. This means that zero blocking is not guaranteed by the structure of the network; it is obtained by means of the control algorithm used to set up connections.

If a multistage network with N incoming and N outgoing trunks is to avoid blocking, there must clearly be at least N links at each intermediate stage. Also, if a network having input switches with n inlets and output switches with n outlets is to be made nonblocking, its intermediate links must permit n simultaneous connections between each input switch and each output switch. The fully interconnected three-stage network of Figure 5.13 and the four-stage network of Figure 3.17 satisfy these conditions; the two-stage network of Figure 5.11 and the partially interconnected three-stage network of Figure 5.14 do not. The condition for a rearrangeable fully interconnected three-stage network to be nonblocking is simply $g_2 \geq n$. It can be shown[2] that the maximum number of existing connections to be moved to enable a new connection to be made is $n - 1$.

Rearrangeable networks are not used in space-division telephone exchanges. The clicks caused by the interruptions of current when connections were rearranged would be objectionable to users. Rearrangement could be used in time-division exchanges, because each connection is made and disconnected eight times every millisecond. A reduction in the number of crosspoints could thus be obtained. However, there is no incentive to obtain this reduction because crosspoints are used very economically by being time shared between a large number of connections, as described in Chapter 6. Moreover, time-division switching networks normally have a high connectivity, which provides very low blocking.

5.10 Strict-sense nonblocking networks

It was shown in Section 5.9 that rearrangeable networks can have zero blocking loss. These networks are nonblocking in the wide sense, because zero blocking is obtained by using a prescribed control algorithm. If a network can never have blocking, no matter what existing connections are present and without any need to rearrange connections, the network is said to be *nonblocking in the strict sense*.

The single-stage network of Figure 5.1 is obviously strictly nonblocking. A connection can always be made to a free outgoing trunk.

In order to make the two-stage network of Figure 5.11 strictly nonblocking, it is necessary to make the number of links from a primary switch to a secondary switch equal to the number of outlets of the secondary switch. If the network has N incoming and outgoing trunks, the primary switches have n inlets and the secondary switches

Switching networks

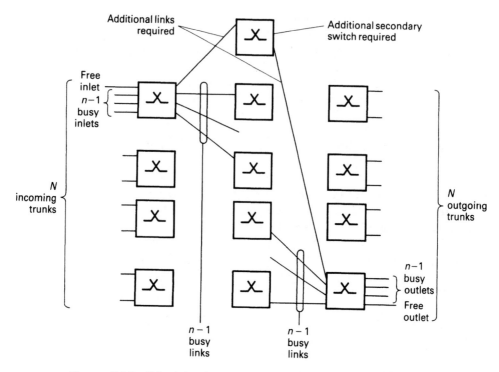

Figure 5.18 Principle of three-stage strictly nonblocking network.

have n outlets (where $n = \sqrt{N}$), then the primary switches need N outlets and the secondary switches need N inlets. Thus, each needs nN crosspoints. The total number of crosspoints is: $C_2 = 2n \times nN = 2N^2$. This is twice as many as needed by a single stage!

If an attempt is made to modify the four-stage network of Figure 3.17 to make it strictly nonblocking, a similar change must be made to each of its switches. Thus, the total number of crosspoints is again greater than that required for a single stage.

Networks having odd numbers of stages can be made strictly nonblocking while requiring fewer crosspoints than a single stage. A method of designing such networks was discovered by Clos[20]. These networks are therefore often called *Clos networks*.

Consider a three-stage network, as shown in Figure 5.18, in which each primary switch has n inlets and each tertiary switch has n outlets. The greatest number of calls that a primary switch can be carrying and still be capable of accepting another is $(n-1)$. These calls will occupy $(n-1)$ links to different secondary switches. Similarly, the selected tertiary switch can be carrying $(n-1)$ calls and occupying $(n-1)$ links from secondary switches. The worst case for blocking will occur when the busy links from the primary switch and the busy links to the tertiary switch terminate on different secondary switches. In order to be able to make the new connection, there must still be one secondary switch to which there are free links. The minimum number of secondary switches, g_2, required is therefore

$$g_2 = (n - 1) + (n - 1) + 1$$
$$= 2n - 1 \qquad (5.31)$$

Then:

No. of primary switches = N/n
No. of secondary switches = $2n - 1$
No. of tertiary switches = N/n
No. of crosspoints per primary switch = $n(2n - 1)$
No. of crosspoints per secondary switch = $(N/n)^2$
No. of crosspoints per tertiary switch = $(2n - 1)n$
and the total no. of crosspoints is:

$$C_3 = (2n - 1)[2N + (N/n)^2] \qquad (5.32)$$

Differentiating equation (5.32) w.r.t. n and equating to zero gives:

$$2n^3 - Nn + N = 0 \qquad (5.33)$$

This has integer roots: $n = 2$, $N = 16$ and $n = 3$, $N = 27$. However, if $n \gg 1$, equation (5.33) approximates to $2n^2 - N = 0$ and $n = \sqrt{N/2}$. Substituting in equation (5.32) gives:

$$C_3 = 2^{5/2} N^{3/2} - 4N \qquad (5.34)$$

(Clos[20] used the nonoptimum value $n = \sqrt{N}$ and obtained $C_3 = 6N^{3/2} - 3N$.)

It follows that a three-stage nonblocking network has fewer crosspoints than a single stage if $N \geq 28$. However, comparing equation (5.34) with equation (5.19) shows that a three-stage nonblocking network contains nearly twice as many crosspoints as a conventional fully interconnected three-stage network.

For networks having large values on N, fewer crosspoints are required if more than three stages are used. A five-stage network[20] can be considered as a three-stage network in which each secondary switch is replaced by a 'level' which itself consists of a three-stage nonblocking network, as shown in Figure 5.19. The number of such levels required is, of course, $2n - 1$. It can be shown that the number of crosspoints (C_5) in this network is a minimum when $n = (2N)^{1/3}$. Then:

$$C_5 = 3 \times 2^{7/3} N^{4/3} - 14N + 2^{5/3} N^{2/3} \qquad (5.35)$$

Similarly, a seven-stage network[20] can be designed as a three-stage network in which each secondary stage is replaced by a level containing five stages. The method can be extended in this way to design nonblocking networks having any odd number of stages.

Nonblocking networks are not used in commercial space-division switching systems because they are uneconomic. However, they have been used in certain military applications where economic considerations are secondary to operational

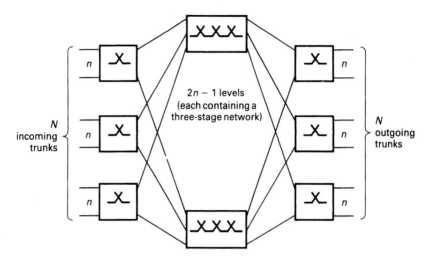

Figure 5.19 Five-stage strictly nonblocking network.

requirements. It can be economic to use nonblocking networks in time-division switching systems, since crosspoints are time-shared between a large number of connections.

Example 5.9

Design a strictly nonblocking network for 100 incoming and 100 outgoing trunks.
  Choose a three-stage network. The minimum number of crosspoints is obtained when

$$n = \sqrt{N/2} = \sqrt{50} = 7.07 \therefore \text{ use } n = 5$$

Thus, the number of secondary switches is $2n-1 = 9$ and the network is as shown in Figure 5.20(a).
  The total number of crosspoints is:

$$C_3 = 20 \times 5 \times 9 + 9 \times 20 \times 20 + 20 \times 9 \times 5 = 5400$$

Example 5.10

Design a strictly nonblocking network for 1000 incoming and 1000 outgoing trunks.
  Choose a five-stage network. The minimum number of crosspoints is obtained when

$$n = (2N)^{1/3} = 2000^{1/3} = 12.599 \therefore \text{ use } n = 10$$

The number of levels needed is $2n-1 = 19$.

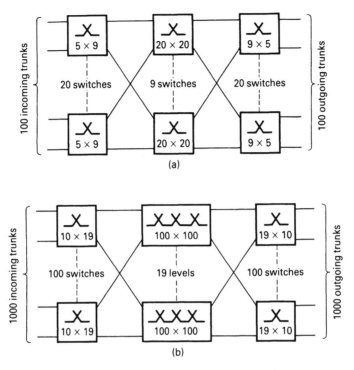

Figure 5.20 Examples of strictly nonblocking networks. (a) Three-stage network with 100 incoming and outgoing trunks. (b) Five-stage network with 1000 incoming and outgoing trunks. (Each level consists of a three-stage network as shown in Figure 5.20(a).)

No. of switches in 1st stage = no. of switches in 5th stage = 1000/10 = 100.

Thus, the network is as shown in Figure 5.20(b). Each level has 100 incoming and 100 outgoing links. Therefore, it can consist of a three-stage network as shown in Figure 5.20(a).

The number of crosspoints is:

$$C_5 = 100 \times 10 \times 19 + 19 \times 5400 + 100 \times 19 \times 10 \\ = 140\,600$$

5.11 Sectionalized switching networks

Since trunks in small groups normally have a lower traffic capacity than trunks in large ones, the division of a large switching network into smaller sections that are not interconnected would appear to be inefficient. However, it is possible to provide an electronic central control to determine which of the networks is best able to deal with

[152] *Switching networks*

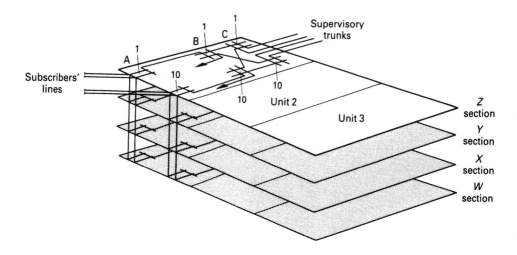

Figure 5.21 Sectionalized switching network.

each call. By doing so, it can coordinate the combined activities of the separate sections as if they were a homogeneous whole. This enables a sectionalized exchange to use fewer crosspoints than the equivalent unsectionalized switching network.[21]

Figure 5.21 shows a sectionalized switching network containing three separate units, each serving up to 1000 subscribers' lines. Each unit consists of a partially interconnected three-stage switching network of the form shown in Figure 5.14, but having concentration. The network contains four sections (shown as planes W, X, Y, Z in the figure). Each section provides only one path between each inlet (subscriber's line) and each outlet (supervisory trunk). It also contains its own marker. The supervisory trunks provide connections between different units in the same plane and between units in different planes.

Call packing is used. In order to set up a connection, the central control instructs the marker in each section which trunks and links to interrogate. The resulting pattern of busy/free conditions is signalled to the central control for choice of the route to be used. This choice is made in accordance with a program of priorities to select the route least likely to cause blocking of subsequent calls. The marker in the selected section then sets up the connection.

The trunking shown in Figure 5.21 can be considered as a number of minimal exchanges connected in parallel. It enables exchanges to be constructed from standard modules, each consisting of one section of one unit. The number of units provided depends on the number of lines and the number of sections (planes) depends on the traffic. Failure of one section results in a degradation of service to all customers, but it does not cause loss of service to any. Thus, the only equipment that needs to be replicated to ensure security is the central control unit.

The sectionalized three-stage network shown in Figure 5.21 is employed in the

British Telecom TXE4 reed–electronic system.[12] It uses from six to eight planes. A fourth switching stage (the D switch), which is shown in Figure 3.24, acts as a mixing stage to improve the occupancy of the supervisory trunks.

Switching networks consisting of separate planes in parallel are now used in some digital time-division switching systems. Usually, there are two planes. If one fails, the other can carry the traffic.

Note

1. This would occur if originating calls were allotted to outgoing trunks by a perfectly random selection process.

Problems

1. Explain briefly the meanings of the following terms applied to gradings: graded group, availability, progressive grading, skipped grading, homogeneous grading.

2. A grading is required to connect 30 outgoing trunks to switches of availability 10.
(a) Design a progressive grading.
(b) Design a homogeneous grading.

3. (a) Estimate the traffic capacity of the above gradings if the grade of service is to be:
(i) 0.001, (ii) 0.01.
(b) If the gradings are each offered 12 E of traffic evenly distributed over the graded groups, how much traffic is carried by the first-choice trunks and the second-choice trunks?

4. Define the following terms: Full availability, link system, blocking, conditional selection, distribution stage, mixing stage.

5. (a) Sketch a two-stage network using switches of size 3 × 4 to connect any of 9 incoming trunks to any one of 16 outgoing trunks. How many crosspoints are needed?
(b) Sketch a fully interconnected three-stage network for the same number of incoming and outgoing trunks, using 3 × 4 switches in the primary stage and 4 × 4 switches in the tertiary stage. How many crosspoints are needed? Comment on this result.
(c) Demonstrate by means of a simple example that the probability of blocking is less with network (b) than with network (a).

6. (a) Design a two-stage network, using switch modules of size 10 × 10, suitable for connecting 200 incoming trunks to 200 outgoing trunks. The outgoing trunks serve ten routes with 20 trunks on each route. How many switch modules are needed?
(b) The occupancies of the incoming and outgoing trunks are each 0.5 E. Estimate the grade of service obtained.

7. (a) A two-stage space-division network acts as a concentrator. It has M incoming trunks, N outgoing trunks and N links between the two switching stages (where $M > N$). Two methods of designing the network are as follows:

(i) To use the same number of switches in each stage, but to have larger switches in the first stage.
(ii) To use switches in the primary stage which have the same number of inlets as the secondary switches have outlets, but to use a different number of switches in each stage.

Which requires the smaller number of crosspoints?
(b) Verify the answer to part (a) by designing

a network of each type for a two-stage concentrator having 400 incoming trunks and 100 outgoing trunks. How many crosspoints does each contain?

(c) The concentrator serves 10 outgoing routes with ten trunks on each route. Each incoming trunk originates 0.1 E of traffic, of which the same proportion is directed to each outgoing route. Estimate the grade of service given by each network and state any assumptions you make. Comment briefly on the results obtained.

8. A three-stage fully interconnected switching network is to connect 600 incoming trunks to 100 outgoing trunks. It is to use switches assembled from blocks of size 5×5. Design a suitable network and determine the number of switch blocks required.

9. A switching network is to have N incoming trunks, N outgoing trunks and not less than N links between stages.

(a) Show that a two-stage network has fewer crosspoints than either a single stage or a partially interconnected three-stage network for $6 \leq N \leq 19$.

(b) The cost per switch is $K(1 + C_1/100)$, where C_1 is its number of crosspoints (e.g. the total cost of a 20×20 switch is 25% more than the cost of its crosspoints). Show that a two-stage network costs less than either a single-stage or a partially interconnected three-stage network for $42 \leq N \leq 400$.

10. Define the following terms: Connectivity, call packing, expansion, quasi-nonblocking network, wide-sense nonblocking network, strict-sense nonblocking network.

11. (a) Design a fully interconnected three-stage network with 200 incoming and 200 outgoing trunks with the minimum number of crosspoints. How many crosspoints does the network contain?

(b) If the occupancy of the trunks is 0.4 E and connections are to be made to particular outgoing trunks, estimate the grade of service. State any assumptions you make.

(c) Redesign the above network to provide a grade of service better than 1 in 1000. How many crosspoints does this network require?

(d) Redesign the network to be strictly nonblocking. How many crosspoints does this network require?

12. A partially interconnected three-stage network as shown in Figure 5.14 consists of switches of size 10×10 and connects 1000 incoming trunks to 1000 outgoing trunks. The outgoing trunks serve ten routes, each having one trunk connected to each tertiary switch.

(a) (i) Compare the number of crosspoints in the network with the number required by a three-stage fully interconnected network with the same number of incoming and outgoing trunks.

(ii) What advantage would the fully interconnected network have over that shown in Figure 5.14?

(b) (i) Assuming that the traffic is evenly distributed over the incoming and outgoing trunks, determine the grade of service when each incoming trunk originates 0.6 E of traffic.

(ii) If the total traffic offered to the network is unchanged but the traffic offered to one primary switch increases by 20%, find the grade of service for these calls.

(iii) If the total traffic offered to the network is unchanged but the traffic offered to one outgoing route increases by 20%, find the grade of service for calls offered to that route.

References

[1] Syski, R. (1986), *Introduction to Congestion Theory in Telephone Systems*, 2nd edn, Oliver & Boyd, Edinburgh.
[2] Benes, V.E. (1965), *Mathematical Theory of Connecting Networks and Telephone Traffic'*,

Academic Press, New York.
[3] Midwinter, J.E., 'Optical switching', chapter 13 in Davies, D.E.N., Hilsum, C. and Rudge, A.W. (eds) (1993), *Communications after AD 2000*, Chapman & Hall, London.
[4] O'Dell, G.F. (1927), 'An outline of the trunking aspects of automatic telephony', *Jour. IEE*, **65**, 185–222.
[5] Leighton, A.G. and Kirby, W. (1972), 'An improved method of grading: the partially-skipped grading', *P.O. Elect. Engrs. Jour.*, **65**, 165–72.
[6] Bear, D. (1988), *Principles of Telecommunication Traffic Engineering*, 3rd edn, Peter Peregrinus, Stevenage.
[7] Atkinson, J. (1950), *Telephony*, Vol. 2, Pitman, London.
[8] Corner, A.C. (1966), 'The 5005 crossbar telephone exchange system', *PO Elect. Engrs. Jour.*, **59**, 170–7.
[9] Feiner, A. and Hayward, W.S. (1964), 'No.1 ESS switching network plan', *Bell Syst. Tech. Jour.*, **43**, 2193–220.
[10] Ronayne, J. (1986), *Introduction to Digital Communications Switching*, Pitman, London.
[11] Long, R.C. and Gorringe, G.E. (1969), 'Electronic telephone exchanges: TXE2 – a small electronic exchange system', *PO Elect. Engrs. Jour.*, **62**, 12–20.
[12] Goodman, J.V. et al. (1976), 'TXE4 electronic exchange system', *PO Elect. Engrs. Jour.*, **68**, 196–203 and **69**, 68–78.
[13] Hills, M.T. (1979), *Telecommunications Switching Principles*, Allen and Unwin, London.
[14] Lee, C.Y. (1955), 'Analysis of switching networks', *Bell Syst. Tech. Jour.*, **34**, 1287–315.
[15] Jacobaeus, C. (1950), 'A study of congestion in link systems', *Ericsson Techniks*, **48**, 1–70.
[16] Seshu, S. and Reed, M.B. (1961), *Linear Graphs and Electric Networks*, Addison-Wesley, Reading, MA.
[17] Takagi, K. (1968), 'Design of multi-stage link systems by means of optimum channel graphs', *Electronics and Communications in Japan*, **51A**, 37–46.
[18] DeBoer, J. (1973), 'Comparison of random selection and selection with a fixed starting position in a multi-stage link network', *Philips Telecom. Rev.*, 31, 148–55.
[19] Benes, V.E. (1962), 'On rearrangeable three-stage switching networks', *Bell. Syst. Tech. Jour.*, **41**, 1481–92.
[20] Clos, C. (1953), 'A study of nonblocking switching networks', *Bell Syst. Tech. Jour.*, **32**, 406–24.
[21] Bear, D. and Warman, J.B. (1966), 'Trunking and traffic aspects of a sectionalised telephone exchange system', *Proc. IEE*, **113**, 1331–43.

CHAPTER 6

Time-division switching

6.1 Introduction

As described in Chapter 3, the first application of digital time-division switching was to provide tandem switching of PCM junction and trunk circuits[1]. Examples of such systems[2] are the Bell ESS No.4 system and the French E 12 system. Since a tandem or trunk exchange is similar to the route switch in a local exchange, local-exchange systems were developed by adding reed-relay space-division concentrators. Examples of such systems[2] include the initial versions of System X and the AXE 10 and E 10 systems. However, developments in the technology of solid-state integrated circuits eventually solved the BORSCHT problem and enabled TDM concentrators to replace space-division concentrators, thus extending digital operation to the customer's line circuit.[2–5]

This has led to the evolution of *integrated digital networks* (IDN), in which compatible digital transmission and switching are used throughout a network. Finally, extension of digital transmission over customers' lines has enabled *integrated-services digital networks* (ISDN) to be introduced. An ISDN can provide the customer with a wide variety of services, based on 64 kbit/s transmission, over a single line from a local exchange.

As an example, the architecture of a System X local exchange[6] is shown in Figure 6.1. The digital switching subsystem (DSS) corresponds to the route switch of Figure 3.20. The line-terminating units of PCM junctions are connected to it directly. Voice-frequency junctions are connected via a signalling interworking subsystem (SIS) and an analog line terminating subsystem (ALTS) that provides analog/digital and digital/analog conversion. Analog and digital customers' lines are connected to the DSS via a concentrator, known as the digital subscribers' switching subsystem (DSSS). These concentrators may be in the main exchange or located remotely. The processor utility subsystem uses software built largely from modules corresponding to the hardware subsystems that they control, as shown in Figure 6.1. The system has a capacity for 60 000 lines and 10 000 E of traffic. A tandem or trunk exchange uses a similar DSS, but it has no concentrators.

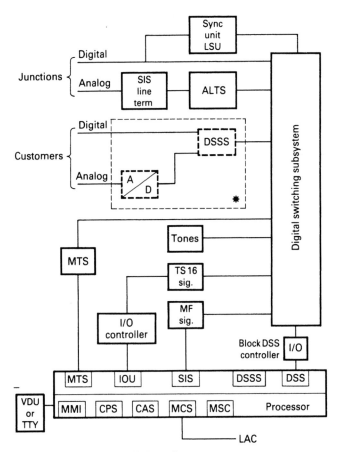

*May be located remotely or in the exchange

Figure 6.1 System X local exchange. ALTS = analog line terminating subsystem, CAS = call-accounting subsystem, CPS = call-processing subsystem, DSS = digital switching subsystem, DSSS = digital subscribers' switching subsystem, LAC = local administration centre, MCS = maintenance control subsystem, MMI = man–machine interface, MSS = management statistics subsystem, MTS = message transmission subsystem, SIS = signalling interworking subsystem, VDU = visual display unit.

A broadly similar architecture is used by other digital telephone-exchange systems.[2,7] Examples of such systems include the AXE-10 system (developed in Sweden), the DMS-10 system (Canada), the E-10 system (France), the No. 5 ESS system (USA), the EWS-D system (Germany) and the NEAX system (Japan).

6.2 Space and time switching

6.2.1 General

A tandem switching centre, or the route switch of a local exchange, must be able to connect any channel on one of its incoming PCM highways to any channel of an outgoing PCM highway. The incoming and outgoing highways are spatially separate, so the connection obviously requires space switching. In general, a connection will occupy different time-slots on the incoming and outgoing highways. Thus, the switching network must be able to receive PCM samples from one time-slot and retransmit them in a different time-slot. This is known as *time-slot interchange*, or simply as *time switching*. Consequently, the switching network of a tandem exchange, or the route switch of a local exchange, must perform both space switching and time switching.

Simple time-division switching networks make connections between channels on highways carrying a primary multiplex group, i.e they operate at 1.5 Mbit/s or 2 Mbit/s. A 2 Mbit/s line system has 32 time-slots. However, it only carries 30 speech channels; time-slot 0 is used for frame alignment and time-slot 16 for signalling. Within an exchange, time-slot 0 is not needed for frame alignment since all switches are driven synchronously from the clock-pulse generator of the exchange. It is also unnecessary to use time-slot 16 for signalling associated with the channels on the highway when this is handled over a separate path (e.g. when common-channel signalling is used, as described in Chapter 8). In this case, all 32 time-slots can be used to switch speech connections. Some systems have large switches operating at multiples of the primary rate (e.g. at 8 Mbit/s) in order to increase traffic capacity by having more time-slots.

6.2.2 Space switches

Connections can be made between incoming and outgoing PCM highways by means of a crosspoint matrix of the form shown in Figure 3.11. However, different channels of an incoming PCM frame may need to be switched by different crosspoints in order to reach different destinations. The crosspoint is therefore a two-input AND gate. One input is connected to the incoming PCM highway and the other to a *connection store* that produces a pulse at the required instants. A group of crosspoint gates can be implemented as an integrated circuit, for example by using a multiplexer chip.

Figure 6.2 shows a space switch with k incoming and m outgoing PCM highways, each carrying n channels. The connection store for each column of crosspoints is a memory with an address location for each time-slot, which stores the number of the crosspoint to be operated in that time-slot. This number is written into the address by the controlling processor in order to set up the connection. The numbers are read out cyclically, in synchronism with the incoming PCM frame. In each time-slot, the number stored at the corresponding store address is read out and decoding logic converts this into a pulse on a single lead to operate the relevent crosspoint.

Since a crosspoint can make a different connection in each of the n time-slots, it is

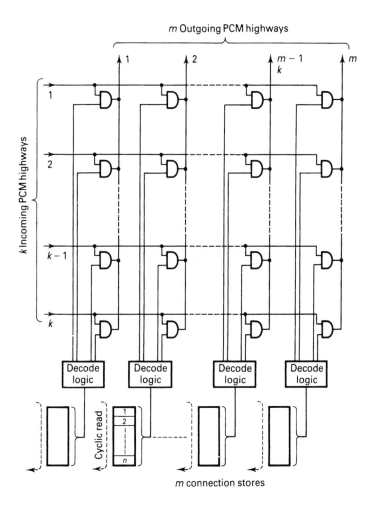

Figure 6.2 Space switch.

equivalent to n crosspoints in a space-division network. The complete space switch is thus equivalent to n separate $k \times m$ switches in a space-division switching network.

6.2.3 Time switches

The principle of a time switch is shown in Figure 6.3(a). It connects an incoming n-channel PCM highway to an outgoing n-channel PCM highway. Since any incoming channel can be connected to any outgoing channel, it is equivalent to a space-division crosspoint matrix with n incoming and n outgoing trunks, as shown in Figure 6.3(b).

Time-slot interchange is carried out by means of two stores, each having a storage address for every channel of the PCM frame. The *speech store* contains the data of each

[160] *Time-division switching*

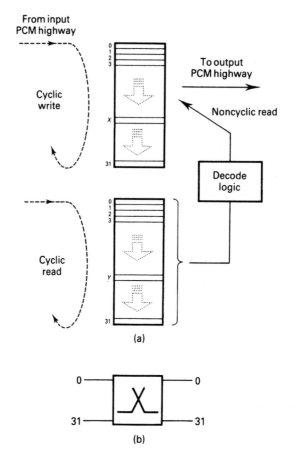

Figure 6.3 Time switching. (a) Time switch. (b) Space-division equivalent.

of the incoming time-slots (i.e. its speech sample) at a corresponding address. Each address of the *connection store* corresponds to a time-slot on the outgoing highway. It contains the number of the time-slot on the incoming highway whose sample is to be retransmitted in that outgoing time-slot. Information is read into the speech store cyclically, in synchronism with the incoming PCM system; however, random-access read-out is used. The connection store has cyclic read-out, but writing in is noncyclic.

To establish a connection, the number (X) of the time-slot of an incoming channel is written into the connection store at the address corresponding to the selected outgoing channel (Y). During each cyclic scan of the speech store, the incoming PCM sample from channel X is written into address X. During each cyclic scan of the connection store, the number X is read out at the beginning of time-slot Y. This is decoded to select address X of the speech store, whose contents are read out and sent over the outgoing highway.

An alternative way of implementing a time switch uses a speech store with

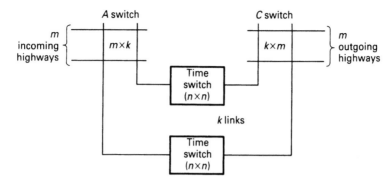

Figure 6.4 Space–time–space (S–T–S) switching network. m = no. of PCM highways, n = no. of time-slots.

random access for writing and cyclic access for reading. In order to transfer data from time-slot X on the incoming highway to time-slot Y on the outgoing highway, the connection store holds Y at address X. This is therefore read out at time X and decoded to write the incoming sample in the speech store at address Y. The cyclic scan of the speech store then reads out the sample at time Y for retransmission on the outgoing highway.

Time switching introduces delay. If $Y > X$, the output sample occurs later in the same frame as the input sample. If $Y < X$, the output sample occurs in the next frame. In a multi-link connection, several such delays occur. Since these are in addition to propagation delay, they adversely affect the echo performance of the connection.

6.3 Time-division switching networks

6.3.1 Basic networks

Figure 6.4 shows a space–time–space (S–T–S) switching network. Each of the m incoming PCM highways can be connected to k links by crosspoints in the A switch, and the other ends of the links are connected to the m outgoing PCM highways by crosspoints in the C switch. Each link contains a time switch. To make a connection between time-slot X of an incoming PCM highway and time-slot Y of an outgoing highway, it is necessary to select a link having address X free in its speech store and address Y free in its connection store. The time switch is then set to produce a shift from X to Y. The connection is completed by operating the appropriate A-switch crosspoint at time X and the appropriate C-switch crosspoint at time Y in each frame.

Figure 6.5 shows a time–space–time (T–S–T) switching network. Each of the m incoming and m outgoing PCM highways is connected to a time switch. The incoming and outgoing time switches are connected by the space switch. To make a connection between time-slot X of an incoming highway and time-slot Y of an outgoing highway, it is necessary to choose a time-slot Z which is free in the connection store of the incoming

[162] *Time-division switching*

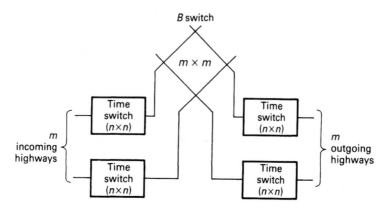

Figure 6.5 Time–space–time (T–S–T) switching network. m = no. of PCM highways, n = no. of time-slots.

highway and the speech store of the outgoing highway. The connection is established by setting the incoming time switch to shift from X to Z, setting the outgoing time switch to shift from Z to Y and operating the appropriate crosspoint at time Z in each frame.

Early designs of digital switching systems used S–T–S networks. This was because storage of speech samples, and hence time switching, was expensive. With the advent of semiconductor memories, time switching is no longer expensive. As a result, most current systems use T–S–T networks. However, a small switching network may need only two stages (e.g. T–S). Large switching networks may have more than three stages. For example, the Bell No.4 ESS toll switching system[8] uses a five-stage (T–S–S–S–T) network.

6.3.2 Bidirectional paths

The switching networks described above provide a connection for only one direction of transmission. Since PCM transmission systems use four-wire circuits, it is necessary to provide separate paths for the 'send' and 'receive' channels. One way of doing this would be to provide a separate switching network for each direction of transmission. However, this may be avoided by connecting the 'send' highways of both incoming and outgoing circuits to one side of the switch and the 'receive' highways to the other side, as shown in Figure 6.6.

In an S–T–S network the same speech-store address in the time switch may be used for each direction of transmission. For a connection between time-slot X on one trunk and channel Y on another, for one direction of transmission, the contents at the address are written at the end of time-slot X and are read at the beginning of time-slot Y. For the opposite direction of transmission, they are written at the end of the same

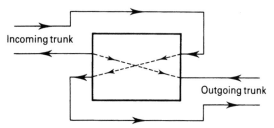

Figure 6.6 Bidirectional transmission through time-division switching network.

time-slot Y and read at the beginning of the next time-slot X. This method cannot be used if both external circuits use the same time-slots; however, this is rarely necessary.

In a T–S–T network, speech in the two directions must be carried through the space switch using different time-slots. In order to simplify control of the switching network, the time-slots for the two directions of transmission have a fixed time difference. Usually, the time-slots have a phase difference of 180°. In a 32-channel system, if time-slot 12 is used for one direction of transmission, then time-slot (12 + 16) = 28 is used for the reverse direction. One advantage of this arrangement is that if one time-slot is found to be free, the associated time-slot will also be free. Since the same time-slot is received from the input highway of a junction and sent to its output highway, the same connection store can be used to control the time switches of both. Speech stores associated with input highways have cyclic read-in and noncyclic read-out (as described in Section 6.2.3). However, speech stores associated with output highways have noncyclic read-in and read out cyclically to the output highways.

6.3.3 More complex switching networks

Many variations are possible on the basic three-stage T–S–T network shown in Figure 6.5. These include:

1. Increasing the size of stores in the time switch, so that each can serve more than one PCM highway.
2. Operating the space switch at a higher bit rate than the incoming and outgoing PCM systems. This enables each crosspoint to serve more than 32 channels, thus effectively increasing the size of the switch.
3. Using parallel instead of serial transmission of PCM words through the space switch. This has the same effect as (1) without an increase in speed; however, it increases the numbers of highways and crosspoint gates.
4. Duplicating, or even triplicating, the switching network to increase security of operation in the presence of faults. This is uneconomic for a space-division network. It is also unnecessary, because failure of individual switches has little effect on the overall grade of service. It is practicable in a time-division switching network because of the cost reduction brought about by time-sharing. It is also

desirable, because failure of an individual unit (e.g. a connection store) has much more serious consequences than in a space-division switching network.

Each of these techniques is used in the Mark 1 digital switching sub-system of System X,[9] shown in Figure 6.7. The receive and send time switches each have a speech store containing 1024 locations and so can serve up to 32 PCM systems. The complete network can contain up to 96 time switches, thus serving up to 3072 PCM systems.

The space switch therefore requires a maximum size of 96×96 and each crosspoint must be able to switch 1024 channels. This is done by using parallel transmission at a digit rate of 8.194 Mbit/s. To minimize problems of pulse distribution, the 1024-channel highways are each split into two highways of 512 channels. The space switch is therefore divided into two segments (A and B), switching odd and even time-slots respectively.

To provide adequate security, the complete network is duplicated. The duplicated systems operate in synchronism and faults are detected by means of a parity check. The PCM words transmitted across the switching network therefore use 9 bits, instead of the normal 8 bits used for transmission.

This Mark 1 digital switching subsystem has now been superseded by a more complex Mark 2 subsystem.[10] It uses a five-stage switching network.

6.3.4 Concentrators

A concentrator connects to a PCM highway a number of customers' line units greater than the number of time-slots on the highway. In a simple concentrator, the customers' codecs are all connected to the common highway and each may use any time-slot. A codec is operated in the required time-slot by means of a connection store. This method is used, for example, in the AXE system[2,4]. However, each 128-line concentrator module gives access to two PCM highways. One is individual to it and, when all its channels are busy, calls overflow to a second highway that is common to 16 modules.

Alternatively, a group of codecs equal to the number of available time-slots (e.g. 24 or 30) uses fixed channel times on a highway. Several such highways are concentrated onto a second highway by means of a time switch. System X uses this method[10]. A concentrator connects up to 2048 customers' lines to eight 32-channel PCM systems (i.e. a concentration of 8 to 1).

Since a concentrator is connected to the route switch by a PCM highway, it may be located at a distance from the main exchange. The concentrator can be controlled by the central processor in the main exchange by means of signals sent over the PCM link (e.g. in time-slot 16 of a 30-channel system). If the PCM link between a remote concentrator unit and the main exchange fails, customers on the concentrator lose all service. Duplicate PCM links are therefore often provided.

The control functions of the concentrator may be enhanced to enable it to connect calls between its own customers (but not with others) if the PCM link fails. Facilities must be added to receive and analyze address signals, generate tones and make cross-switch connections between customers' lines. The unit is then known as a *remote switching unit*.

Time-division switching networks [165]

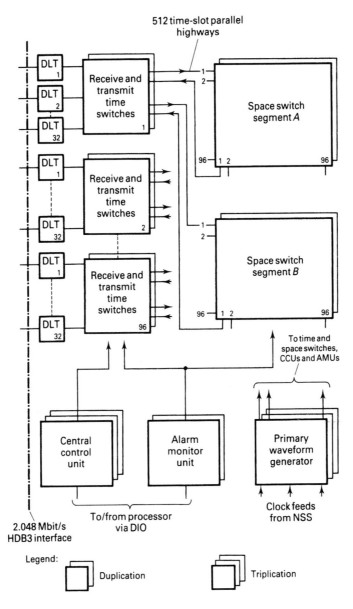

Figure 6.7 Mark 1 digital switching subsystem of System X. DLT = digital line terminating unit.

[166] *Time-division switching*

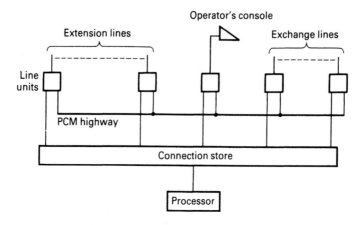

Figure 6.8   Trunking of a digital PBX.

6.3.5 PBX switches

A large PBX may use a switching network similar to that of a public exchange. However, a small PBX may only generate sufficient traffic for all its connections to be made over a single highway.[12] All its ports, i.e. those for extension lines, exchange lines and the operator's position, have codecs connected to a common highway, as shown in Figure 6.8. The codecs are operated in the required time slots by a connection store. In order to increase the line capacity of the PBX, the number of time-slots on the common highway may be increased by using 8-bit parallel transmission instead of serial transmission. To provide two-way communication over the same highway, each connection uses two time-slots, as described in Section 6.3.2.

6.3.6 Digital cross-connect units

For a telephone call, a connection is made through a digital switching network at the start of a call and cleared down as soon as the call ends. However, a similar digital switching network may be used for semi-permanent connections. It is controlled manually from an operating terminal instead of automatically by the processor of an exchange. Such a switching network is called a *digital cross-connect unit*. It performs a function for digital circuits similar to that of a distribution frame for analog circuits. It is sometimes called a 'slow switch', in contrast to a 'fast switch' used to connect telephone calls and the connections made by a digital cross-connect unit are sometimes called 'nailed-up time-slots'.

Two functions that can be performed by digital cross-connect units are *grooming* and *consolidation*. In grooming, 64 kbit/s channels on a common PCM bearer are separated for routing to different destinations. For example, a line from a customer's PBX may carry a mixture of PCM channels, some to the public exchange and some to other PBXs in the customer's private network. In consolidation, channels on several PCM bearers that are not fully loaded are combined onto a smaller number of bearers, thereby improving the utilization of the PCM systems.

6.4 Grades of service of time-division switching networks

In the S–T–S network of Figure 6.4 each crosspoint of the space switch is time shared by n channels. It is therefore equivalent to n separate crosspoints in a space-division switch. Thus, the A switch is equivalent to n space-division switches of size $m \times k$ and the C switch is equivalent to n space-division switches of size $k \times m$. Each of the k time switches is equivalent to a space-division switch of size $n \times n$, as shown in Figure 6.3(b). The S–T–S network of Figure 6.4 thus corresponds to the equivalent three-stage space-division network of Figure 6.9.

In the T–S–T network of Figure 6.5, each time switch is equivalent to a space-division switch of size $n \times n$ and there are m of them associated with incoming highways and m associated with outgoing highways. The space switch is equivalent to n space-division switches of size $m \times m$. The T–S–T network of Figure 6.5 thus corresponds to the equivalent three-stage space-division network of Figure 6.10.

As a result, it is unnecessary to invent new traffic theory to determine grades of service for time-division switching systems. The loss probability obtained for a given traffic offered to a time-division switching network can be determined by studying the equivalent space-division network.

Example 6.1

An S–T–S network has 16 incoming and 16 outgoing highways, each of which conveys 24 PCM channels. Between the incoming and outgoing space switches there are 20 links containing time switches. During the busy hour, the network is offered 300 E of

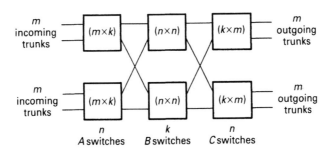

Figure 6.9 Space-division equivalent of S–T–S switch: m = no. of PCM highways, n = no. of time-slots, k = no. of time-switch links.

Time-division switching

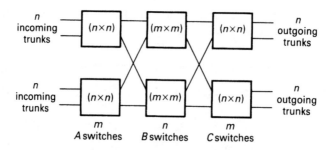

Figure 6.10 Space-division equivalent of T–S–T switch. m = no. of PCM highways, n = no. of time-slots.

traffic and it can be assumed that this is evenly distributed over the outgoing channels. Estimate the grade of service obtained if:

1. Connection is required to a particular free channel on a selected outgoing highway (i.e. mode 1).
2. Connection is required to a particular outgoing highway, but any free channel on it may be used (i.e. mode 2).

1. For the equivalent space-division network shown in Figure 6.9: $m = 16$, $n = 24$, $k = 20$. The occupancy of a link is $b = 300/(24 \times 20) = 0.625$ E.
   From equation (5.26):
   $$B_1 = [1 - (1 - b)^2]^k = [1 - (1 - 0.625)^2]^{20}$$
   $$= 0.859^{20} = \underline{0.048}$$

2. The occupancy of a highway is
   $$c = 300/(24 \times 16) = 0.781 \text{ E}$$
   From equation (5.27):
   $$B_2 = [B_1 + c(1 - B_1)]^n = [0.048 + 0.781(1 - 0.048)]^{24}$$
   $$= 0.792^{24} = \underline{0.0037}$$

Example 6.2

A T–S–T network has 20 incoming and 20 outgoing PCM highways, each conveying 30 channels. The required grade of service is 0.01. Find the traffic capacity of the network if:

1. Connection is required to a particular free channel on a selected outgoing highway (i.e. mode 1).
2. Connection is required to a particular outgoing highway, but any free channel on it may be used (i.e. mode 2).

1. For the equivalent space-division network shown in Figure 6.10: $m = 20$, $n = 30$.
   Let the occupancies of the mn links and trunks be b.
   From equation (5.26):

   $$B^1 = [1 - (1-b)^2]^n = [1 - (1-b)^2]^{30} = 0.01$$
   $$\therefore \quad 1 - (1-b)^2 = 0.01^{0.0333} = 0.858$$
   $$(1-b)^2 = 0.142 \text{ and } b = 0.623.$$

   $\therefore$ Total traffic capacity of network is
   $0.623 \times 20 \times 30 = \underline{374 \text{ E}}$

2. All the channels on a route are provided by the same outgoing time switch; therefore all the trunks on a route are connected to the same C switch in the equivalent space-division network.

   The probability of blocking for a connection to this C switch is B_1.
   The probability that all trunks outgoing from the C switch are busy is approximately b^n.
   $\therefore$ The probability that connection can be made to a free outgoing trunk is $(1-B_1)(1-b^n)$ and the loss probability is $B_2 = 1 - (1-B_1)(1-b^n)$.

   However, if n is large, b^n is very small (e.g. if $b = 0.623$, $b^{30} = 6.8 \times 10^{-7}$).

   $$\therefore \quad B_2 \doteqdot B_1$$

Thus, approximately the same loss probability is obtained in either mode and the traffic capacity of the network for $B_2 = 0.01$ is again 374 E.

6.5 Nonblocking networks

Time-division switching networks often have large values of connectivity and are therefore quasi-nonblocking. For example, the network described in Section 6.3.3 has $n = 1024$ and it can carry up to 0.95 E per time-slot.[9] Time-division networks can also be strictly nonblocking.

A time-division switching network will be nonblocking in the strict sense if its equivalent space-division network is strictly nonblocking. It is shown in Section 5.10 that a three-stage space-division network whose primary switches have n inlets and tertiary switches have n outlets is strictly nonblocking if there are $2n-1$ secondary switches.

To make the S–T–S switch of Figure 6.4 strictly non-blocking, the number of B switches in the equivalent space-division network of Figure 6.9 must be at least $2m-1$. This can be done by providing $2m-1$ time-shifting links.

To make the T–S–T switch of Figure 6.5 strictly nonblocking, the number of B switches in the equivalent space-division network of Figure 6.10 must be at least $2n-1$. This can be done by operating the space switch at a higher bit rate than the external highways in order to provide more time-slots (e.g. by doubling the speed to provide 64 time-slots instead of 32).

6.6 Synchronization

6.6.1 Frame alignment

For correct operation of a time-division switching network the PCM frames on all the incoming highways must be exactly aligned. However, since incoming PCM junctions come from different places, their signals are subjected to different delays. Thus, even if all exchange clock-pulse generators are in perfect synchronism, there will be time differences between the starting instants of different PCM frames entering a digital exchange.

To solve this problem, the line-terminating unit of a PCM junction stores the incoming digits in a *frame-alignment buffer*, as shown in Figure 6.11. Digits are read into this buffer at the rate, f_a, of the incoming line, beginning at the start of each frame. They are then read out at the rate, f_b, of the exchange clock, beginning at the start of the PCM frame of the exchange. To cater for the maximum amount of misalignment between a digital line system and the exchange, the aligner must have a buffer capacity of at least one frame (e.g. 256 bits for a 2 Mbit/s PCM system). This introduces delay additional to that caused by time switching.

A frame-alignment buffer caters perfectly for a constant misalignment. The fill of the buffer is constant and its level depends on the phase difference between the incoming line system and the exchange. It will also cope with a misalignment that changes slowly between limits (e.g. due to temperature changes in cables). However, if the exchanges at the two ends of a line have slightly different clock frequencies, the contents of the buffer will change until it either overflows or empties. If the buffer overflows, its contents are erased so that it can start refilling. If the buffer empties completely, the contents of the previous frame are repeated to refill it. In either case, a complete frame is in error. This is known as a *frame slip*. Of course, slips can also arise

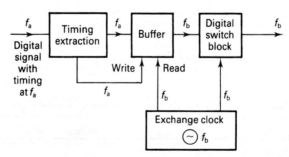

Figure 6.11 Frame alignment of PCM signals entering a digital exchange.

from malfunctions in transmission or switching systems. These are called *uncontrolled slips*, whereas a slip made deliberately to regain frame alignment is called a *controlled slip*.

A digital network may be plesiochronous (i.e. each exchange has an independent clock) or it may be synchronous (i.e. all exchange clocks are controlled by a single master clock). If plesiochronous working is employed, with crystal-controlled clocks having a frequency accuracy of 1 part in 10^7, about 68 frame slips per day will occur in an exchange.[5] For a connection of seven inter-exchange links in tandem, this would cause about 20 slips per hour. For telephony, only about one slip in 25 results in an audible click,[13] so this is tolerable. However, regular slips would be much more serious for data transmission. Either frequent adjustment of clock frequencies would be needed or clocks of atomic standard would be required in all exchanges. Consequently, all the exchanges in an integrated digital network are usually synchronized by a common master clock, as described below.

6.6.2 Synchronization networks

In a synchronous digital network just one or two atomic reference clocks control the frequencies of the clocks of all the exchanges in the network.[14,15] This is sometimes called *despotic control*. For this purpose, a synchronizing network is added to the PSTN in order to link the exchange clocks to the national reference standard. Under normal conditions the network will be free of slip, whereas a plesiochronous network will always experience some slips.

The local clock in each exchange is provided by a crystal oscillator whose frequency can be adjusted by a control voltage. This control voltage is derived from the incoming digit stream on a synchronizing link, which is used to determine whether the exchange clock rate should be increased, decreased or left unchanged. Adjustments are made periodically, as a single quantum increase or decrease. This ensures that exchanges maintain the same long-term average frequency, although short-term deviations may occur. This is known as *mesochronous working*.

Synchronizing links may be unilateral or bilateral. In the first case, there is a 'master–slave' relationship; the clock frequency of the exchange at one end of the link is controlled solely by the exchange at the other end. In the second case, there is a mutual relationship; each exchange influences the frequency of the other. The principles of these methods are shown in Figure 6.12.

A unilateral sync system is shown in Figure 6.12(a). Exchange A is the 'master' and exchange B is the 'slave'. Exchange B determines the phase difference between its own clock and that of exchange A by the fill of the aligner buffer on the incoming link. A change in phase causes a step increase or decrease in clock frequency lasting for a few milliseconds. If there is more than one sync link into exchange B, its correction is based on a majority decision.

In a single-ended bilateral sync link, as shown in Figure 6.12(b), the above decision process is made at each end of the link. As a result, both exchange clocks achieve the same average frequency. In a mesh of such sync nodes, the exchanges would

[172] *Time-division switching*

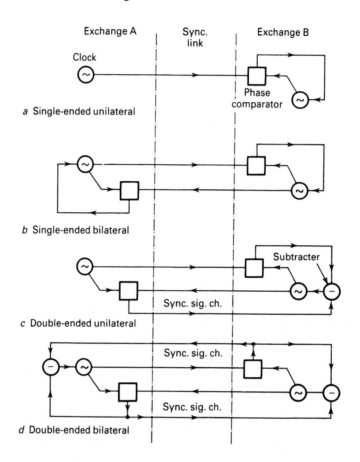

Figure 6.12 Exchange synchronization systems. (a) Single-ended unilateral system. (b) Single-ended bilateral system. (c) Double-ended unilateral system. (d) Double-ended bilateral system.

mutually agree on a common frequency without being controlled by an overall master clock.

A disadvantage of the single-ended unilateral and bilateral sync systems is that the phase comparators are unable to distinguish between phase changes due to frequency drift and those due to changes in propagation time (e.g. caused by cable temperature changes). The former necessitates frequency adjustment, but the latter does not. This disadvantage of the single-ended unilateral and bilateral sync systems is overcome by the double-ended systems shown in Figure 6.12(c) and (d). These eliminate the influence of propagation-delay variations by subtracting the change in phase determined at one end of the link from that determined at the other end.

Let the phase error detected at exchange A be $\delta(\phi_A - \phi_B) + \delta\phi_T$, where $\delta(\phi_A - \phi_B)$ is the phase change due to discrepancy between the clocks and $\delta\phi_T$ is that due to a

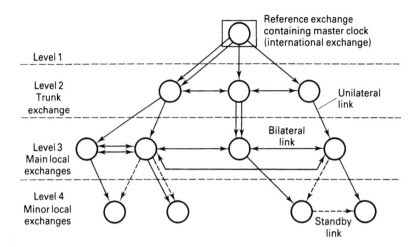

Figure 6.13 Synchronization hierarchy of an integrated digital network.

change in propagation time. Then the phase change detected at exchange B is $\delta(\phi_B - \phi_A) + \delta\phi_T$. Since $\delta(\phi_B - \phi_A) = -\delta(\phi_B - \phi_B)$, the difference between the two measurements is $2\delta(\phi_A - \phi_B)$ and $\delta\phi_T$ is eliminated.

A signalling channel is required to carry the result of the phase comparison to the other end of the link in order to make the subtraction. For a unilateral link, as shown in Figure 6.12(c), this channel is needed in only one direction. For a bilateral link, a signalling channel is needed in each direction, as shown in Figure 6.12(d). If the network uses 2 Mbit/s PCM systems, this signalling information can be carried in the spare capacity of time-slot 0 of the 32 time-slot frame.[5]

A synchronizing network for an IDN is shown in Figure 6.13. Since this auxiliary network must link all exchanges in the IDN, the sync network has the same nodes and the same hierarchical structure as the parent PSTN. The sync links are provided by PCM systems that carry normal traffic between the exchanges.

Frequency control is exerted downwards from the national reference standard by unilateral links from each exchange to those in the next lower level. However, bilateral links are used between exchanges in the same level of the hierarchy. Thus, if there is an incomplete network, or failure of the sync signal from the master source, exchanges at that level mutually determine their own clock frequency and synchronize lower-level exchanges to it.

Problems

1. (a) Sketch an S–T–S network to connect m incoming highways to m outgoing highways, each carrying n PCM channels and having k time-switch links. Explain briefly how it works.

(b) An S–T–S network has ten incoming highways, ten outgoing highways and ten time-switch links. The highways convey 32 PCM time-slots. The average occupancy of incoming PCM channels is 0.7 E.

[174] *Time-division switching*

(i) Derive an equivalent space-division network.
(ii) Estimate the blocking probability.
(iii) Estimate the grade of service when an incoming call must be connected to a selected outgoing highway but may use any free channel on it.

2. (a) Sketch a T–S–T network to connect m incoming lines to m outgoing lines, each carrying n PCM channels. Explain briefly how it operates.
(b) A T–S–T network has ten incoming highways and ten outgoing highways, each carrying 32 PCM channels. The average occupancy of the incoming channels is 0.6 E.

(i) Derive an equivalent space-division network.
(ii) Estimate the blocking probability.
(iii) Estimate the grade of service when an incoming call must be connected to a selected outgoing highway but may use any free channel on it.

3. Redesign the S–T–S network of question 1(b) and the T–S–T network of question 2(b) to be strictly nonblocking.

4. A T–S–T and S–T–S switch both have 32 incoming and outgoing highways, each having 32 PCM channels. The S–T–S network has 32 time-switch links, so both networks give the same blocking probability.
Compare the numbers of crosspoints and bytes of storage required for these two networks.

5. Since bits of storage are very much cheaper than crosspoints, an approximate measure that has been proposed for comparing the relative costs of networks is the 'complexity', defined as:

Complexity = $N_c + N_b/100$, where
N_c = number of space-stage crosspoints
N_b = number of bits of memory.

Compare the complexities of the S–T–S network in question 1(b) and the T–S–T network in question 2(b).

6. A two-stage digital switching network is to make connections between m incoming PCM highways and m outgoing PCM highways, each having n channels. Each call from an incoming PCM highway is to be connected to a selected outgoing PCM highway but may use any free channel on it.

(a) The network may use either space-time switching or time-space switching. Draw an equivalent space-division network for each. Hence, state which is preferable and explain why.
(b) For each network, determine the grade of service when the network has ten incoming and ten outgoing 30-channel PCM systems, each channel having an average occupancy of 0.4 E.

7. A digital PBX has all its ports (serving extensions, exchange lines and the operator) connected to a common bus, as shown in Figure 6.8. The bus carries 128 time-slots and can be used for making a connection between any two ports.
In the busy hour, the average both-way traffic per extension is 0.1 E and 10% of this constitutes traffic to and from the public exchange. The occupancy of the operator's position is 0.3 E. Since extension users can dial outside calls directly, traffic between extensions and the operator is very small.
If the required grade of service is 0.01, how many extensions can the PBX accommodate and how many lines are needed to the public exchange? (Table 4.1 may be used.)

8. In the digital switching subsystem of System X, as shown in Figure 6.7, the sending and receiving time switches each have a speech store containing 1024 storage locations and so can serve 32 PCM systems (each having 32 time-slots). The complete switching network contains up to 96 time switches, serving up to 3072 PCM systems. The space switch is thus of size 96 × 96 and handles 1024 channels. However, to minimize problems of pulse distribution, the 1024-channel highways are each split into two highways of 512 channels each. For security, the complete network is duplicated.
If the switch is fully equipped and the average occupancy of incoming channels is

less than 0.45 E, show that connections to outgoing routes (which may use any free time-slot) have negligible loss probabilty.

9. (a) What are the principal requirements for the interface between a PCM line system and an incoming time switch?
(b) Show the application of this in designing a time switch capable of handling 32 PCM systems.

10. A plesiochronous digital network uses 2.048 Mbit/s line systems whose frames contain 256 bits. The exchange clocks have a frequency stability of 1 part in 10^7.

(a) Prove that a connection of seven inter-exchange links can have about 20 slips per hour.
(b) On average, what proportion of bits are then received in error?
(c) In what percentage of 3-minute telephone calls on such connections will clicks be audible?

References

[1] Chapuis, R.J. and Joel, A.E. (1990), *Electronics, Computers and Telephone Switching: a book of technological history*, North-Holland, Amsterdam.
[2] Joel, A.E. (ed.) (1982), *Electronic Switching: digital central office systems of the world*, IEEE Press, New York.
[3] McDonald, J.C. (ed.) (1983), *Fundamentals of Digital Switching*, Plenum Press, New York.
[4] Ronayne, J. (1986), *Introduction to Digital Switching*, Pitman, London.
[5] Redmill, F.J. and Valdar, A.R. (1990), *SPC Digital Telephone Exchanges*, Peter Peregrinus, Stevenage.
[6] Tippler, J. (1979), 'Architecture of System X: Part I', *PO Elect. Engrs. Jour.*, **72**, 138–41.
[7] 'The No.5 ESS switching system', (1985), Special Issue of *ATT Bell Labs, Tech. Jour.*, **70**, Part 2.
[8] 'The 4ESS switch', (1977), Special Issue of *Bell Syst. Tech. Jour.*, **56**, Part 7.
[9] Risbridger, J.N.A. (1979), 'System X subsystems, Part 1: the digital switching subsystem', *PO Elec. Engrs. Jour.*, **73**, 19–26.
[10] Maddern, T.S. and Ozdmar, M. (1989), 'Advances in digital circuit switching architecture', *Proc. 2nd IEE National Conf. on Telecommunications*, pp. 303–8.
[11] Foxton, M.C. (1980), 'System X subsystems, Part 5: the subscribers' switching subsystem', *PO Elec. Engrs. Jour.*, **73**, 216–22.
[12] Rayfield, D.A.T. and Gracie, S.J. (1979), 'The Plessey PDX: a new digital PABX', *Int. Conf. on Private Electronic Switching Systems*, London, IEE Conf. Pub. no. 163, pp. 211–14.
[13] McLintock, R.W. (1991), 'Transmission performance', Chapter 4 in Flood, J.E. and Cochrane, P. (eds), *Transmission Systems*, Peter Peregrinus, Stevenage.
[14] Boulter, R.A. and Bunn, W. (1977), 'Network synchronisation', *PO Elect. Engrs. Jour.*, **70**, 21–8.
[15] Abate, J., Cooper, C., Pan, J. and Shapiro, I. (1979), 'Synchronization considerations for the switched digital network', *Int. Commun. Conf. Rec.*, June.

CHAPTER 7

Control of switching systems

7.1 Introduction

As described in Chapter 3, switching systems have evolved from being manually controlled to being controlled by relays and then electronically. The change from the manual system to the Strowger step-by-step system brought about a change from centralized to distributed control. However, as systems developed and offered more services to customers, it became economic to perform particular functions in specialized equipment that was associated with connections only when required; thus, common control was introduced.

Later, the development of digital-computer technology enabled different functions to be performed by the same hardware by using different programs; thus switching systems entered the era of stored-program control (SPC). An SPC public exchange can offer a much wider range of services than electromechanical exchanges provide. Moreover, since the processor's stored data can be altered electronically, some of these services can be controlled by customers. Examples include: optional call barring, repeat last call, reminder calls, call diversion and three-way calls.

In the first generation of SPC systems[1,2] the controlling processors were virtually mainframe computers and were large and expensive. Consequently, it was uneconomic to provide more than one (with a standby to provide service if a fault occurred). Thus, centralized control was reintroduced.

Subsequently, the cost of call processing was reduced by the advent of the microprocessor. This enabled various tasks to be devolved from the central processor to small processors associated with different parts of the system. For example, in the system shown in Figure 6.1, the digital switching subsystem (DSS), each of the customers' concentrators (DSSS) and the message transfer system (MTS) has its own controlling processor and the central processor provides overall coordination. Thus, there has now been an evolution away from fully centralized control to a more distributed form of control.

7.2 Call-processing functions

7.2.1 Sequence of operations

Similar basic processes must be performed by switching systems in any network, whether circuit switched or packet switched. They were summarized for telephony in Section 3.5. They will now be described in more detail for a simple telephone call between two customers whose lines terminate on the same exchange. A sequence of operations takes place in which the calling and called customers' lines and the connections to them change from one state to another as described below.

Idle state

Initially, the customer's handset is in the 'on-hook' condition. The line is idle, waiting for calls to be originated or received (state 0). Meanwhile, the exchange is monitoring the state of the line, ready to detect a calling condition.

Call request signal

The customer sends a signal to the exchange to request a call. For a telephone call, this is done by lifting the handset, which causes current to flow in the line. The calling signal is also sometimes known as a *seize signal*, since it obtains a resource from the exchange.

Calling line identification

The exchange detects the calling condition and this identifies the line which originated it. In general, this signal appears on a termination associated with the customer's equipment number (EN). Equipment-number to directory-number (EN-to-DN) translation must therefore be performed in order to charge for the call.

Determination of originating class of service

The originating class of service (COS) corresponds to the range of services available to the calling customer. It must therefore be determined before a connection can be set up. In electromechanical exchanges it was necessary for lines with different classes of service to be segregated into different groups in order to distinguish between them. For example, ordinary lines and payphone lines in a Strowger exchange must be connected to different first selectors. In an SPC exchange, a customer's COS forms part of the data stored for that customer in the *line store*; it is sometimes called a *class mark*. Many more different classes of service can be provided and these are alterable electronically, some under the control of the customer (e.g. optional call barring).

Identification of calling party

If the originating COS indicates a multi-party line, it is necessary to ensure that the correct party is billed for the call. In a manual exchange, party identification was carried out by the operator asking the caller. In an automatic exchange, customers on two-party lines can be distinguished by the calling condition placing an earth on one line wire or the other instead of looping the line.[3]

Connection to the calling line

The exchange now makes a connection to the calling line.

Proceed to send signal

The exchange sends a signal (dial tone) back to the caller to indicate that it is ready to receive the identity of the line termination to which connection is to be made. The exchange is now waiting for this information (state 1).

Address signal

The caller now sends a signal to the exchange to instruct it to route the call to the required destination. In a telephone exchange this is done by dialling or by sending tone pulses from a push-button telephone.

Selection of outgoing line termination

The exchange determines the required outgoing line termination from the address information that it has received. Since the caller dialled the directory number (DN) of the called customer, in general, this involves DN-to-EN translation.

Determination of terminating class of service

The exchange needs to determine the terminating COS of the called line, since this affects the procedure for handling the call. For example, if the call is to a customer having a PBX, the customer will have a group of lines and any of these may be used for the connection. (This also applies to junction or trunk calls; any free circuit on the required outgoing route may be used.) If the called customer shares a party line, it is also necessary to determine which party is to be alerted.

Testing called line termination

The called line may be unavailable, either because it is busy or it is out of service. Therefore, the exchange tests the state of the line before making connection to it. In the case of a call to a PBX (or to an outgoing junction), the exchange tests each termination until either it finds a free one or all are found busy.

Status signal

A *status signal*, sometimes called a *call-progress signal*, is now sent back to inform the caller of the progress of the call. This is usually an audio-frequency tone. However, it may be a recorded announcement (e.g. for trunk-route congestion or a customer's changed number). If the signal is busy tone or number-unobtainable tone, the caller replaces the handset and the connection is released. Then the idle state (state 0) is resumed.

Connection to called line termination

Having discovered that the called line is obtainable and free, the exchange makes a connection to it.

Alerting called customer

The exchange sends a signal to the called line to alert the customer to receive the call. In a telephone exchange, this is done by sending ringing current to the line. In the case of a two-party line, either customer can be alerted by ringing applied between one line or the other and earth.[3] For a four-party line, the ringing may also be superimposed on either a positive or negative voltage.

At the same time, ringing tone is sent back to the caller as a call-progress signal. The exchange is now waiting for an answer (state 2).

Answer signal

When the called customer answers by lifting the handset the line is looped and current flows. This provides an 'answer' signal to the exchange, causing it to cease sending the alerting signal. The exchange also removes ringing tone from the calling line, thus providing an answer signal to the caller.

If the called customer does not answer, the caller replaces the handset. This causes the exchange to release the connections to the calling and called line terminations. The caller's line is now again in the idle state (state 0).

Completion of the connection

Receipt of the answer signal from the called customer causes the exchange to complete the connection between the line terminations of the calling and called customers.

Conversational state

The connection has now been completed between the lines of the two customers and they can converse for as long as they wish (state 3). The exchange supervises the connection to detect the end of the call and to charge for it (by metering or ticketing).

[180] *Control of switching systems*

Clear signals

When each customer replaces the handset, line current ceases and so provides a 'clear' signal to the exchange.

Release of connection

The exchange then clears down the connection between the calling and called line terminations. Thus, the idle state (state 0) is resumed.

There are theoretically four ways of controlling release of the connection, as follows:

- Release when the calling party clears
- Release when the called party clears
- Release when the first party clears
- Release when the last party clears.

Since it is normally the calling party who pays for the call, it is reasonable for the caller to decide how long it shall last. Thus, the connection is normally released when the calling party clears down.

In the above example, everything has proceeded normally. However, the system also needs to be able to cope with abnormal situations. Difficulties arise when one party clears but the other does not. If the called customer replaces the handset but the caller does not, then the called line is still held and the customer is unable to originate or receive further calls. This is known as the *called subscriber held* (CSH) condition. If the caller replaces the handset but the called customer does not, the connection is released, but the loop still present on the line looks to the exchange like a calling condition. A new connection will be made to it, but there will be no response to dial tone. This is called the *permanent loop condition*. It is also known as the *permanent glow condition* (PG), because it resulted in a lamp remaining alight on a Strowger selector. The PG condition is particularly harmful if a large number arises from earth faults on lines when a cable is damaged. This can result in congestion in the exchange, e.g. all the registers can be held and genuine calls prevented from being made.

In older systems, PG and CSH conditions had to be removed through manual disconnection by a maintenance technician. In more modern systems, a *time-out* process is used. If the condition persists for more than a certain period of time (e.g. a few minutes for a connection or a fraction of a minute for a register), the connection is automatically released. However, this alone is insufficient. If a permanent loop engages a register and this releases itself by timing out, the line will only seize another register. When a line is forcibly released, it is therefore necessary to place the line termination in a state which the exchange can recognize as not being a calling condition, but which the customer can remove by eventually replacing the handset. This is called the *parked condition*.

In an electromechanical system the customer's line circuit has two relays: the line relay L (which operates to the calling signal) and the cut-off relay K (whose operation removes the calling condition when connection is made to the line). In an SPC system these states correspond to two bits in the line store. The combination can provide $2^2 = 4$ states. However, in normal operation, only three are used, as follows:

Line state	Relay states	
	L	K
Line idle	0	0
Line calling (but not yet connected)	1	0
Line connected	0	1

Thus, the fourth state (1 1) can be used to indicate that the line is parked.

7.2.2 Signal exchanges

The processes described above for a local call involve a number of actions taken in response to signals. The relative times of these signals are shown in Figure 7.1. Signals sent in the direction away from the caller (and towards the called line) are called *forward signals* and those sent towards the caller (and away from the called line) are called *backward signals*.

Forward signals pass from the caller to the exchange and from the exchange to the called customer; backward signals pass from the exchange to the caller and from the called customer to the exchange, as shown in Figure 7.2. It will be seen from this figure that there is a 'handshake' protocol. Each signal should produce a response in the opposite direction, thus verifying correct operation, as follows:

1. The call request signal is answered by the proceed-to-send signal.
2. The address signal is answered by a call-status signal.
3. The answer signal is a response to the alerting signal.

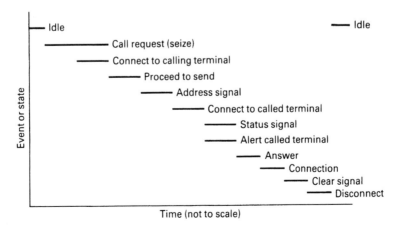

Figure 7.1 Timing of signals exchanged for a local call.

[182] *Control of switching systems*

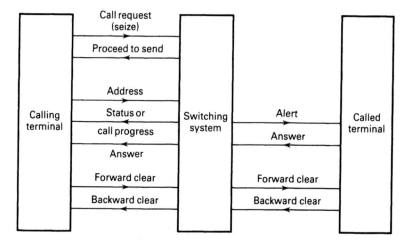

Figure 7.2 Signal exchange diagram for a local call.

4. The caller responds to the answer signal by commencing the conversation.
5. The backward clear signal is a respose to the forward clear signal (or vice versa).

For a call over a junction between two exchanges, the actions between the customer's calling signal and connection to an outgoing line occur at the originating exchange. This exchange then sends a seize signal to the terminating exchange. After the originating exchange has sent address information to the terminating exchange, the actions from receipt of the address information to alerting the called customer take place at the terminating exchange. When the called customer sends the answer signal, this is extended back to the originating exchange, in order that its supervision can commence. At the end of the call, the clear signal originated by the caller is extended forward to the terminating exchange and that from the called customer is sent back to the originating exchange. Both exchanges then release their connections.

For a long-distance call the charging rate is normally determined at the trunk exchange as part of the translation of the area code of the called customer. When metering is employed it is therefore necessary for the supervisory apparatus at the trunk exchange to send meter pulses back over the junction to the originating local exchange. When ticketing is employed the calling-line identification (CLI) must be sent forward from the local exchange at the start of the call, instead of meter pulses being sent back during it.

In a network of electromechanical exchanges, signals between exchanges are sent over the same inter-exchange circuits as the calls to which they relate. This is called *channel-associated signalling*. In a network of stored-program-controlled exchanges, inter-exchange signals are generated by the processor in one exchange and acted upon by the processor in the other. The signals can therefore be transferred over a high-speed data link directly between the two processors, instead of being transmitted over the speech connection. This is called *common-channel signalling*. Methods used for both types of signalling are described in Chapter 8.

7.2.3 State transition diagrams

As described in Section 7.2.1, a call progresses from one state to another in response to events. One event is the arrival of a signal. This may result in the performance of an action, often called a *task*. The combination of the present state and a new event defines a task and performing this results in the next state. Sometimes there can be more than one possible next state, the choice depending on external information (e.g. depending on whether a called line is free or busy).

This sequence of operations can be clearly described by means of a *state transition diagram* (STD). An international standard for such diagrams[4] has been produced by the CCITT; it is known as the *Specification and Description Language* (SDL). The basic symbols defined for use in STDs are shown in Figure 7.3(a). They are:

- *State boxes* These are labelled with a number and a title. Additional information may also be included in the box if required.
- *Event boxes* These have an indented arrow, indicating whether the event corresponds to the receipt of a forward or backward signal.
- *Action boxes* These are rectangular boxes, except when the action is the sending of a signal. The box then has a protruding arrow, indicating whether the signal is sent forward or backward.
- *Decision boxes* The basic symbol is the diamond-shaped box used in computing flow charts. However, if more than two decisions are possible, the modified symbol shown in Figure 7.3(b) is used.

For a complex system, the STD may extend over several pages. These can be joined by the connector symbol shown in Figure 7.3(a).

Example 7.1

Draw a state transition diagram, using the SDL symbols, for a local telephone call. To simplify the diagram, omit events and states arising from misoperation (e.g. time-outs and parking).

The diagram is shown in Figure 7.4. The events and states shown are those described in Section 7.2.1.

State transition diagrams may be drawn for a complete system, for a subsystem or for functional units within it. The amount of detail shown by the STD may be expected to increase at the lower levels of system implementation. For example, Figure 7.4 shows a single action resulting from receipt of the address signal. However, the STD for a register, or for the register task in a central processor, would show receipt of the address information digit by digit and the action performed after each digit is received.

An STD can be used to describe actions performed by a hardware functional unit

[184] *Control of switching systems*

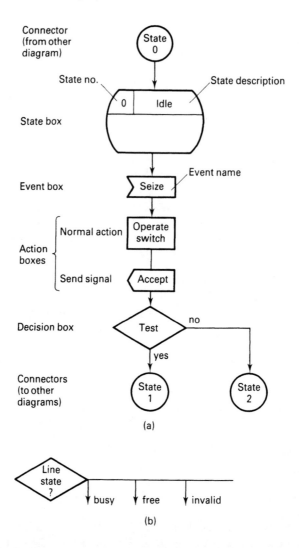

Figure 7.3 Symbols used in state transition diagrams (CCITT Recommendation Z 101). (a) Basic symbols. (b) Multiple decision box.

or a software program. In principle, the STD for a system or subsystem is independent of whether its actions are performed by hardware or software. In practice, however, the choice of either a hardware or software solution will often affect the way in which a system is divided into functional units and thus result in different STDs for these functional units.

The state transition diagram can specify precisely and unambiguously the functions to be performed by a system and its subsystems. It is therefore an important tool that is useful at every stage in the life cycle of a switching system, as follows:

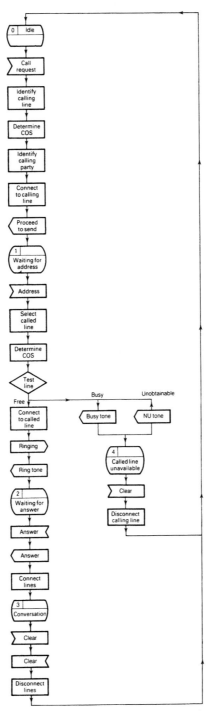

Figure 7.4 Simplified state transition diagram for a local call.

- Initial specification
- Design (of both hardware and software)
- Laboratory testing
- Manufacturing testing
- Installation and commissioning
- Acceptance for service
- Operation and maintenance
- Training of staff.

The graphical representation of SDL described above is not its only form. SDL supports a text form which can be machine processed. This has a formal syntax and semantics that assign unambiguous meanings to SDL constructs.

7.3 Common control

A common control performs a specific call-processing function. Thus, the control of the switching system employs *functional division*. A common control is brought into a connection only when required and released when it has performed its task. Thus, its actions serve different calls on a time-division basis. If several common controls are required to operate concurrently in order to handle the call-processing traffic load, then there is also space division.

Switching networks are lost-call systems. However, when common controls are busy, calls offered to them are not normally lost; they are delayed. For example, if all the registers in an exchange are busy, a caller merely has to wait longer for dial tone. If a central processor is used, tasks for processing queue until the processor can handle them. The traffic performance of common controls can therefore be analyzed using queueing theory, as described in Section 4.7.

The output of a control unit is a function of both its present input and previous inputs. It is thus a sequential logic circuit and must possess both combinatorial logic and memory. Common control units have been designed using relays, electronic digital circuits and stored-program control (SPC). The use of SPC enables the same logic circuits to perform different tasks under the control of different programs. An SPC common control can therefore be more flexible than one using wired logic. In the limit, a single SPC control can perform all the call processing needed for an exchange, leading to centralized control.

There is extensive literature on the design of both hardware[5,6] and software[7,8] for digital logic systems, so no further consideration will be given here to this subject. Instead, the interconnection of common controls and the trunks of switching networks in order to exchange signals will now be discussed. The principal methods that have been employed are as follows:

- Auxiliary switching networks
- Tree networks

- Use of the main switching network
- Buses
- Scanning.

A separate small network using switches similar to those in the main switching network can be used to connect common controls to trunks as required. For example, as shown in Figure 3.9, uniselectors were added to the step-by-step system to connect registers to the train of selectors. In crossbar systems, as shown in Figure 3.19, crossbar switches are used to connect registers and senders to trunks.

Since a number of trunks may request the use of a common control at the same time, contention can arise. A circuit in a common control used to resolve this contention is called a *one-only selector*, an *allotter*, or an *arbiter*.[9,10] Its operation is as follows:

1. If the arbiter is free, then an incoming seize signal is accepted by it for the common control.
2. If it is busy, incoming seize signals are rejected.
3. If it is free, but two or more seize signals are received simultaneously, only one is accepted and the others are rejected.
4. As an alternative to 2 and 3, incoming requests are queued. Seize signals are stored and, when the common control gives a clear signal, a previously received seize signal is accepted.

If a function is provided by several common controls, then a further one-only selector is needed to select a particular free control to respond to an incoming request.

If there is only a single common control, or a small number, it may be more economic to provide each with a selecting tree[5] than to use a more complex switching network. A tree consisting of n relays or electronic elements can connect a common control to 2^n trunks.

Instead of using an auxiliary switching network, the main switching network may be employed for connecting common controls by using the crank-back principle. In the Bell No.5 crossbar system, shown in Figure 3.19, an initial connection is made from a calling line to a subscribers' register through the main switching network. After the outgoing line has been selected, this connection is replaced by one between the calling and called lines. Crank-back is also used in the Bell ESS No.1 system.[2]

This principle is extensively used in the TXE4 system, where it is called *serial trunking*. In this system, as shown in Figure 3.24, registers are connected to the same side of the switching network as customers' lines and junctions. Thus, all connections pass through the switching network in both directions.

The connections used in making an own-exchange call are shown in Figure 7.5(a). When a customer calls, an initial connection (path 1) is made to a register via a through link (i.e. a link which does not contain a transmission bridge). When the register has received the address information, a connection is then made between the calling and called lines (path 2). This uses a link that contains a transmission bridge and provides supervisory facilities. When the second connection has been tested for

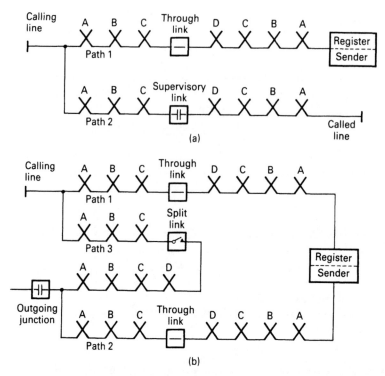

Figure 7.5 Serial trunking in the TXE 4 system. (a) Local call. (b) Outgoing junction call.

continuity, the first connection is released. The connection to the register is held until this has been done, so that a second attempt can be made if the test fails.

The connections made for an outgoing junction call are shown in Figure 7.5(b). An initial connection (path 1) is made from a calling line to a register, via a through link, as before. When the address information has been obtained, a connection (path 2) is made from the sender side of the register to an outgoing junction via another through link. The register then sends out the required routing digits over the junction. A final connection (path 3) is then made between the calling line and the junction. This uses a split link, which does not complete the connection until the register has finished sending. When sending is complete and the continuity of path 3 has been checked, path 3 is switched through at the link and paths 1 and 2 are released. Supervision of the call is provided by the outgoing junction terminal unit.

In a space-division switching system, call-progress tones are usually applied by the supervisory unit in each connection. In a digital time-division switching system, tones may be supplied from a common unit over a PCM highway, to which connections are made through the main switching network. An example of this is shown in Figure 6.1. The method is practicable because the tone signals are synthesized in the form of trains of PCM samples at 64 kbit/s.[11]

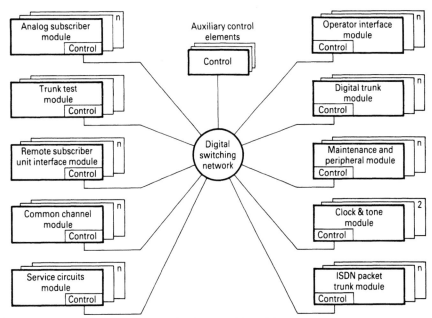

Figure 7.6 System 12 architecture.

Another example of the use of the main switching network is the Alcatel System 12, shown in Figure 7.6. Control of the exchange is distributed among a number of processors, each performing a specific function.[11,12] These communicate with each other and with customers' connections by means of temporary connections made through the main switching network, as required.

When it is necessary to exchange signals between a number of functional units, but the signal exchanges need only take place one at a time, the units can be connected by a common bus. If information is transmitted in serial mode, the bus can be a single wire. If parallel transmission is used, a wire is needed for each digit (e.g. eight wires to transmit 8-bit bytes). An example is the use of a bus to connect a processor unit, or units, with stores and input and output ports, as shown in Figure 7.7.

In scanning, electronic gates, forming the equivalent of a rotating switch, connect a common control to each trunk in turn. Since, at any time, only one trunk can communicate with the common control, there can be no contention. However, scanning presents the following requirements, which may conflict:

1. In every cycle the scanner must connect the common control to each trunk for a sufficiently long period to exchange the required signals.
2. The period of a complete scanning cycle must be sufficiently short for the common control to detect every change that occurs in the states of the trunks being scanned.

If these requirements do conflict, it may be possible to use start–stop scanning.[10] If the states of most trunks are unchanged since the previous cycle, the scanner can step

[190] Control of switching systems

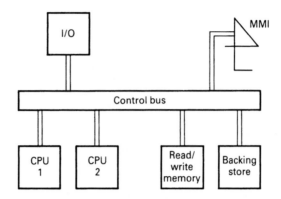

Figure 7.7 Functional units interconnected by a bus. CPU = central processor unit, I/O = input/output ports, MMI = man–machine interface.

over them rapidly and then stop for a longer period on a trunk whose state has changed, in order to allow sufficient time for a signal exchange. The requirements for a scanning system are considered in more detail in Example 7.2 below.

Scanning is used in the TXE 4 system shown in Figure 3.24, in the Bell No.1 ESS system[2,13] and in other electronic switching systems.[2] Scanners serving customers' lines and junctions enable the controlling processors to detect their calling and clearing signals.

Scanning may be inherent when time-division switching is employed, making it unnecessary to provide separate scanning equipment. If analog circuits are multiplexed in a 2 Mbit/s PCM system, their signalling states are scanned by the primary multiplexer and transmitted in time-slot 16. Thus, an exchange can determine the states of these circuits simply by monitoring time-slot 16. For a 1.5 Mbit/s PCM system, the exchange needs to monitor bit no. 8 for each channel.

Example 7.2

1. A continuous cyclic scanner is to be used to signal the loop/disconnect states of customers' lines to a common control, to enable it to detect call and clear signals. It is required to signal a change in the state of a line to the common control within 100 ms. To reduce the chance of misoperation due to a voltage transient on a line, the common control does not act on a change of state unless the new state is present during two out of three consecutive scans. If the scanner is to serve 500 lines, what is its output digit rate?

   Time taken for 3 scans = 100 ms
   ∴ Scanning rate = 30 scans/second
   500 lines are scanned during each period
   ∴ Output digit rate = 30 × 500 bit/s = <u>15 kbit/s</u>

2. An alternative design is to use a 500-line start–stop scanner. The common control causes this to step rapidly over lines whose state has not changed since the previous scan, but to pause on a line whose state has changed in order to register the new state reliably. A scanning period of 100 ms can now be used. This must not be overrun when up to 10% of lines have changed state since the previous scan. If the output digit rate is to be the same as for the previous scanner, how long is available for testing a line whose state has changed?

> 450 lines have not changed state.
> Time taken to scan these = 450/15 = 30 ms.
> ∴ 100 − 30 = 70 ms is available to test the remaining 50 lines.
> ∴ Time per line = 70/50 = 1.4 ms

3. Calling lines are connected to trunks which are scanned continuously, in order for a common control to detect dialled digits. A change in state of a trunk will not be registered unless the new state is present in two out of three successive scans. Customers' dials generate a nominal 10 pulses per second, with a tolerance of from 7 to 12 pulses per second. The dial pulses have a nominal break/make ratio of 67% to 33%, with a tolerance on the break period from 63% to 72%. If the output digit rate is to be the same as before, how many trunks can the scanner serve?

> The scanner must detect the minimum make period.
> This is 28% × 1/12 second = 23.333 ms.
> Three scans are used, so the scanning period is
> 23.333/3 = 7.78 ms.
> At 15 kbit/s, the number of trunks scanned in 7.78ms is
> 15 × 7.78 = 116 trunks.

7.4 Reliability, availability and security

Switching systems must be reliable. The Strowger system was very fault tolerant because of the distributed nature of its control. Many faults occured in individual items of apparatus, but these had little effect on the overall quality of service. Complete failures of entire exchanges hardly ever happened. In contrast, the use of common control, and particularly centralized control, makes the operation of an exchange critically dependent on a small number of equipments. These must be designed to high standards of reliability[14] to obtain a long mean time between failures (MTBF). In an SPC system, the software must also be very reliable.

Once an equipment has failed, the fault must be diagnosed and rectified. The longer the MTTF and the shorter the mean time to repair (MTTR), the greater the proportion of time for which an equipment provides service. This proportion is called the *availability* of the equipment.[1] Thus:

$$\text{Availability} = \frac{\text{MTTF}}{\text{MTTF} + \text{MTTR}}$$

Control of switching systems

Table 7.1 CCITT Unavailability objectives for telephone exchanges

Entire system	$< 1.5 \times 10^{-5}$
	(e.g. $<$ 6 hours outage in 50 years)
Customer's line	$< 10^{-4}$
	(e.g. $<$ 1 day in 25 years)
Inter-exchange circuit	$< 10^{-4}$
Emergency calls	$< 1.5 \times 10^{-5}$
Basic telephony service	$< 10^{-4}$
Supplementary services	$< 10^{-3}$
Charging	$< 10^{-4}$
Traffic measurement	$< 10^{-3}$
Admin. operations	$< 10^{-2}$

Source: CCITT.[15]

The availability gives the probability that the equipment will operate correctly when required. The probability that the equipment will not operate is called its *unavailability*. Thus:

$$\text{Unavailability} = 1 - \text{availability} = \frac{\text{MTTR}}{\text{MTTF} + \text{MTTR}}$$

The failure of a complete exchange is a very serious matter. It should not happen more often than once every 50 years or more; the availability required therefore approaches unity. A fault on a customer's line may be serious to that customer, but it does not affect others. The failure of a single call is usually only an inconvenience; it can be redialled. The failure of a central control to carry out administrative functions causes problems to the operating company, but it does not inconvenience customers. As a result of these considerations, the CCITT has recommended objectives for unavailability.[15] and these are listed in Table 7.1.

If common control or central control is employed, an exchange that had the minimum possible configuration of equipment would be unlikely to obtain adequate availability, even if constructed with the most reliable components that can be obtained. Measures need to be taken to provide *security*, that is to ensure correct operation even when faults are present. This entails providing additional equipment. The security measures that are used are as follows:

- Line circuits: none
- Switching network: none (or partial duplication)
- Common controls: 1 in n sparing
- Central processors: replication.

A terminating unit for a customer's line or a junction[9,11,16] contains relatively few components. Thus, it should suffer faults less frequently than the line itself. No additional measures are needed to obtain the required availability.

A space-division switching network normally provides a choice of many different paths for each connection. Consequently, a fault affecting one trunk in the network causes only a very small degradation in the grade of service and this can be tolerated. In a time-division switching network, the paths are time shared; thus a failure of one has a greater effect. However, time sharing reduces the amount of equipment, compared with a space-division network, so it is less costly to provide redundant equipment and this is sometimes done. For example, as shown in Figure 6.7, the space switch in the T–S–T network of System X is duplicated. This also has the advantage that, if the exchange needs to be extended, one switch can carry the traffic while the other is being augmented.

For common-control equipments, such as registers, 1-in-n sparing is used. If n equipments are sufficient to handle the traffic, $n+1$ are provided. Failure of one of them has little effect on service and the probability of another failing before the first can be repaired is very small.

If all calls had be handled by a single central processor, its failure would put the entire exchange out of service, which would be intolerable. Therefore, except in small PBXs, at least two processors are provided. This also has the advantage that, if the software of the exchange is to be upgraded, one processor can carry the traffic while the other is being reloaded. The security methods used are discussed in more detail in Section 7.5.1.

7.5 Stored-program control

7.5.1 Processor architecture

In order to obtain adequate security a switching system with central processors requires a minimum of two. If two processors are used, each must carry the full traffic load if the other fails, so each must have sufficient processing capacity to do this. Two processors may be configured to operate in the following ways:

- Worker and standby
- Load sharing
- Synchronous operation.

In a system with a worker and a standby, either *cold standby* or *hot standby* may be used. The term 'cold standby' is really a misnomer; the spare processor must be switched on and working. However, its memories are not updated, so there will be disruption of calls when change-over takes place. In hot standby, the spare processor is constantly updated with details of all calls; it can therefore assume control without disruption.

In load sharing, both processors are working independently. Thus, at any time, they will be performing different tasks for different calls. This makes it difficult for the remaining processor to take over from the failed one without disruption of traffic. Contention between the processors must be prevented and each must update the other

[194] Control of switching systems

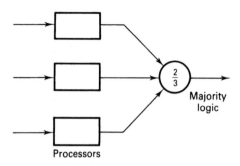

Figure 7.8 Triplication with majority logic.

as calls are set up and cleared down. Load sharing can be extended beyond two processors. Additional processors can be added as traffic grows; this leads to a multiprocessor configuration, as described below. If there are more than two processors, 1-in-n sparing can be used instead of full duplication.

In synchronous dual operation both processors receive identical inputs and operate in synchronism to produce the same output at the same time. Comparison of the outputs, to check if they are identical, gives immediate fault detection. A discrepancy leads to the running of a test program to discover which is the faulty unit. This is then taken out of service while the other proceeds with the task. However, this method cannot detect a software fault. Since both processors are executing the same program, they produce the same error and loss of service will result. Synchronous dual processors are used, for example, in the No.1 ESS system[12,13] and the AXE system.[11,12]

An extension of this principle is to use three processors with majority-decision logic at their outputs, as shown in Figure 7.8. If one of the processors fails, the majority logic still gives the correct output and the faulty processor is identified as the one which disagreed with the other two. This principle is used by the main control units of the TXE 4 system[2] and by the control units of the digital switching subsystem of System X (as shown in Figure 6.7).

In a large exchange the traffic will require more than a single processor to handle it. Either a *multicomputer* or a *multiprocessor* architecture may be used. In the former the unit that is replicated is a complete computer (with its memory). The load is usually divided between them on a geographical basis, each handling processing for a different part of the exchange. This represents a reversion towards progressive control. An example[11,12] is System 12, shown in Figure 7.6.

In a multiprocessor system the processors share program and data stores; however, each can control any part of the exchange, using any program. Each processor repeatedly runs diagnostic programs to check peripherals, input/output channels, memories, etc., and the processors check each other by exchanging signals according to strict protocols. System X uses clusters of up to four processors.[11] A

Stored-program control [195]

small exchange may have a single cluster of only two and a large exchange may have up to eight clusters of four processors.

Failure of a processor system may be due to a hardware fault or a software fault. A hardware fault may occur at any time because of failure of a component which has hitherto been healthy. However, a software fault is usually due to an error which has been present in a program since the day when it was written.

In a system with replicated processors a hardware fault does not cause failure of the system. The fault is detected by the software that is running or by a routine-testing program. Recovery is by *reconfiguration*; the faulty unit is automatically taken out of service and, if necessary, replaced by equipment which has been on standby. A diagnostic program is then run to find the location of the fault within the unit.

When processing encounters a fault in a program or data, the program being run cannot proceed any further. Recovery is obtained by means of a *roll-back*. The processor is loaded with data that represent the state of the system a short time prior to the failure. For this to be possible, data representing the states of the parts of the system must be regularly dumped to a backing store. When the failure is detected, an error message is sent that contains sufficient data to enable the fault to be subsequently analyzed and corrected. The roll-back causes loss of calls which are being set up, but not of connections that have already been established.

Roll-back is usually effective, because 'bugs' that lie on the most common processing paths have been eliminated during testing while the system was being developed. Consequently, failures during service should occur only from rare combinations of circumstances. Rolling back provides a set of circumstances different from that present when the fault occurred. However, if another failure occurs within a short time after the roll-back, this shows that the system has not recovered. More roll-backs are then made, each going back further in time, until recovery is obtained. The last resort is a complete initialization of the system.

Example 7.3

An SPC switching system has two central processors, which set up all connections. Normally, the load is shared evenly between them; however, if one fails, the other handles all the traffic.

During the busy hour, the total traffic offered to the exchange is 1000 E and the average holding time of calls is 3 minutes. The processing times for calls can be assumed to have an exponential distribution, with a mean time of 162 ms.

1. Find the percentage of tasks which must queue for processing during the busy hour when both processors are operating.
2. Find the percentage of tasks which must queue for processing during the busy hour when one processor is out of service.
3. What will happen if the number of calls offered to the exchange during the busy hour increases by 25% when one processor is out of service?

Traffic on the exchange is 1000 E = Ch/T

$$\therefore C = 10^3 \times 60/3 = 20\,000 \text{ calls per hour.}$$

Traffic offered to processors is

$$A = Ch/T = 20 \times 10^3 \times 162 \times 10^{-3}/3600 = 0.9 \text{ E}$$

1. From equation (4.13):

$$\frac{1}{P(0)} = \sum_{x=0}^{1} \frac{A^x}{x!} + \frac{2A^2}{2!(2-A)} = 1 + 0.9 + 0.9^2/1.1$$

$$= 2.64$$

From equation (4.15):

$$\text{Probability of delay} = E_{2,2}(A) = \frac{A^2}{2!} \frac{2}{2-A} P(0)$$

$$= \frac{0.9^2}{2} \times \frac{2}{1.1} \times \frac{1}{2.64} = 0.279$$

i.e. 27.9% of tasks queue for processing.

2. $E_{2,1}(A) = A = 0.9$
i.e. 90% of tasks queue for processing.
3. $A = 0.9 \times 1.25 = 1.125$
More than one erlang is offered to a single server, so the queue length increases continually and the system crashes.

7.5.2 Distributed processing

The reduction in the cost of processing brought about by the microprocessor has enabled the control of switching systems to be decentralized. Instead of all processing being performed by a central processor, routine tasks associated with parts of the system (e.g. the route switch, concentrators, etc.) or particular functions (e.g. line scanning, digit reception, signalling, etc.) are delegated to separate small processors. Such processors are sometimes called *regional processors*.

Since a connection involves a number of different functions and passes through different parts of the system, a central processor is still required to direct the regional processors and to perform the more complex tasks. When there is such a hierarchy of processors, messages may be exchanged between central and regional processors over a bus[11,12] (as in the AXE system), through an auxiliary switch provided for this purpose[11,12] (as in the No.5 ESS system), or through the main switching network,[11,12] either by semi-permanent connections (as in the EWS-D system and System X) or by switched connections (as in System 12).

In early SPC systems, fully centralized processing placed a restriction on the amount of directly addressable memory that was available. Thus, the dimensioning of common storage was a critical issue. Nowadays the use of distributed microprocessors, each with its own extensive random-access memory (RAM), has removed this limitation.

Since regional processors relieve the central processor of tasks such as line scanning and digit reception, the complexity of its call-processing software is reduced and reliability and maintainability are improved. As a result of data being stored at the regional processors, no processor requires access to data that is beyond its direct, addressing capability. This eliminates the need for frequent disk access and increases the speed of call processing. However, copies of all software still need to be held in a backing store for reloading to RAM if the contents of a memory become corrupted.

7.5.2 Software

Like any digital computer, an exchange processor uses stored programs for processing input data to give outputs. Many different calls must be processed concurrently in real time. A processor therefore performs successive tasks for one call interspersed with those for other calls. Each time a processor returns to a call, it needs to determine its state by reading data from memory. The store areas that are accessed include:

1. *The line store* In addition to class of service, the status of the line is stored (e.g. free, busy or parked).
2. *The call record* Data stored for each call include: time of origin, equipment number of line, equipment numbers of other equipments associated with the connection, actions already taken, switch paths used, last signal received, current state of the call, address digits received, time of answer signal, time of clear down.
3. *Translation tables* Translations are required for address codes, for EN-to-DN translation and for DN-to-EN translation.
4. *A map of the switching network* If the 'map in memory' technique is used, the memory must contain a bit for each link in the switching network to indicate whether it is free or busy. The call set-up program looks at these in order to select a suitable path for each connection. (However, if the 'map in network' technique is used, the P wires of the links are scanned to determine which are free and which are busy.)

The complete software for an exchange consists of application programs and an operating system. The application programs each deal with part of the system operation. They form a set of modules, for example as shown in Figure 6.1. The operating system provides the environment in which the applications programs run. Its functions include:

1. *Control of timing* to ensure that processes are executed at specified times (e.g. for alarm calls) or periodically (e.g. for scanning).
2. *Scheduling* to ensure that processes are carried out with a predetermined order and timing.
3. *Interrupt handling* to ensure that high-priority processes are given precedence.

4. *Interprocess communications* to facilitate communication between software processes and, if necessary, between processors.
5. *Input/output control* to allow communication between the processor and the system it controls.
6. *Storage management* to control the storage and accessing of exchange, customer and call data.
7. *Human–machine communication* to provide the protocols for communication between terminals and the processor.

In addition to call processing, a central processor performs many administrative tasks. These include: database management, fault diagnosis and alarm processing, automatic testing, traffic recording, changing customers' facilities and directory numbers, changing routings and routing codes and the generation of exchange management statistics. These tasks require a large amount of software. However, since they are performed infrequently, they use much less processing time than call processing. For example, results published by Tokita et al.[17] for the HDX-10 system showed that administration programs comprised 71% of the total software, but used only 0.1% of the total processing time.

Software for SPC switching systems is usually written in a high-level language, such as CHILL (the CCITT high-level language[18]). The source code is then compiled into a final object code which is stored in the program memory of the processor. However, some programs for time-critical applications may be written in assembler language. For further information on software design, the reader is referred to the literature.[7,8,11,19,20]

Call-processing programs can be derived directly from state-transition diagrams in SDL.[11,21,22] The SDL description, in text form, is machine read and stored in memory in the form of data structures of linked lists and translation tables. An interpreter program is written to access the lists and tables and to process the call by interpreting the data within them. The call processing is then said to be data driven.

A switching system is required to be suitable for many different kinds of exchange: large and small exchanges, local and trunk exchanges, etc. Many system functions, e.g. basic call processing, are required for all of these. Other functions are required in certain applications, but not others. For example, the provision of customer facilities is not required for a trunk or international exchange. The use of modular software enbles a manufacturer to supply a telecommunications operator economically with just what is required for each exchange.

The first step towards the *software build* for a particular exchange is to select the appropriate program modules from a software library, e.g. for the required customer facilities, types of line termination and signalling system, etc. These are then linked by a link loader program to form the system software for a standard exchange of the type required.

The next step is to link this generic software with data that differ from exchange to exchange. The exchange-dependent data include the number of customer lines and their classes of service, junction quantities, calling rates and traffic loads. These result in different switch and processor configurations, different equipment quantities and different requirements for memory capacity. The result is the complete software build for an exchange, ready for commissioning and service in the network.

7.5.4 Overload control

A processor system has only finite capacity, so there is a limit to the number of calls per hour that an SPC exchange can handle successfully. The capacity of an exchange is therefore specified in terms of its number of lines, the traffic in erlangs and the number of *busy-hour call attempts* (BHCA). For example, the System X local-exchange system[23] shown in Figure 6.1 can accommodate up to 60 000 line terminations, 10 000 E of traffic and 500 000 BHCA.

If the processors in an ideal system could handle N calls per hour, the throughput of calls per hour would equal the number of calls offered until this reaches N. As the number of calls offered per hour increases further, the throughput of successful calls would remain constant at N, as shown by curve (a) in Figure 7.9.

In practice, some processing time is used by unsuccessful calls as well as by successful ones. If the number of offered calls exceeds N, some are unsuccessful. Since these use processing time, the number of successful calls is less than N. As the number of offered calls increases further, the number of unsuccessful calls increases and the throughput of successful calls decreases, as shown by curve (b) in Figure 7.9.

In order to prevent this unwanted behaviour, *overload control* is introduced. This restricts the load on the processors to no more than can be handled by rejecting some of the traffic offered to the exchange. Consequently, as shown by curve (c) in Figure 7.9, the throughput of the exchange remains at its maximum level when the number of offered calls increases beyond N per hour.

The overload-control program monitors the queues of tasks awaiting processing and takes progressively more drastic actions as these increase in length beyond

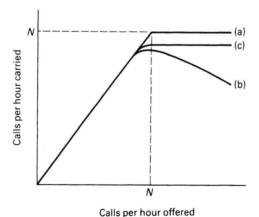

Figure 7.9 Overload control. (a) Overload characteristic of ideal system. (b) Overload characteristic of system without overload control. (c) Overload characteristic of system with overload control.

[200] Control of switching systems

predetermined thresholds. These actions are not discontinued until the queue lengths have subsided below the original thresholds. The first action is to discontinue 'background' tasks, such as routine maintenance programs and the output of data other than for billing. Next, a proportion of new calls is rejected. Finally, all calls are rejected, except any having priority (e.g. to emergency services). The processing of existing connections is not discontinued. If existing calls were not cleared down on termination, the overload would not be overcome; moreover, customers would be overcharged.

The exchange of human–machine messages also continues when overload control is operating, since these may be concerned with human actions designed to cure the problem. A traffic overload in one exchange may be part of a more widespread overload in the network. A network management centre may then intervene with the overload controls in several exchanges, as described in Section 10.10, in order to reduce the congestion in the network.

Note

1. Use of the term 'availability' in this context should not be confused with use of the same word in teletraffic engineering to denote the outlet capacity of a switch.

Problems

1. The control circuit for operating a crossbar switch must meet the following requirements:

(a) Action is initiated by closure of a relay contact which remains closed only long enough for the select magnet to operate fully.
(b) The circuit shall make sure that the select magnet has operated before starting timing for hold magnet operation.
(c) A delay (to allow the select finger to settle) shall intervene between completion of select magnet operation and energization of the hold magnet.
(d) The circuit shall make sure that the hold magnet has locked operated to a separate ground before releasing the select magnet and restoring to normal.

(The hold and select magnets are each provided with a single off-normal make contact to signal to the control circuit that they have operated.)

Draw a state transition diagram to specify the operation of the control circuit.

2. Draw a state transition diagram for the functions of a customers' register–sender in a city telephone exchange.

The city has a seven-digit numbering scheme. The first three digits are an exchange code and the last four denote a customer's number on the exchange. Code '999' provides access to emergency services and codes of the form '1xx' provide access to other services.

An initial digit '0' is the trunk prefix, which causes calls to be routed to a trunk exchange. This prefix digit is followed by a 10-digit customer's national number, or up to 14 digits for an international call.

To simplify the diagram, only functions required for calls to customers in the city area need be shown.

3. In the TXE2 system, when a customer calls, operation of the line relay applies a single pulse of current to the calling-number generator. This initiates a connection and provides EN-to-DN translation. There are two calling-number generators; at any time, one is waiting for calls and the other is on standby. Provided that two calls arrive with an interval between them of more than 100 μs, both numbers will be generated satisfactorily. The system is designed to handle a total traffic of 250 E. For an average holding time, this corresponds to 7500 calls in the busy hour.

What is the probability that a call attempted during the busy hour will be unsuccessful due to a clash of calling pulses entering the calling number generator?

4. (a) A switching network has connectivity k. During the busy hour, the blocking probability of any path through the network is b. Show that the average number of paths, $\bar{x}$, tested in order to establish a connection to a selected outgoing trunk is

$$\bar{x} = \frac{1-b^k}{1-b}$$

It can be assumed that paths being busy are independent random events. For $1 < x < k$, the control unit selects the xth choice if its previous $x-1$ choices are busy and the xth is free. However, the search must stop at the kth choice, whether it is free or busy.

(b) A fully interconnected three-stage space-division switching network has 20 secondary switches. Find the average number of paths tested to make a connection to a particular outgoing trunk when the link occupancy is (i) 0.3 E, (ii) 0.5 E, (iii) 0.7 E, (iv) 0.9 E.

5. A central-processor system contains two identical units, each of which can carry the full load. The mean time to failure (MTTF) of each unit is 1000 hours. It can be assumed that failures of the units are independent random events.

Estimate the MTTF of the system if the mean time to repair (MTTR) for a unit is (a) 10 hours, (b) 1 hour.

6. A small telephone exchange has an originating traffic of 90 E during the busy hour and the average holding time of calls is 3 minutes. It has nine registers. The holding times of the registers have a negative exponential distribution with a mean of 10 seconds.

(a) On average, how many calls originate in the busy hour and how much traffic is offered to the registers?
(b) From Figure 4.10, estimate the percentage of calls that encounter delay in obtaining a register.
(c) What proportion of calls will encounter delay in obtaining a register if one register is out of service?
(d) What will happen if the calling rate increases by 75% while one register is out of service?

7. A telephone exchange has two markers, operating on a basis of load sharing. Dialled digits are received by registers which attempt to seize a marker after having analyzed the received address digits. If both markers are busy, registers queue. Marker holding times have a negative exponential distribution with a mean time of 100 ms. In the busy hour, the average number of call attempts is 30 000.

For the busy hour, find the following:
(a) The proportion of calls that must wait for a marker.
(b) The average delay.
(c) The proportion of calls for which the delay exceeds 0.5 second.
(d) The average delay when one marker is inoperative due to a fault.
(e) The proportion of calls for which the delay exceeds 0.5 second when one marker is inoperative.
(f) What will happen if the number of busy-hour call attempts increases by 40% while one marker is inoperative.

8. An SPC exchange has two central processors that share the processing load. Tasks awaiting execution are dealt with in their order of arrival. Various tasks have different processing times and it can be assumed that they have a negative exponential distribution.

On average, 14 400 calls are made in a busy hour. On average, 20 processing tasks are performed for each call and its total processing time is 200 ms. Find:

(a) The mean value of the total of the delays encountered for these tasks for each call during the busy hour.
(b) The probability that this total delay will be zero.
(c) The mean value of the total of the delays encountered for these tasks for each call during the busy hour if one processor is out of service.

9. (a) In order to improve the security of the above exchange against processor faults, the manufacturer designs a Mark 2 version of the system. This has three processors that share the processing load evenly. The number of processing tasks per call and their average duration remain the same.

When the exchange has a full busy-hour load of 14 400 BHCA, find the percentage of tasks performed without delay, both under normal conditions and when one processor is out of service.

(b) The manufacturer designs a Mark 3 system which uses distributed processors for many of the routine tasks. These share the total processing operations with two load-sharing central processors. As a result, the average number of tasks performed for each call by a central processor is reduced to ten and their average total duration to 150 ms.

If the probability of tasks being delayed is to be the same as for the Mark 1 system, how many more calls can the Mark 3 system serve in the busy hour?

(c) Comment briefly on the relative performances of the three systems.

10. The processors of an SPC telephone exchange can successfully handle N calls per hour. The processor time used for a successful call is T. If the exchange is overloaded, each unsuccessful call uses processor time τ.

(a) Show that, without overload control, the throughput (in calls per hour) rises linearly until the number of attempted calls is $x = N$ and then falls linearly to zero at $x = NT/\tau$.

(b) If $\tau = 0.2T$, what percentage of calls is lost at 50% overload?

References

[1] Chapuis, R.J. and Joel, A.E. (1990), *Electronics, Computers and Telephone Switching: a book of technological history*, North-Holland, Amsterdam.
[2] Joel, A.E. (ed.) (1976), *Electronic Switching: central offices of the world*, IEEE Press, New York.
[3] Atkinson, J. (1950), *Telephony*, Vol.2, Pitman, London.
[4] CCITT Recommendation Z101, *Specification and Description Language*.
[5] Keister, W., Ritchie, A.E. and Washburn, S.H. (1952), *The Design of Switching Circuits*, Van Nostrand, New York.
[6] Lewin, D. and Prothero, D. (1992), *Design of Logic Systems*, 2nd edn, Chapman & Hall, London.
[7] Norris, M. and Rigby. P. (1992), *Software Engineering Explained*, Wiley, Chichester.
[8] Sommerville, I. (1992), *Software Engineering* 4th edn, Addison-Wesley, Reading, MA.
[9] Hills, M.T. (1979), *Telecommunications Switching Principles*, Allen and Unwin, London.
[10] Flowers, T.H. (1976), *Introduction to Exchange Systems*, Wiley, Chichester.
[11] Redmill, F.J. and Valdar, A.R. (1990), *SPC Digital Telephone Exchanges*, Peter Peregrinus, Stevenage.
[12] Joel, A.E. (ed.) (1982), *Electronic Switching: digital central offices of the world*, IEEE Press, New York.
[13] Talley, D. (1982), *Basic Electronic Switching for Telephone Exchanges*, Hayden, Rochelle Park, NJ.
[14] Smith, D.J. (1972), *Reliability Engineering*, Pitman, London.

[15] CCITT (1981), 'Economic and technical aspects of the choice of telephone switching systems', GAS 6 ITU, Geneva.
[16] McDonald, J.C. (ed.) (1983), *Fundamentals of Digital Switching*, Plenum Press, New York.
[17] Tokita, Y., Suzuki, T., Shoda, A. and Hiyama, K. (1978), 'ESS software architecture for multi-processor system HDX-10', *Proc. Conf. on Software Engineering for Telecommunications Switching*, Helsinki.
[18] CCITT Recommendation Z 200, *Communications high-level language*'.
[19] Hills, M.T. and Kano, S. (1976) *Programming Electronic Switching Systems*, Peter Peregrinus, Stevenage.
[20] Kawashima, H. and Nakajima, N. (1979), *Software Design for Electronic Switching Systems*, Peter Peregrinus, Stevenage.
[21] Saracco, R., Smith, J.R.W. and Reed, R. (1989), *Telecomunications Systems Engineering Using SDL*, North-Holland, Amsterdam.
[22] Belina, F., Hogrefe, D. and Sarma, A. (1991), *SDL with Applications from Protocol Specification*, Prentice Hall, Englewood Cliffs, NJ.
[23] Tippler, J. (1979), 'Architecture of System X: Part 1, an introduction to the System X family', *PO Elec. Engrs. Jour.*, **72**, 138–41.

CHAPTER 8

Signalling

8.1 Introduction

In a telecommunications network, signalling systems are as essential as switching systems and transmission systems. For a multilink connection, it is necessary to send signals in both directions between the the caller and the originating exchange, between the called customer and the terminating exchange and between exchanges.

   Signalling systems must obviously be compatible with the switching systems in a network. They must be able to transmit all the signals required to operate the switches. They must also be compatible with the transmission systems in the network in order to reach the exchanges that they control. Thus, the design of signalling systems is directly influenced by both switching and transmission requirements and the evolution of signalling has followed developments in switching and transmission.[1] The signalling methods used with different types of transmission system are described in Sections 8.3 to 8.5.

   Transmitted signals may be either *continuous signals* or *pulse signals*. An example of a continuous signal is the DC off-hook signal on a customer's line. A pulse signal may be either a single pulse or a coded group of pulses. An example of the latter is a decimal digit sent by loop/disconnect pulsing.

   Transmitted signals may be either *unacknowledged signals* or *acknowledged signals*. Address digits sent by customers are normally unacknowledged. When an acknowledgement signal is returned, it confirms receipt of the signal that was sent. Acknowledgements may be continuous or pulse signals. If pulse signalling is used, a signal may be repeated until it is acknowledged. To obtain more rapid signalling, receipt of a group of pulse signals may be confirmed by a single acknowledgement signal. When continuous signalling is used, a signal is sent until the acknowledgement is received and the acknowledgement signal persists until the original signal has been removed. This is called *compelled signalling* and it is the most reliable method. However, when a circuit has a long propagation time, compelled signalling is slow. Four propagation times elapse before the sending equipment has detected the end of the acknowledgement and can send another signal. Fully compelled signalling is therefore not used over satellite circuits.

Most inter-exchange circuits have used one-way working. On a route between exchanges A and B, a separate group of circuits is used for calls from A to B and from B to A. However, as shown in Section 4.6.2, fewer circuits would be required if traffic from A to B and from B to A were combined on a single group of trunks. Both-way working has therefore been used when circuits are long and expensive and the traffic level is not very high. Both-way signalling systems[1] are more complex than systems for one-way working. Identical apparatus is needed at each end of a circuit, since it may be seized from either exchange. Contention can arise, since the circuit may be seized simultaneously at each end. When this happens, the circuit is released and both calls are lost.

Traditionally, exchanges have sent signals over the same circuits in the network as the connections which they control. This is known as *channel-associated signalling*. For a simple telephone call only the following basic signals are required between exchanges:

- Call request or seize (forward)
- Address signal (forward)
- Answer (backward)
- Clear signals (forward and backward).

The introduction of stored-program control (SPC) enabled customers to be provided with a wider range of services than those available with electromechanical switching systems. It is desirable for customers to be able to use these enhanced services right across a network instead of only on their own exchange. For example, a customer wishing to divert incoming calls to another location should be able to make the diversion to any telephone in the network rather than only to telephones on the same exchange. These services require more signals to be transmitted between exchanges than have traditionally been provided. Since the signals are generated by the central processor in one exchange and sent to the processor of another exchange, they can be transmitted directly between the processors over a separate data channel. This is known as *common-channel signalling* (CCS) and is described in Sections 8.7 to 8.9.

Common-channel signalling is now widely used in public telecommunications networks, both nationally and internationally. It is also used in private networks for signalling between digital PBXs.

8.2 Customer line signalling

In a local telephone network loop/disconnect signalling is used for sending customers' call and clear signals to the exchange. Since there is a minimum line current that the exchange can detect, there is a maximum permissible line resistance. This limits the maximum length of line and the size of the area served by the exchange. (In addition to being limited by DC resistance, the length of lines is also limited by the permissible attenuation at voice frequencies. Ideally, both limits should be the same.)

[206] *Signalling*

Hz \ Hz	1209	1336	1477	1633
697	Digit 1	2	3	Spare
770	4	5	6	Spare
852	7	8	9	Spare
941	*	0	#	Spare

Figure 8.1 Frequency coding used by pushbutton (keyphone or touch tone) telephone sets.

When dial telephones are used, customers send address information by decadic pulsing. For each digit, the dial makes and breaks the circuit to send a train of up to 10 loop/disconnect pulses at approximately 10 pulses per second. The exchange is able to detect the end of each pulse train because the minimum pause between digits (e.g. 400 ms to 500 ms) results in a loop state significantly longer than the 'makes' during pulsing (e.g. 33 ms.)

A relay circuit to receive dial pulses[2] was required in every selector in a Strowger exchange. However, the introduction of registers reduced the number of dial-pulse receivers needed, so these could be more complex. This led to the introduction of push-button telephones, which send voice-frequency pulses and thus provide faster signalling.

A push-button (keyphone or touchtone) telephone uses *dual-tone multifrequency signalling* (DTMF). It sends each digit by means of a combination of two frequencies, one from each of two groups of four frequencies,[1] as shown in Figure 8.1. This is done to reduce the risk of *signal imitation*. Since each digit uses two frequencies, and these are not harmonically related, there is much less chance of the combination being produced by speech or room noise picked up by the telephone transmitter than if a single frequency were used.

In addition to the digits '1' to '0', the telephone keypad has buttons with the symbols '*' and '#'. These are used by SPC exchanges for facilities that are under the control of customers. For example, a customer who wishes incoming calls to be diverted to another telephone may key '*', followed by the appropriate instruction digits, before leaving. On returning, an instruction prefixed by '#' is keyed to remove the call diversion.

When a line serves a payphone additional signals must be transmitted for coin collection and return. For example, the exchange can make a polarity reversal for coin collection and apply a higher voltage for coin return. Some coin boxes require a coin to be inserted before dial tone is sent. These use a *ground start signal* instead of a loop seize signal. An earth is applied to the negative (R) wire of the line. A ground start signal may also be used to call a PBX from the main exchange. Otherwise, the line could be seized by an outgoing call from the PBX before the first burst of ringing current from the main exchange.

In an integrated-services digital network (ISDN), digital transmission is used for customer access. The signalling methods used are described in Section 8.10.

8.3 Audio-frequency junctions and trunk circuits

A two-wire junction circuit can transmit loop/disconnect signals, as used on customers' lines. On an amplified four-wire circuit, centre taps of line transformers can be connected, as shown in Figure 8.2, to provide a two-wire phantom circuit for signalling.

In customer-line signalling, the signals to and from the caller and those to and from the called customer are sent over different lines, as shown in Figure 7.2. However, for a junction call, the forward seize and clear signals and the backward answer and clear signals must be sent between the exchanges over the same line. This is accomplished as follows. The originating exchange uses the loop and disconnect states to send its forward seize and clear signals and to send dial pulses. The terminating exchange sends back its answer signal by interchanging the battery and earth connections to the line, thus reversing the direction of the line current. The originating exchange recognizes this reversal by means of a polarized relay.[1,3] The terminating exchange sends its backward clear signal by removing the polarity reversal.

The repertoire of different signals may be increased by using pulse signals in addition to continuous signals. When signals are exchanged between registers it is necessary to delay sending from the originating exchange until a register has been connected at the receiving exchange. This can be achieved by not sending digits until a

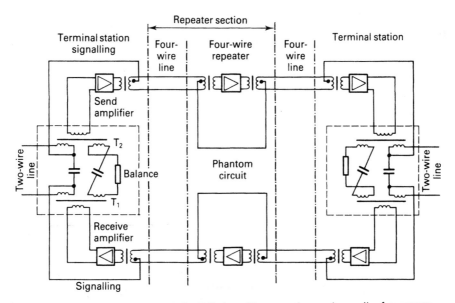

Figure 8.2 Phantom circuit for DC signalling over four-wire audio-frequency line.

proceed-to-send signal has been received. In North America a short pulse of line reversal has been used as a proceed-to-send signal. When periodic pulse metering was introduced in the UK it became necessary to send back meter pulses from the trunk exchange to the local exchange. Line reversals were used for this purpose. Since these pulses occur during a conversation, interference with speech was avoided by causing the pulses to have a slow rise and fall.

Multi-level signals have also been employed. For example, in North America higher-voltage pulses were sent back from a tandem office to an end office to operate customers' message registers for remote-zone registration.[4] Multi-level signals have also been used to enable address information to be sent from an automatic exchange to a manual exchange more rapidly than by loop/disconnect pulsing. Each decimal digit was sent as a group of 'heavy' and light' positive and negative pulses. Receiving equipment decoded the digits and displayed them to the operator. Examples were the panel call-indicator (PCI) system[4] used in North America and the coded call-indicator (CCI) system[2] used in the UK.

As an alternative to sending loop/disconnect signals forward and reversals backward over a balanced pair, the two wires of a junction may be used to provide a separate one-wire signalling circuit with earth return in each direction. An earth is applied at one end when the other end has a relay connected to battery (e.g. for a ground start signal) or battery is applied when the other end has a relay connected to earth. These methods have a lower immunity than loop signalling to induced interference and earth potential differences. Hence, loop signalling is preferable.

Old power-drive systems,[2] which used revertive pulsing to control the switches in an exchange, also used revertive pulsing to send address signals over junctions between exchanges. A higher signalling speed could be obtained because the pulses were generated by the motion of a selector, instead of being used to step a selector (as in a network of Strowger exchanges). In North America revertive impulsing was used between panel offices, then between panel offices and crossbar offices and, finally, over trunks between crossbar offices.

When address information is sent between exchanges by loop/disconnect pulsing over a long line, distortion of the pulses can cause errors in the received information.[1,2] As shown in Figure 8.3, the capacitance of the line causes a slow decay of the pulse waveform, so the output break from the receiving relay is shorter than that originally sent. On a multilink connection, when all the switches are directly controlled by the originating exchange (as in a network of step-by-step exchanges), the distortion is cumulative. However, when register–senders are used the pulses are regenerated at each exchange and the conditions are thus less onerous.

The nonsymmetrical waveform shown in Figure 8.3 occurs with loop/disconnect pulsing because the sending impedance of the circuit is zero in the loop state and infinite in the disconnect state. The pulse distortion can be reduced, and the signalling range increased, by using a waveform that is symmetrical. Long-distance direct-current (LDDC) signalling systems were therefore designed which obtain symmetrical waveforms by using *double-current working*. Pulses are sent by polarity reversals instead of by makes and breaks.[1,2,4] The sending impedance is thereby kept constant and the

Audio-frequency junctions and trunk circuits [209]

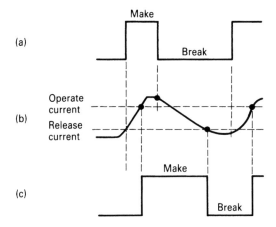

Figure 8.3 Distortion of loop/disconnect pulses caused by distributed capacitance of a long line. (a) Input pulse at sending end. (b) Arrival current waveform. (c) Output pulse from receiving relay.

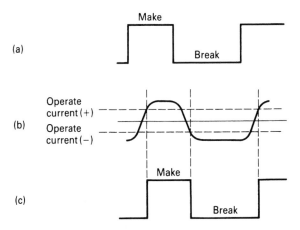

Figure 8.4 Pulsing by current reversals (double current working). (a) Input pulse at sending end. (b) Arrival current waveform. (c) Output pulse from receiving polarized relay.

duration of the received pulse (between the crossing points of a zero-voltage reference) is equal to that of the transmitted pulse, as shown by Figure 8.4. The pulses are received by a polarized relay. The signalling range is increased by the sensitivity of the polarized relay as well as by the reduction in pulse distortion.

Long-distance DC signalling is largely of historical interest, since most long-distance circuits are now provided in high-capacity multiplex transmission systems. The signalling methods used with multiplex systems are described in Sections 8.4 and 8.5.

8.4 FDM carrier systems

8.4.1 Outband signalling

In frequency-division multiplex (FDM) systems the carriers are spaced at intervals of 4 kHz and the baseband is from 300 Hz to 3.4 kHz. By using channel filters with a sharp cut-off it is possible to insert a narrow-band signalling channel above the speech band (i.e. between 3.4 kHz and 4 kHz). Signal frequencies of 3.7 kHz and 3.85 kHz have been used. This is known as *outband signalling*.

An outband signalling system is shown in Figure 8.5. A DC signal on the input lead M at one terminal causes the signal frequency to be sent over the transmission channel. This is detected at the other terminal to give a corresponding DC signal on the output lead E. If the repeater station containing the FDM channelling equipment is adjacent to the switching equipment, it is simpler for the latter to send and receive signals over separate E and M wires than to extract them from and re-insert them into the speech circuit. E and M wire signalling has been widely used in North America,[4] not only for outband signalling but also for DC and inband signalling systems. The E lead always carries signals from the signalling apparatus to the switching equipment

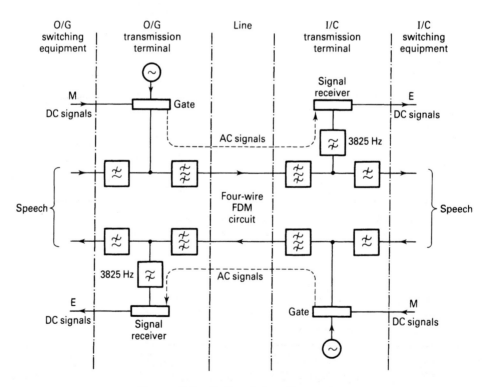

Figure 8.5 Outband signalling system.

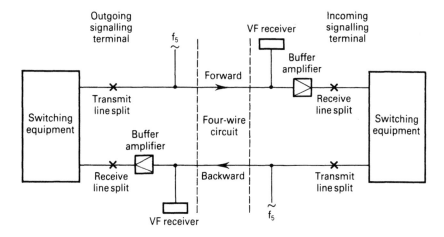

Figure 8.6 Voice-frequency (VF) signalling system.

and the M lead carries signals from the switching equipment to the signalling apparatus.

To use outband signalling successfully in a network, all routes must use FDM systems with built-in outband signalling. In practice, however, a route may contain a section with audio-frequency transmission or a carrier system not equipped with outband signalling. It is then necessary to use *inband signalling*, as described in Section 8.4.2.

8.4.2 Inband (VF) signalling

Systems that transmit signals within the baseband of FDM systems are known as *inband signalling systems* or *voice-frequency* (VF) *signalling systems*. They have the important advantages that they are independent of the transmission systems used and they will function over any circuit that provides satisfactory speech transmission.

A VF signalling system is shown in Figure 8.6. The line is split when the signal tone is transmitted in order to confine it to the link concerned. The line split at the receiving end is operated by the signal receiver. Consequently, the tone spills over before the receiver has operated; however, this is ignored because the duration of the spill-over (e.g. 20 ms) is less than the length of the signals used. The unity-gain buffer amplifier at the receiving end (where the signal level may be low) prevents transients produced by electromechanical switching equipment from reaching the VF receiver.

Since voice-frequency signals are used, there is a possibility of signal imitation, i.e. the receiver may be operated because the signal frequency is present in transmitted speech. This is obviously undesirable; for example, it could clear down a connection before the users have finished their conversation. The following measures are taken[1] to make the probability of signal imitation almost negligible:

[212] *Signalling*

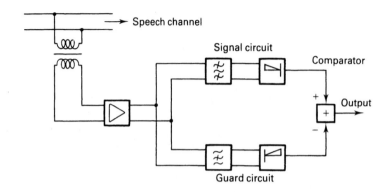

Figure 8.7 Block schematic diagram of voice-frequency receiver.

1. A signal frequency is chosen at which the energy in speech is low (i.e. above 2 kHz). For example, 2280 Hz is used in the UK[1] and 2600 Hz in North America.[5]
2. The durations of signals are made longer than the period for which the speech frequency is likely to persist in speech (e.g. ≥ 50 ms).
3. Use is made of the fact that the signal frequency is unlikely to be produced in speech without other frequencies also being present.

A block diagram of a VF receiver is shown in Figure 8.7. In order to make use of measure (3), the receiver contains a signal circuit with a band-pass filter to accept the signal frequency and a guard circuit with a band-stop filter to accept all other frequencies and reject the signal frequency. The outputs of both circuits are rectified and compared. If the output from the signal circuit exceeds that from the guard circuit, the receiver operates; if the output from the guard circuit exceeds that from the signal circuit, the receiver gives no output signal.

To avoid interfering with speech, VF signals must not be transmitted while a conversation is in progress. To avoid this, two signalling methods have been employed:

- Tone-on-idle signalling
- Pulse signalling.

Tone-on-idle signalling has been extensively used in North America.[5] In principle, it is very simple. An idle four-wire circuit transmits tone continuously in both directions and this is switched off in one direction by the forward seize and in the other by the backward answer. Clear signals switch on the signalling tone again. In practice, however, there are complications. Since the tone is present on every idle circuit, it must be transmitted at a low level to avoid overloading the FDM system. Also, the receiver must guard against short interruptions of the tone causing false seizures.

In a pulse-signalling system, the seize, answer and clear signals (and the breaks in loop/disconnect pulsing) are all sent by tone pulses with durations typically between 50 ms and 700 ms. (The clear signals are of long duration to minimize the chance of connections being released in error.) Since the same pulse of tone means a different signal at different stages in a call, the control circuit operated by the VF receiver must have a memory in order to produce the correct responses. Pulse-type VF signalling systems have been used by many operating organizations. An example is the British AC 9 system.[1]

8.5 PCM signalling

PCM primary multiplexers were designed from the outset to incorporate signalling. The DC signals associated with the audio-frequency baseband circuits in each direction are sampled and the signal samples are transmitted within the frame of PCM channels. It is therefore unnecessary to use VF signalling.

The 2 Mbit/s system has 32 8-bit time-slots, but it only provides 30 channels. Time-slot zero is used for frame alignment and time-slot 16 is used for signalling, as shown in Figure 2.11. The eight bits of channel 16 are shared between the 30 channels by a process of *multiframing*. As shown in Figure 8.8, 16 successive appearances of channel 16 form a multiframe of 8-bit time-slots. The first contains a multiframe alignment signal and each of the subsequent 15 time-slots contains four bits for each of two channels. Thus, every speech circuit can have, in each direction, either a single signalling channel operating at 2 kbit/s or four independent signalling channels at 500 bit/s. This enables a much larger number of signals to be exchanged than is possible with the DC signalling methods described in Section 8.3.

The North American 1.5 Mbit/s PCM system has a frame of 193 bits. This contains 24 8-bit time-slots, all of which are used for speech, and the 193rd bit in each frame is used for the frame-alignment signal. In every sixth frame, the eighth bit of each channel

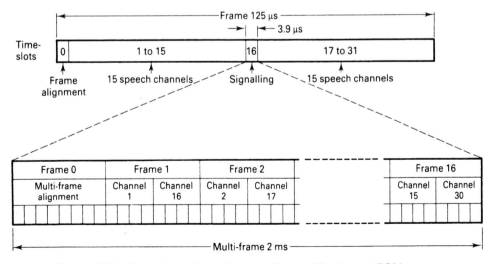

Figure 8.8 Use of multiframe for signalling in 30-channel PCM system.

is used for signalling instead of for speech and this has been found to cause a negligible increase in quantization distortion. A 12-frame multiframe is used. Thus, associated with each speech channel are two independent signalling channels at 666 bit/s or a single signalling channel at 1.3 kbit/s. The 193rd bit of the PCM frame is used on alternate frames for multiframe alignment instead of for frame alignment.

The methods described above provide channel-associated signalling, but PCM systems can also be used for common-channel signalling. Multiframing is then not required. In the 30-channel system, time-slot 16 is used to provide a common signalling channel at 64 kbit/s. In the 24-channel system, the 193rd bit in alternate frames can be used to provide a common signalling channel at 4 kbit/s. However, if a 64 kbit/s channel is required, it is necessary to sacrifice one of the speech channels (i.e. channel 24).

8.6 Inter-register signalling

For a multilink connection in a network of register-controlled exchanges, a register in the originating exchange receives address information from the calling customer and sends out routing digits. Each succeeding register both receives and sends out routing digits, until the terminating exchange is reached. This sequence of operations introduces post-dialling delay. To minimize this delay, a more rapid method of sending routing information than loop/disconnect pulsing is needed and *inband multifrequency (MF) signalling systems* have been developed for this purpose.[1,6,7]

Multifrequency signalling has also been used for keypulsing by operators in manual exchanges.[3,4] This enables an operator to send address information over a junction to an automatic exchange more rapidly than by dialling.

An inter-register signalling system cannot be used for seize, answer and clear signals. No register is connected when an incoming seize signal is received, since it is this signal which initiates the connection to a register. The register is released after it has set up a connection through its exchange and sent out routing digits; therefore, it cannot receive answer and clear signals. Consequently, *line signalling*, using one of the methods described in Sections 8.3 to 8.5, is required in addition to inter-register signalling.

Either *en-bloc* or *overlap signalling* may be used. In en-bloc signalling, the complete address information is transferred from one register to the next as a single string of digits. Thus, no signal is sent out until the complete address information has been received. In overlap signalling, digits are sent out as soon as possible. Thus, some digits may be sent before the complete address has been received and signalling may take place simultaneously on two links (i.e. the signals overlap). This enables subsequent registers to start digit analysis earlier than is possible with en-bloc signalling and this reduces post-dialling delay.

Either *link-by-link signalling* or *end-to-end signalling* may be employed. In link-by-link signalling, information is exchanged only between adjacent registers in a multilink connection, as shown by Figure 8.9(a). This has the following advantages:

Inter-register signalling [215]

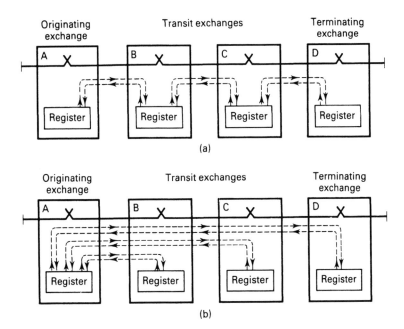

Figure 8.9 Link-by-link and end-to-end signalling between registers. (a) Link-by-link signalling. (b) End-to-end signalling. (Note, backward sigalling is not always provided.)

1. Signals suffer only the transmission impairments (e.g. attenuation, distortion and noise) of a single link.
2. Different signalling systems may be used on different links. Thus, if a network is being modernized, all registers do not need to be modified simultaneously.

However, each transit register must receive, store and retransmit the complete address information. Consequently, link-by-link working has the disadvantage that the holding times of registers and the post-dialling delay are long, particularly if backward signalling is used in addition to forward signalling.

In end-to-end signalling, the originating register controls the setting up of a connection until it reaches its final destination, as shown in Figure 8.9(b). Each transit register receives only the address information required to select the outgoing route to the next exchange in the connection. Having performed its task, it is released and the originating register signals to the next register. End-to-end signalling has the disadvantage that all registers must be compatible with the originating register. Also, signals can suffer the transmission impairments of several links in tandem. Thus, the VF receivers used must have a greater dynamic range than is needed for link-by-link signalling. However, each transit register needs to receive and retransmit only part of the address information, so register holding times and the post-dialling delay are

[216] *Signalling*

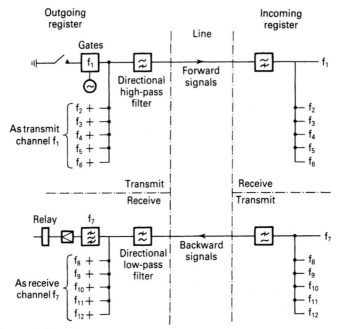

Figure 8.10 Multifrequency (MF) inter-register signalling system.

shorter than for link-by-link working. For this reason, end-to-end working has been used in most networks.

The arrangements used for multifrequency (MF) inter-register signalling are shown in Figure 8.10. Each digit is sent by a combination of two frequencies out of six (2/6 MF). Since two frequencies are required to represent a digit, this gives an error-detection capability. Combinations of two frequencies out of six give 15 possible digit values, so five extra signals are available in addition to the digits '1' to '0'. The frequencies are spaced at intervals of 120 Hz or 200 Hz. Since inter-register signalling precedes a conversation, signal imitation cannot occur. Thus, the VF receivers do not need speech immunity and this enables short signal pulses to be used. The directional filters shown in Figure 8.10 enable the system to be used, if required, for signalling over two-wire circuits.

The Bell R1 system[1,7] (which corresponds to the CCITT Regional Signalling System No.1) uses six signal frequencies spaced at 200 Hz between 700 Hz and 1700 Hz. Its signals are unacknowledged, so no frequencies are provided for backward signalling. The system transmits 12 forward signals as follows:

    KP (start of pulsing)
    Digits 1 to 0
    ST (end of pulsing)

The system employs link-by-link signalling.

Inter-register signalling [217]

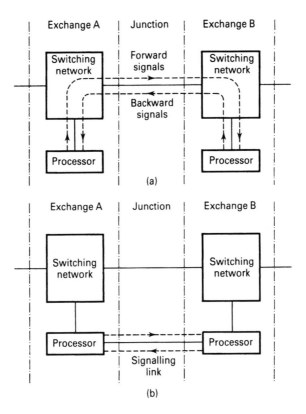

Figure 8.11 Principle of common-channel signalling (CCS). (a) Channel-associated signalling between central processors. (b) Common-channel signalling between central processors.

The CCITT R2 system[1,8] (Regional Signalling System No.2) provides both forward and backward signals. It normally uses continuous compelled signalling. However, a version with pulse signalling is used for satellite circuits because of their long delay. End-to-end working is employed, i.e. a connection is controlled by the originating register.

The R2 system uses signal frequencies spaced at 120 Hz, as follows:

Forward: 1380, 1500, 1620, 1740, 1860, 1980 Hz
Backward: 540, 660, 780, 900, 1020, 1140 Hz

In this system more than 15 signals can be sent in each direction, because one signal frequency combination is used as a shift signal to enable each of the others to have two meanings. Thus, a very wide repertoire of signals is available. Some of the additional signals are specified for use on international calls. However, since the system was defined as a regional one, operating administrations were free to choose how to use the other signals. Consequently, there are different national versions of the R2 system. An example is the British MF2 system.[1]

8.7 Common-channel signalling principles

8.7.1 General

In a network of SPC exchanges, a connection that is made through two exchanges requires call processing by the central processor in each exchange. If channel-associated signalling is used for calls from exchange A to exchange B, as shown in Figure 8.11(a), it is necessary for the central processor of exchange A to send its outgoing forward signals to the individual speech circuit for transmission to exchange B. At this exchange, the signals must be detected on the speech circuit and passed to the central processor. Similarly, backward signals from processor B must be sent over the speech circuit, detected at exchange A and extended to its processor. This is an inefficient arrangement for signalling between the two processors!

If a high-speed data link is employed between the processors, as shown in Figure 8.11(b), it can provide a channel for all signals between exchanges A and B. This is known as *common-channel signalling* (CCS). It gives the following advantages:

1. Information can be exchanged between the processors much more rapidly than when channel-associated signalling is used.
2. As a result, a much wider repertoire of signals can be used and this enables more services to be provided to customers.
3. Signals can be added or changed by software modification to provide new services.
4. There is no longer any need for line signalling equipment on every junction, which results in a considerable cost saving.
5. Since there is no line signalling, the junctions can be used for calls from B to A in addition to calls from A to B. Both-way working requires fewer circuits to carry the traffic than if separate groups of junctions are provided from A to B and from B to A.
6. Signals relating to a call can be sent while the call is in progress. This enables customers to alter connections after they have been set up. For example, a customer can transfer a call elsewhere, or request a third party to be connected into an existing connection.
7. Signals can be exchanged between processors for functions other than call processing, for example for maintenance or network-management purposes.

The error rate for common-channel signalling must be very low and the reliability required is much greater than for channel-associated signalling. Failure of the data link in Figure 8.11(b) would prevent any calls from being made between exchanges A and B, whereas failure of a line signalling equipment, or even of an inter-register signalling system, would result only in the loss of a small fraction of the traffic. When channel-associated signalling is used, the successful exchange of signals over a circuit proves that the circuit is working. CCS does not inherently provide this checking facility, so a separate means (e.g. automatic routine testing) must be provided to ensure the integrity of the speech circuits.

CCS systems use *message-based signalling*. Successive messages exchanged between the processors in Figure 8.11(b) usually relate to different calls. Each message must therefore contain a label, called the *circuit identity code*, that indicates to which speech circuit, and thus to which call, it belongs. Since messages pass directly between central processors, no connection is required to an incoming junction before an address signal is received. The address signal can therefore be the first message sent and there is no need for a seize signal. In a multilink connection, signalling takes place from one transit exchange to the next without involving the originating exchange. Thus, link-by-link signalling is inherent with CCS.

In a CCS system messages from a processor queue for transmission over the signalling link. The number of speech circuits that can be handled by a CCS system is therefore determined by the acceptable delay. A signalling link operating at 64 kbit/s normally provides signalling for up to 1000 or 1500 speech circuits.[1,9] However, more may be handled (with increased delays) when the load of a link that has failed is added to the existing load on a back-up link.

The use of CCS for inter-exchange signalling has been followed by its application to customers' lines in integrated-services digital networks (ISDN). Basic-rate ISDN access provides two 64 kbit/s channels and a 16 kbit/s common signalling channel in each direction. Primary-rate access provides a 64 kbit/s common signalling channel for either 23 channels (in 1.5 Mbit/s-based networks) or 30 channels (in 2 Mbit/s-based networks). The signalling methods used are described in Section 8.10.

8.7.2 Signalling networks

Figure 8.11(b) shows a direct CCS link between two exchanges. This is known as *associated signalling*. In a multi-exchange network, there will be many CCS links between exchanges and these form a *signalling network*. In principle, CCS signals can follow different routes from the connections which they control and they can pass through several intermediate nodes in the signalling network. This is called *nonassociated signalling* and is shown in Figure 8.12(a). Since, in general, signal messages entering the network may be destined for any other exchange, the messages must include labels containing their destinations. The network used for nonassociated signalling is thus a form of packet-switched network.

In practice, CCS messages are usually only routed through one intermediate node, as shown in Figure 8.12(b). This is known as *quasi-associated signalling* and the intermediate node is called a *signal-transfer point* (STP). In Figure 8.12(b), the CCS equipment in exchange C handles signalling for connections between exchanges A and B, in addition to signalling for connections between C and A and between C and B.

Since CCS signals may be routed via an STP, each message contains a *destination-point code* to enable it to be routed to the correct exchange. It also contains

[220] *Signalling*

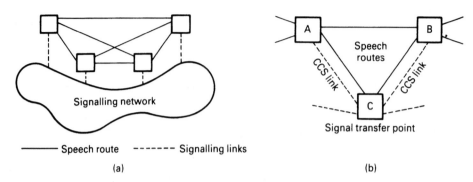

Figure 8.12 Signalling networks. (a) Use of non-associated signalling. (b) Use of quasi-associated signalling. (Associated signalling is shown in Figure 8.11(b).)

an *originating-point code* to enable the messages sent back to be correctly routed. If the CCS system in an exchange recognizes the destination-point code of an incoming message as its own, the message is accepted and passed to the central processor. If the code is that of another exchange, the CCS system looks up a translation table to determine the route for onward transmission of the message.

Quasi-associated signalling is used when there are few circuits between A and B and thus little signalling traffic between them. It is then economic to share a single signalling link from A to C between the route from A to B and routes from A to other exchanges. When there are many circuits between A and B, and thus much signalling traffic, it is economic to use associated signalling. However, an alternative route via an STP is normally provided in case the associated-signalling link fails.

The transmission bearers used for a CCS network are channels in the main transmission bearer network. The first generation of CCS systems (CCITT no.6) used modems to transmit at 2.4 kbit/s or 4.8 kbit/s over analog telephone channels. A 4 kbit/s channel could also be provided over a 1.5 Mbit/s PCM system, as described in Section 8.5. Current CCS systems (CCITT no. 7) use a 64 kbit/s channel provided by time-slot 16 in a 2 Mbit/s PCM system or time-slot 24 in a 1.5 Mbit/s system.

8.8 CCITT signalling system no.6

The first common-channel signalling system to be standardized internationally was the CCITT no.6 system.[1,10,11] This was designed for use in analog networks and used bit rates of 2.4 kbit/s and 4.8 kbit/s. It used fixed-size signal units of 28 bits (20 information bits + 8 parity-check bits). A later version for use in digital networks added four padding bits to be compatible with 8-bit PCM time-slots. However, this pioneer system has now been superseded by CCITT signalling system no.7, which is described in Section 8.9.

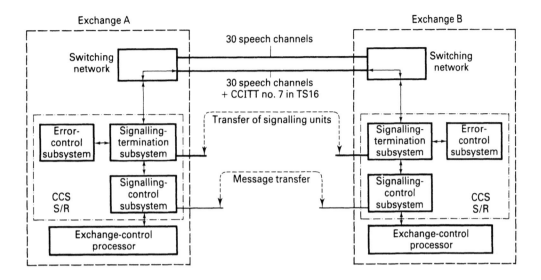

Figure 8.13 Block schematic diagram of CCITT no.7 signalling system.

8.9 CCITT signalling system no.7

8.9.1 General

A block schematic diagram of the CCITT no.7 signalling system[9,10,12] is shown in Figure 8.13. Signal messages are passed from the central processor of the sending exchange to the CCS system. This consists of three microprocessor-based subsystems: the signalling-control subsystem, the signalling-termination subsystem and the error-control subsystem. The signalling-control subsystem structures the messages in the appropriate format and queues them for transmission. When there are no messages to send, it generates filler messages to keep the link active. Messages then pass to the signalling-termination subsystem, where complete *signal units* (SU) are assembled using sequence numbers and check bits generated by the error-control subsystem. At the receiving terminal, the reverse sequence is carried out.

The system can be modelled as a stack of protocols like the ISO seven-layer model described in Section 1.7. However, the system was specified before the ISO model was published and the layers are referred to as *levels* in its literature. The levels are as follows:

>   *Level* 1: The physical level
>   *Level* 2: The data-link level
>   *Level* 3: The signalling-network level
>   *Level* 4: The user part.

[222] *Signalling*

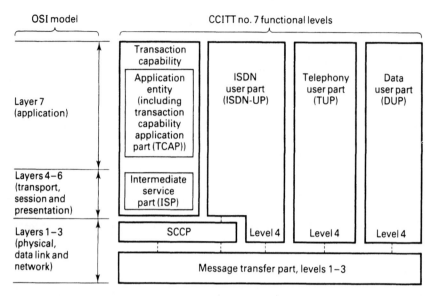

Figure 8.14 Relationship between CCITT no.7 functional levels and layers of OSI seven-layer model.

The relationship between these levels and the layers of the OSI model is shown in Figure 8.14. The user part encompasses layers 4 to 7 of the OSI model.

Level 1 is the means of sending bit streams over a physical path. It uses time-slot 16 of a 2 Mbit/s PCM system or time-slot 24 of a 1.5 Mbit/s system.

Level 2 performs the functions of error control, link initialization, error-rate monitoring, flow control and delineation of messages.

Level 3 provides the functions required for a signalling network. Each node in the network has a *signal-point code*, which is a 14-bit address. Every message contains the point codes of the originating and terminating nodes for that message.

Levels 1 to 3 form the *message-transfer part* (MTP) of CCITT no.7. Level 4 is the *user part*. This consists of the processes for handling the service being supported by the signalling system. The message-transfer part is capable of supporting many different user parts. So far, three have been defined: the telephone-user part (TUP), the data-user part (DUP) and the ISDN-user part (ISDN-UP).

CCS systems are now being used for messages that are not associated with calls, for example to interrogate a remote database (as described in Section 10.6), for traffic management (as described in Section 10.10) and for operations, maintenance and administration. This led to the specification of the part known as the *transaction-capabilities* (TC). Since this was designed after publication of the OSI model, its protocols were specified to conform with it.[10] As shown in Figure 8.14, a *signalling-connection control part* (SCCP) has been added to level 3 to make it fully compatible with layer 4 of the OSI model. The *intermediate service part* (ISP) performs

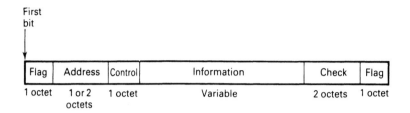

Figure 8.15 Frame structure for high-level data-link control (HDLC) protocol.

the functions of layers 4 to 6 of the OSI model and the *transaction-capabilities application part* (TCAP) provides for layer 7.

8.9.2 The high-level data-link control protocol

The level 2 protocol used in the CCITT no.7 signalling system uses the international standard known as *high-level data-link control* (HDLC).[13] Messages are sent by packets contained within frames having the format shown in Figure 8.15.

The beginning and end of each HDLC message is indicated by a unique combination of digits (01111110) known as a *flag*. Of course, this sequence of digits can occur in messages and must then be prevented from being interpreted as a flag. This is done by a technique known as *zero-bit insertion and deletion*, which is also called *bit stuffing and unstuffing*. When sending digits of a message between the two flags, the sending terminal inserts a '0' after every sequence of five consecutive '1's. The receiving terminal deletes any '0' which occurs after five consecutive '1's and so restores the original message.

The opening flag is followed by bit fields for address and control information. These are followed by the data field containing the message information. Between this and the closing field there is an error-check field, which enables the receiving system to detect that a frame is in error and request retransmission. The error-check field contains 16 bits generated as a cyclic redundancy check (CRC) code.[14]

8.9.3 Signal units

Information to be sent is structured by the signalling-control unit (level 2) into a *signal unit* (SU). The SU is based on the high-level data-link control (HDLC) protocol described in Section 8.9.2.

Signal units are of three types:

1. *The message signal unit (MSU)* This transfers information supplied by a user part (level 4) via the signalling network level (level 3).
2. *The link-status signal unit (LSSU)* This is used for link initialization and flow control.

[224] *Signalling*

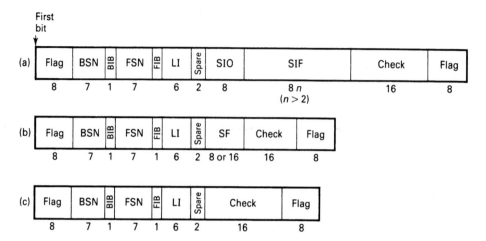

Figure 8.16 Formats of signal units in CCITT no.7 signalling system. (a) Message signalling unit. (b) Link-status signalling unit. (c) Fill-in unit. BIB = backward indicator bit, BSN = backward sequence number, FIB = forward indicator bit, FSN = forward sequence number, LI = length indicator, SF = status field, SIF = signalling information field, SIO = service information octet.

3. *The fill-in signal unit (FISU)* This is sent to maintain alignment when there is no signal traffic.

The format of of the MSU is shown in Figure 8.16(a). Messages are of variable length and are sent in 8-bit bytes (octets) as follows:

1. Opening and closing flags are used to delimit signals. They have the code pattern '01111110'.
2. The forward indicator bit (FIB), backward indicator bit (BIB), forward sequence number (FSN) and backward sequence number (BSN) are used for error correction, as described below.
3. The length indicator (LI) gives the length of the signal unit. A value of LI greater than two indicates that the SU is a message signal unit.
4. The service-information octet (SIO) indicates the user part appropriate to the message (e.g. telephone, data or ISDN).
5. The signalling-information field (SIF) may consist of up to 272 octets and contains the information to be transmitted. Its format is defined by the user part, as described in Section 8.9.4.
6. The error-check field is immediately before the closing flag. It contains 16 bits generated as a cyclic redundancy check code.[14]

The basic method of error correction[10] is described as a noncompelled positive/negative-acknowledgement retransmission error-correction system. When the signalling link is normal, the FIB and BIB in the transmitted SU are set to the same value (e.g. zero). The transmitted FSN is set to one more than for the previous message. Level 2 of the receiving CCS system analyzes the check bits and discards the message if it detects an error. If the message is not discarded, the received FSN is compared with the expected value (i.e. one more than the previous FSN) and the message is passed to level 3. The FSN is sent back as a BSN and the BIB provides a positive acknowledgement.

If there has been an error and the message has been discarded, the comparison of FSNs at the receiving end shows a discrepancy. The BIB is then inverted to provide a negative acknowledgement and the BSN has the value of the last correctly received MSU. On receipt of the negative acknowledgement, the sending CCS system interrupts transmission of further SUs and sends previous MSUs in their original order.

The link-status and fill-in signalling units are more simple. Their formats are shown in Figure 8.16 (b) and (c).

8.9.4 The signalling information field

A message from the user part occupies the signal information field (SIF) of the MSU shown in Figure 8.16(a), i.e. it commences after the SIO. An example of a TUP message is the initial address message (IAM). This is the first message to be sent, since a separate seize signal is not required. The format of the IAM is shown in Figure 8.17. Its fields, in order of transmission, are:

1. A label of 40 bits (5 octets) containing:
   the destination point code (14 bits)
   the originating point code (14 bits)
   the circuit identity code (12 bits).
2. The H0/H1 octet
   The H0 field (4 bits) identifies a general category of message (e.g. forward address message).
   The H1 field completes the definition of the message (e.g. variable-length IAM).
3. The calling party category (e.g. ordinary customer or operator).
4. The message indicator
   This indicates any special requirements (e.g. whether a satellite link or echo suppressors will be used).
5. The number of address signals (4 bits)
   This gives the number of address digits in the IAM.
6. The address signal.

This signalling information field is then followed by the check bits and the closing flag to complete the MSU.

[226] *Signalling*

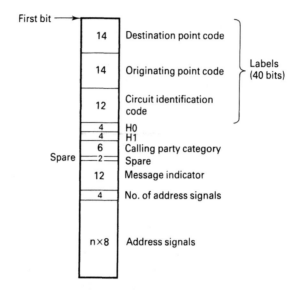

Figure 8.17 Format of TUP initial address message in CCITT no.7 signalling system.

8.10 Digital customer line signalling

Digital transmission is used on customers' lines to provide access to an ISDN.[15,16] Each line may give access to several terminals on the customer's premises and common-channel signalling is used to serve them. Basic-rate access[17] provides, in each direction, two B channels at 64 kbit/s and a D channel at 16 kbit/s for signalling. Primary-rate access[18] provides, in each direction, 30 channels (in a 2 Mbit/s-based network) or 23 channels (in a 1.5 Mbit/s-based network), together with a 64 kbit/s signalling D channel.

The CCITT has defined the Digital Subscriber Signalling System no.1 (DSS1) for signalling over the D channel.[10,16] The transfer of information in each direction between the customer's premises and the exchange is by messages, called *frames*, which are similar to the CCITT no.7 signal units described in Section 8.9.3. The form of HDLC protocol used is known as LAP-D (link access protocol for the D channel).

The format of a frame[10,19] is shown in Figure 8.15. Each frame begins and ends with an 8-bit flag and there is a 16-bit cyclic-redundancy check field. The address field is not used for routing in the network; it selects ports at either end of the line. Its first octet contains the *service-access-point identifier*, which indicates the exchange terminal to be used (e.g. for a circuit-switched call or a packet-switched call). The second octet contains the *terminal-end-point identifier*, which identifies the equipment on the customer's premises to which the message refers. Each octet begins with an *extension bit*. If this is '0', it indicates that another octet is to follow; if it is '1', it indicates that the octet is the last. Normally, the extension bit of the first octet is '0' and that of the second octet is '1'. However, the latter can be changed to '0' to add a third octet, or more, if required. The second bit in the address field is the *command/response bit* (C/R). If the frame is a command, it indicates the receiver (i.e. either the exchange or the customer). If the frame is a response, it indicates the sender.

The control field (1 or 2 octets) indicates the type of frame being transmitted. There are three types of frame:

- I format for information transfer
- S format for supervisory functions
- U format for un-numbered transfers.

The I frame and the S frame contain sequence numbers to ensure that frames are not lost. The U frame does not.

The information field contains the layer-3 information to be transfered between customer and exchange. Signals to be transmitted for establishing and controlling connections have been recommended by the CCITT.[20] For example, I.451 messages are for basic call control[21] and I.452 messages are to control supplementary services.[22]

Primary-rate access with common-channel signalling is used between digital PBXs and public exchanges. Common-channel signalling based on similar principles can also be used between PBXs in a digital private network. A system[23] known as DPNSS (digital private network signalling system) is widely used in the UK and other countries. When this system is applied to a private network, its PBXs appear to users as a single large PBX, with all its supplementary services available wherever the user is located.

The DPNSS system was developed from a system known as DASS2, which was used for primary-rate access signalling in the UK until supplanted by I.421 in the early 1990s. The two forms of signalling can co-exist in channel 16 and inter-PBX private circuits can be mixed with circuits connected to the PSTN on the same 30-channel multiplex.

The lack of compatibility between DPNSS and I.421 is inconvenient and a successor is clearly needed. The European Computer Manufacturers Association (ECMA) has taken the lead in this, with a signalling system called QSig. This has now been offered for international standardization.

Problems

1. Loop/disconnect signalling is used to transmit customers' dial pulses to an electromechanical telephone exchange. Customers' dials generate pulses at 10 pulses/second with a break/make ratio of 67% to 33%.

Pulses are received by a relay which is connected to the 50 V exchange battery and has 400 Ω resistance and 4 H inductance. The relay operates at a current of 30 mA and releases at 20 mA and it has equal operate and release lags. The effects of saturation and eddy currents in the iron core of the relay may be ignored. As a further simplification, it may also be assumed that an exponential rise or decay of current has reached steady state after three times the time-constant of the circuit.

(a) Find the break/make ratio of the output

[228] *Signalling*

pulses from the relay when the resistance of a customer's line is: (i) zero, (ii) 1000Ω.
(b) Discuss the effects on the performance of the circuit of:
(i) A spark-quench capacitor across the contacts of the dial,
(ii) Distributed capacitance of a long line.

2. Explain the meanings of the following terms applied to inter-register signalling: en-bloc signalling, overlap signalling, compelled signalling, link-by-link signalling, end-to-end signalling.

3. A number of analog transmission circuits are equipped with various types of line signalling system, e.g. loop/disconnect DC, long-distance DC, voice-frequency signalling, etc. These circuits are to be multiplexed on a 30-channel PCM system to access a digital network via an SPC digital exchange. The digital network uses common-channel signalling.
Describe, with the aid of a diagram, suitable arrangements for signalling over the multiplexer.

4. The introduction of stored-program-controlled exchanges has led to channel-associated signalling (CAS) being replaced by common-channel signalling (CCS). Explain why.
What advantages does CCS have over CAS? What disadvantages does it have and how are these overcome?

5. An integrated digital network is equipped with 64 kbit/s common-channel signalling systems. The network is free from bit errors most of the time. However, it is considered necessary to incorporate some form of error control in the common-channel signalling system.
Describe a suitable form of error-control method for the common-channel signalling. It is required that the switching stored-program control shall have no function in the signalling error-control process.

6. An integrated services digital private branch exchange (ISPBX) is to access an integrated services digital network (ISDN) via a digital circuit group. Describe, in general terms, the digital access signalling system you would adopt. Explain:
(i) The formatting principle and the protocol for an initial address signal message.
(ii) The error-control method.

7. (a) Explain why it is necessary in digital inter-exchange common-channel signalling to adopt arrangements to ensure signalling service in the event of failure of a signalling link.
(b) The following inter-exchange common-channel signalling arrangements apply in a national digital network:
(i) Several inter-exchange signalling links between trunk exchanges.
(ii) A single inter-exchange link between local exchanges.
Describe the arrangements you would adopt in each case to maintain the signalling service in the event of signalling link failure.

8. (a) Explain what you understand by the term 'signal transfer point' in relation to inter-exchange digital common-channel signalling in an integrated services digital telecommunications network.
(b) Discuss the network conditions in which the signal transfer point facility would be used. Describe the arrangements included in the signal messages to enable the facility to be performed.
(c) Assuming a common-channel signalling relationship to include signal transfer, state, with reasons, any constraints on the number of signal transfer points which may apply in that signalling relationship.

9. The signal units of the CCITT no.7 signalling system provide address messages. The LAP-D frame of the CCITT Digital Subscriber Signalling System contains an address field. These serve different purposes. Explain what each is for.

10. A common-channel signalling system uses a 64 kbit/s data link which serves a group of 1500 speech circuits on a route between two telephone exchanges. The busy-hour traffic is 1000 E and the average call duration is 2 minutes. On average, each call requires the transmission of ten messages (five signals plus five responses) and the average message length is 20 octets.

What percentage of messages encounter delay and what is the mean delay for these messages?

References

[1] Welch, S. (1979), *Signalling in Telecommunications Networks*, Peter Peregrinus, Stevenage.
[2] Atkinson, J. (1950), *Telephony*, Vol. 2, Pitman, London.
[3] Atkinson, J. (1948), *Telephony*, Vol. 1, Pitman, London.
[4] Breen, C. and Dahlbom, C.A. (1960), 'Signaling systems for control of telephone switching', *Bell Syst. Tech. Jour.*, **39**, 1381–444.
[5] Newell, N.A. and Weaver, A. (1951), 'Single-frequency signaling system for supervision and dialing over long-distance telephone trunks', *AIEE Trans.*, **70**, 489–95.
[6] Gazanion, H.. and LeGare, R. (1963), 'Systèmes de signalisation Socotel', *Commutation et Electron.*, **4**, 32.
[7] Dahlbom, C.A., Horton, A.W. and Moody, D.L. (1949), 'Multi-frequency pulsing in switching', *Electr. Eng. (USA)*, **68**, 505–10.
[8] CCITT Recommendations Q.350–Q.368, 'Specifications of signalling system R2'.
[9] Redmill, F.J. and Valdar, A.R. (1990), *SPC Digital Telephone Exchanges*, Peter Peregrinus, Stevenage.
[10] Manterfield, R.J. (1991), *Common-channel Signalling*, Peter Peregrinus, Stevenage.
[11] CCITT Recommendations Q.251–Q.300, 'Specifications of signalling system no.6'.
[12] CCITT Recommendations Q.700–Q.775, 'Specifications of CCITT signalling system no.7'.
[13] ISO Standard 4335, 'Data communication, high-level data-link control procedure'.
[14] Brewster, R.L. (1989), *Communication Systems and Computer Networks*, Ellis Horwood, Chichester.
[15] Griffiths, J. (ed.) (1992), *ISDN Explained*, 2nd edn, Wiley, Chichester.
[16] Stallings, W. (1992), *ISDN and Broadband ISDN*, 2nd edn, Macmillan, New York.
[17] CCITT Recommendation I.420, 'Basic user-network interface'.
[18] CCITT Recommendation I.421, 'Primary rate user-network interface'.
[19] CCITT Recommendation Q.921, 'ISDN user-network interface, data-link layer specification'.
[20] CCITT Recommendation I.450 (also Q.930), 'ISDN user-network interface layer 3 – general aspects'.
[21] CCITT Recommendation I.451 (also Q.931), 'ISDN user-network interface layer 3 – specification for basic call control'.
[22] CCITT Recommendation I.452 (also Q.932), 'Generic procedures for the control of supplementary services'.
[23] Hiett, A.E. and Dangerfield, W. (1988), 'Private network signalling', *Computer Communications*, **11**, 191–6.

CHAPTER 9

Packet switching

9.1 Introduction

In message switching, each switching centre stores incoming messages until the required outgoing circuit becomes free and then retransmits them. No calls are lost because of congestion, but delays are incurred. As shown in Figure 9.1, message switching enables circuits to attain an occupancy approaching 1.0 erlang during the busy hour. Thus, it is more efficient than circuit switching for types of traffic, such as telegraph traffic, for which delays can be tolerated.

Simple message switching is not very suitable for data traffic, because of the very large variation of holding times. Messages can vary in length from a single character from a keyboard to very long streams of data. The operator of a VDU who is interrogating a distant computer needs a quick response. It will not be obtained if the message has to wait in a queue while a large file is being exchanged between two mainframe computers.

For this reason, data networks use a modified form of message switching called *packet switching*. Long messages are split into a number of short ones, called *packets*, which are transmitted separately, as shown in Figure 9.2. Thus, the single packet from the VDU operator is sent between packets of the large computer file, instead of waiting until its transmission is complete and the delay is minimized.

The format of a typical packet is shown in Figure 9.3. Since each packet is handled as a complete message, its data must be preceded by a *header*, which contains the destination address of the message. It is possible for packets sometimes to arrive at their destination in a different order from that in which they are sent. Each header therefore contains a *sequence number*, to enable packets to be reassembled in the correct order at the receiving terminal. Other bits are added to the header for control purposes, e.g. to indicate whether a packet contains a message or is being sent to control the network. The packet ends with bits added for error detection and correction. The most popular error-detection technique uses a cyclic redundancy check code (CRC).[1]

If the receiving terminal is to obtain correction of errors by requesting retransmission of a packet, then its header must contain the address of the sending

Introduction [231]

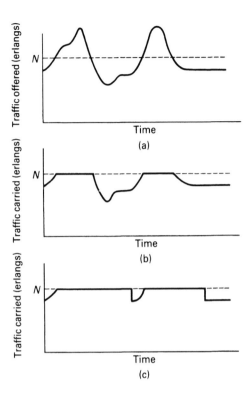

Figure 9.1 Traffic capacity of a circuit-switched and a packet-switched system. (a) Traffic offered. (b) Traffic carried by circuit-switched system. (c) Traffic carried by packet-switched system. N = number of trunks.

terminal. If packets can vary in length, the receiving terminal must be able to know when it has received the complete packet. The length of the packet is therefore also included in the header. However, some systems use a *flag*, i.e. a predetermined unique combination of digits, to indicate the end of a packet, instead of sending a length count.

The data network and its terminals handle the packets by procedures known as *protocols*. These operate at a number of different levels, from the the physical one of sending bits over a data link to the protocols pertaining to the particular application for which the system is used. The protocols can therefore be defined in terms of the OSI 7-layer model[2,3] described in Section 1.7.

Packet switching was first developed for use in private data networks. It is widely used in *local-area networks* (LANs), for data transmission within a single site or building.[1,2,4,5] Packet switching is also used in *wide-area networks* (WANs), for data transmission between different sites of an organization. A WAN may link LANs at different locations. If two networks use the same protocols, they may be linked by a simple apparatus called a *bridge*. If they use different protocols, a more complex

Packet switching

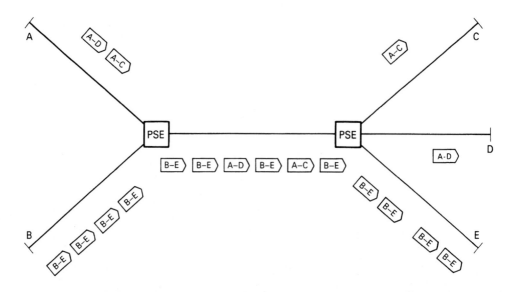

Figure 9.2 Principle of packet switching. PSE = packet-switched exchange.

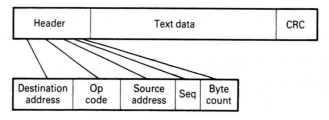

Figure 9.3 Typical packet format.

equipment known as a *gateway* is required to translate from one protocol to the other. LANs and WANs are described in Section 9.3.

The early success of packet switching in such relatively small-scale networks led to the introduction of large packet-switched public data networks (PSPDNs). These are described in Section 9.4.

More recently, packet switching has received attention as a means of handling variable-bit-rate traffic in future broadband networks. This development is discussed in Section 9.5.

9.2 Statistical multiplexing

A conventional multiplex transmission system provides a permanent channel from each input terminal to each output terminal. These channels are used inefficiently if the

Statistical multiplexing

traffic is 'bursty', i.e. each terminal sends or receives information only intermittently. The number of terminals served by a data link can be increased by sending information in packets instead of allocating each terminal a fixed time-slot in a synchronous TDM frame. A multiplexer operating in this mode is called a *statistical multiplexer* (sometimes abbreviated to 'statmux'), or an *intelligent multiplexer*. In effect, it is a packet switch that concentrates the traffic of the terminals onto a single trunk.

Although each terminal generates information only intermittently, several may attempt to send at the same time. A statmux must therefore include a buffer store to queue messages for transmission in turn. The permissible delay therefore limits the number of terminals that can be served. A statmux is a single-server queueing system, so the total of its packets must amount to less than one erlang of traffic. If the buffer becomes full, the multiplexer sends a signal to its terminals to prevent them from sending.

If a data link transmits k bit/s and it serves terminals that generate m bit/s when active, then a synchronous TDM system can cater for $n = k/m$ terminals (ignoring the 'overhead' bits required for a frame sync signal). If each terminal sends its data in packets and it is active for a fraction, a, of the time, then the mean digit rate per terminal is am (ignoring the overhead bits required for packet headers). If the permissible delays limit the occupancy of the data link to b, then $k = nm\,a/b$ and $n = (k/m)(b/a)$. Thus, if the activity, a, is low, a statistical multiplexer can serve many more terminals than a synchronous multiplexer.

The mean holding time, h, of the single trunk is the time taken to send a packet.

$$\therefore \quad h = p/k$$

where p is the average number of bits per packet.

If packets arrive at rate λ, occupancy of the trunk is

$$A = \lambda p/k \text{ erlang}$$

From Section 4.7.5, the mean queueing delay is

$$\bar{T}_q = \frac{Ah}{1-A} = \frac{\lambda(p/k)^2}{1-\lambda p/k}$$

The mean interval between a packet arriving from a terminal and its transmission being completed is

$$\bar{T} = \bar{T}_q + h = \frac{\lambda(p/k)^2}{1-\lambda p/k} + \frac{p}{k}$$

$$= \frac{1}{(k/p) - \lambda} \tag{9.1}$$

Example 9.1

1. A terminal is connected to a computer by a data link transmitting at 2.4 kbit/s in each direction. On average, the terminal user issues a request to the computer

[234] Packet switching

once per minute. The average length of a user's message is ten 8-bit characters and the average length of the computer's reply is 100 8-bit characters. The time taken by the computer to process the information and the propagation time of the link are negligible. Find:

  (i) The average delay before the terminal has received a complete reply.
  (ii) The occupancy of the data channel.

2. It is decided to share the data link between several terminals by installing a statistical multiplexer. This adds a two-character header to each message. It is required that the average delay before a user receives a reply shall not exceed 0.5 second. (It can be assumed that the lengths of messages have a negative exponential distribution.)

  (i) How many terminals can the data link then serve?
  (ii) If all the terminals are in use, what percentage of requests have to wait one second or more for a reply?
  (iii) For the same number of terminals, what data rate would the link need to transmit if a synchronous multiplexer were used instead of a statmux?

1. (i) The data link transmits 2.4/8 = 300 characters/s.

  ∴ Average time taken for computer to send reply is 100/30 = 1/3rd second

  (ii) Occupancy of link from terminal to computer is

  $10/(60 \times 300) = \underline{5.55 \times 10^{-4}}$ E.

  Occupancy of link from computer to terminal is

  $100/(60 \times 300) = \underline{5.55 \times 10^{-3}}$ E.

2. (i) The packet length is $p = 8 \times (100 + 2) = 816$ bits.
  The transmission speed is $k = 2.4 \times 10^3$ bit/s.
  From equation (9.1):

  $$\lambda = \frac{k}{p} - \frac{1}{T} = \frac{2.4 \times 10^3}{816} - 2 = 0.941 \text{ packets/second}$$

  If there are n terminals, each receiving a packet once every 60 seconds, then $n = 0.941 \times 60 = 56.5$.
  i.e. the mux can serve up to 56 terminals.

  (ii) The holding time is $h = 816/2.4 \times 10^3 = 0.34$ second.
  The occupancy is $A = 0.34 \times 56/60 = 0.32$ E.
  If total delay is 1 second, queueing delay is
  $T_a = 1 - 0.34 = 0.66$ second
  From Section 4.7.5:
  Mean queueing delay when there is delay is

$$\overline{T'_q} = \frac{h}{1-A} = \frac{0.34}{1-0.32} = 0.5 \text{ second}$$
$$P(T \geq t) = A \ e^{-t/\overline{T'_q}} = 0.32 \ e^{-0.66/0.5} = 0.0854$$

i.e. 8.5% of requests wait at least 1 second for a reply.

(iii) For syncmux serving 56 terminals, the required data rate is $56 \times 2.4 = \underline{134.4 \text{ kbit/s}}$.

9.3 Local-area and wide-area networks

9.3.1 Bus networks

One commonly used architecture for LANs and WANs is the bus structure shown in Figure 1.2(b). For a LAN, the bus may consist of a cable running round a building. For a WAN, it may be constructed from circuits leased from a public telecommunications operator, or even from the operators in several different countries. Satellite links are also employed.[6]

A bus network is a single-server queuing system. The bus must be able to transmit the total traffic of all the nodes. As this increases towards 1.0 erlang, the delay tends to infinity. Thus, the total traffic must not exceed a fraction of an erlang if long delays are to be avoided. (See Figure 4.11.) Consequently, the bus must transmit data at a much higher speed than the rates at which data are generated at its nodes.

Two principal modes of operation are used. In the first, one node is usually a mainframe computer and communication takes place only between it and data terminals at the other nodes. All communication is completely controlled by the central computer and this is called *polling*. In the second mode all nodes are of equal status and send traffic independently, i.e. *random access* is provided.

In a polling network, the central computer invites the nodes to communicate in turn according to a predetermined rule. Two types of polling are in use: *roll-call polling* and *hub polling*. In roll-call polling the terminals are normally in the 'listening' mode, waiting to receive signals broadcast by the central computer. Each broadcast message includes the address of one terminal. When a terminal receives its address, it either receives a message or responds to an invitation to send data. If it has no message waiting, it replies to that effect and the central computer polls the next node. It is not necessary for all terminals to be polled sequentially. Some may be given priority and polled more frequently than others. The central computer may change the roll-call list if priorities change.

The disadvantage of roll-call polling is that, for very wide areas, there can be significant delays between polling actions because of the long end-to-end propagation time of the network. This can be reduced by *hub polling*. The interrogation is initiated by terminals, instead of by the central computer. When a terminal has sent or received a message, or if it has no message to send, it passes the polling invitation to the next terminal. This greatly reduces the time required for polling on long-distance networks.

[236] Packet switching

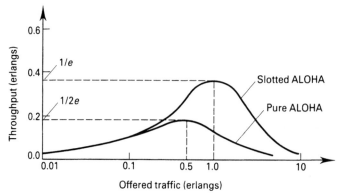

Figure 9.4 Throughput characteristics for pure and slotted ALOHA.

However, the terminal equipment is more complex and it is difficult to modify the polling sequence to allow priorities.

Polling networks meet the needs of organizations having a number of branches where terminals need to communicate with a mainframe computer at headquarters.[7] Examples include the centralized accounting systems of banks and the centralized seat-reservation systems of airlines. The bus may be hundreds or even thousands of kilometres in length. It is usually provided by means of circuits leased from a public telecommunications operating organization, or even from those of several different countries.

One of the simplest random access methods is the *ALOHA protocol* developed by the University of Hawaii for a network using satellite radio transmission.[8,9] It allows each node to send a packet as soon as it is formed and relies on a positive acknowledgement to indicate that the packet is received without error. If each packet takes P seconds to send and the maximum propagation time is t_p, then an acknowledgement should be received within time $P + 2t_p$. If not, the packet is sent again.

A packet is transmitted successfully only if another station does not send within P seconds before or after it. The period vulnerable to contention is thus $2P$. When a collision occurs, both stations retransmit, so further collisions could occur. The probability of this is greatly reduced by introducing a random delay before retransmission. Thus, the two contending stations retransmit at different times and a further collision is less likely. If it does occur, the process is repeated until success is achieved.

The pure ALOHA protocol works well as long as the throughput of data is small. If a large number of calls is generated, collisions are frequent and retransmissions cause further collisions. As a result, the throughput of data actually falls when the traffic increases beyond a certain level,[9,10] as shown in Figure 9.4.

The probability of successful transmission of a packet is the probability, $P(0)$, that no other packet is sent during the interval $2P$. If packets have a mean arrival rate of λ per second and a Poisson distribution, then:

$$P(0) = e^{-2P\lambda}.$$

Therefore, the rate at which packets are transmitted successfully, S, is given by:

$$S = \lambda e^{-2P\lambda}.$$

Differentiating with respect to λ, the maximum throughput is found to be:

$$S_{max} = \frac{1}{2P} e^{-1} \text{ packets/second, when } \lambda = 1/2P.$$

But P is the packet duration, so the traffic offered is $\lambda P = 0.5$ E and the traffic carried is $1/2e$ erlangs. Thus, the throughput obtainable is only 18% of the theoretical capacity of the system.

The throughput of an ALOHA system can be improved significantly if all the transmissions are synchronized. The time frame is divided into time-slots of duration P and each station may only start sending at the beginning of a time-slot. This known as *slotted ALOHA*. When collisions occur, the packets overlap completely. Thus, the vulnerable period is reduced from $2P$ to P and collisions are fewer. The throughput obtainable is increased to $1/e$ erlangs, i.e. 36.8% of the theoretical system capacity.

Example 9.2

1. A pure ALOHA system uses a 56 kbit/s channel. On average, each terminal originates a 1024-bit packet every 30 seconds. How many terminals can the system accommodate?
2. How many terminals could the system accommodate using the slotted ALOHA protocol?

1. Duration of packet $= 1024/56 = 18.3$ ms.
   $\therefore$ Traffic per terminal $= (18.3 \times 10^{-3})/30$
   $= 6.1 \times 10^{-4}$ E

   and the total traffic is $6.1n \times 10^{-4}$ E, where n is the number of stations. But the maximum throughput of the system is $1/2e$ E.

   $\therefore \quad n = 10^4/(6.1 \times 2e) = 301$

2. The maximum throughput $= 1/e$ E.

   $\therefore \quad n = 10^4/6.1e = 603.$

An alternative to the ALOHA protocol is to arrange for terminals to 'listen' to the highway and send data only when they sense that it is free. Thus, collisions are avoided, instead of being resolved by retransmission. This technique[11] is called *carrier-sense multiple access with collision detection* (CSMA/CD). If a terminal senses that the highway is busy it waits for a short period and then checks it again. The waiting period is necessary because several terminals could find the highway free and start to send simultaneously. The probability of a further collision is minimized by making the waiting times random. However, after the random waiting time, the terminal must monitor the bus for a period equal to the longest propagation delay, t_p, to prevent transmission from starting before the preceding packet has reached its destination.

Because collisions are avoided, CSMA/CD is more efficient than ALOHA, The

Sync.	Destination address	Source address	Packet type	Data	Check-sum
32 bits	48 bits	48 bits	16 bits	Variable 45–1500 bytes	32 bits

Figure 9.5 Format of Ethernet frame.

vulnerable period is reduced from P or $2P$ to only $2t_p$. For a LAN, t_p is short, so the throughput can approach 1.0 E. Because collisions are avoided, there is no critical load beyond which the throughput collapses, as in ALOHA. However, the available throughput decreases as t_p increases. Hence, CSMA/CD is more suitable for LANs than WANs, unless packets are very long and data rates are low.

The best-known CSMA/CD network is the Ethernet network.[10,12] This transmits data at 10 Mbit/s, so it can support a large number of relatively slow data terminals. The bus is passive, so the packet header starts with a *preamble* of 32 bits in length to achieve synchronization, as shown in Figure 9.5. The maximum length of a single Ethernet bus is about 500 m. For lengths greater than this, segments of bus are joined by repeaters. Each segment can accommodate up to 100 data terminals.

An alternative to the CSMA/CD bus is the *token bus*. The *token* is a unique pattern of digits which is circulated among the terminals. Each node waits to receive the token. If it has data waiting, it sends the data and then passes the token to the next node. If a node has no data to send, it simply passes on the token. Thus, although the the structure of the network is a bus, its nodes form a logical ring.

The token-bus technique is less efficient than CSMA/CD at low traffic levels, because each terminal must wait for the token even if the bus is free. However, its performance is superior at heavy traffic levels because contention and the time taken to resolve it are entirely eliminated.

Example 9.3

An Ethernet operates at 10 Mbit/s. It is 1 km in length and the velocity of propagation is 2×10^8 m/s. Data packets consist of 512 bits, including a 64-bit overhead. A receiving terminal takes the time of one bit to access the channel in order to send an acknowledgement signal, which consists of an empty packet.

If there are no collisions, at what rate can the system convey data?
The maximum propagation time is $10^3/(2 \times 10^8)$ second $= 5\mu s$

Time taken to send a full packet $= 512/10 = 51.2\mu s$.
Time taken to send an empty packet $= 64/10 = 6.4\mu s$.
∴ Time taken to send packet and receive acknowledgment is

$$51.2 + 0.1 + 2 \times 5 + 6.4 = 67.7\mu s$$

The data conveyed is $512 - 64 = 448$ bits.
∴ Effective data rate is $448/67.6 = \underline{6.62 \text{ Mbit/s}}$

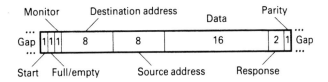

Figure 9.6 Slot bit pattern of Cambridge ring.

9.3.2 Ring networks

A ring network is shown in Figure 1.2(c). When data are transmitted between two nodes, the signal is received at each intermediate node and then retransmitted. Thus, a ring network is an active network, whereas a bus LAN can be passive. A ring network is also a single-server queueing system, so the total traffic from all nodes must be only a fraction of an erlang if long delays are to be avoided. The principal types of ring network are the *slotted ring* and the *token ring*.

The cable between nodes of a ring has a propagation delay (usually about 5 ns/m) and the nodes act as regenerative repeaters. Thus, provided that the duration of a block of data is less than the propagation time round the ring, it can circulate until it is deleted by a node. In a slotted ring, which is often called a *Cambridge ring*, this block of data is formatted into one or more *slots*. Since the length of the ring may be short, the slot must also be short. In the original form of the Cambridge ring,[13] it is only 38 bits, as shown in Figure 9.6. Each slot contains a 'full/empty' bit, the source and destination addresses, the data and some 'housekeeping' bits. A node that is ready to send data waits to receive the slot. If it recognizes that the slot is empty, it inserts a packet of data. As the full slot circulates, each node monitors the destination address. If it matches its own, the data are read off and response bits are inserted in the slot to indicate correct receipt of the data. When the slot returns to the sending node, it is marked empty and passed on for re-use.

A slot cannot be immediately re-used by a node if it still has data waiting. It must wait until the slot has traversed a complete circuit. Thus, use of the ring is shared fairly among all the terminals. If a node has more than 38 bits to send, it will take several revolutions of the ring to transmit the message. Larger packets can be accommodated by operating the ring at a higher bit rate. The *Cambridge Fast Ring* operates at 70 Mbit/s and accommodates 32 octets of data and 16-bit address fields.[14]

In the token ring[15] a *token* (e.g. the bit pattern '11111111') circulates continuously while the ring is idle. To prevent this pattern from occurring during data, bit stuffing (as described in Section 8.9.2) is used. If a node is ready to send data, it *captures* the token. In some token rings this involves removing the token. In others, a bit is changed in the token (e.g. the token may be changed from 11111111 to 11111110) and the modified bit pattern is called a *connector*. Having captured the token, the node may send a packet or several packets (up to a prescribed maximum). The receiving node recognizes its address, reads off the message and sets the response bit before

[240] *Packet switching*

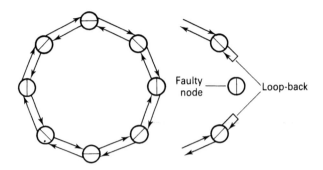

Figure 9.7 Duplicated ring with self-healing capability.

retransmitting. When the message returns to the sending node it resets the token for use by other nodes.

There must be sufficient delay around the ring to accommodate the complete token. Otherwise, the first bit of the token would return to the sending node before the last bit had left it. Each node delays the token by at least one bit in deciding whether to change it. This delay is called the *node latency*.

The token ring accommodates much larger packets than the slotted ring, typically a few thousand octets. Thus, the proportion of time occupied by address and control information is much smaller and the throughput of data is larger. However, the delay encountered by a node in accessing the ring is the sum of the service times of all nodes upstream of it. Since the packets are longer than for a slotted ring, so are these delays.

Correct operation of a basic ring network requires the correct operation of every node and link. Thus, failure of any node or link causes complete failure of the entire system. This can be combatted by duplicating the ring. The additional ring may be simply used as a standby; alternatively, both may carry traffic until a fault occurs. It is possible to protect against failure of both main and standby units in a single node or link by using two rings transmitting in opposite directions, as shown in Figure 9.7. A faulty node or link can now be isolated by looping back at the nodes at opposite sides of the fault to form a single complete path. Automatic techniques for detecting and isolating faulty units result in a *self-healing network*.

Example 9.4

A token ring operates at 10 Mbit/s. It is 1 km in length and the propagation velocity is 2×10^8 m/s. Fifty terminals are spaced around the ring and the node latency is one bit. Packets are 512 bits long, including 64 overhead bits. The token consists of eight bits.

Compare the effective data rate when the ring is fully loaded with that of the Ethernet in Example 9.3.

Time taken to send connector = 8/10 = 0.8 μs
Time taken to send packet + token = (512 + 8)/10 = 52 μs
Total latency of 50 nodes = 50/10 = 5 μs
Time taken for packet plus connector and token to circulate round ring and return to originating node is

52 + 0.8 + 5 + 5 = 62.8 μs

No. of data bits in packet = 512 − 64 = 448
∴ Effective data rate = 448/62.8 = 7.13 Mbit/s

In this case, the effective data rate of the token ring is about 8% more than that of the Ethernet.

9.3.3 Comparison of bus and ring networks

The relative advantages and disadvantages of bus and ring networks are as follows:

1. A bus network can be more reliable than a ring network because its highway can be passive. Thus, failure of one or more nodes does not interrupt service for the others.
2. Additional nodes can be connected to a bus without disrupting the network. Adding nodes to a ring network necessitates taking the network out of service.
3. A bus network uses twisted-pair or coaxial cable because of the need to transmit signals in both directions. The highway of an optical-fiber bus network would require to be tapped by bidirectional optical couplers.
4. Bus networks can suffer from signal reflections at impedance irregularities. Care is therefore required to avoid impedance mismatches at nodes and at the ends of the bus.
5. Fault isolation is difficult in bus networks, but it is straightforward in ring networks.

9.3.4 Optical-fiber networks

Optical fibers have low attenuation and can transmit data at very high bit rates. They are also immune to electrical interference. This makes the use of optical fibers attractive for LANs. In order to avoid the need for bidirectional optical tappings to construct a bus network, optical-fiber LANs use ring configurations. The fibers simply provide unidirectional transmission links between the nodes.

The first use of optical-fiber LANs was for the connection of mainframe computers to high-speed peripheral equipments. This led to the development of a standard for a *fiber-optic distributed data interface* (FDDI). An FDDI network[4,16,17] uses a dual ring, transmitting data at 100 Mbit/s. Loop-back, as described in Section 9.3.2, is used to obtain fault tolerance. It is also possible to use optical fibers in a network with a star configuration, as shown in Figure 1.2(d).

Recently, telecommunications operating organizations have considered introducing a high-speed data service to link together the LANs of customers across a city or even a wider area. Such a network is called a *metropolitan-area network* (MAN). A network called a *distributed-queue dual-bus* (DQDB) network has been proposed and standards for it have been developed.[4,17] Although the DQDB network is logically a bus, it is implemented physically as a ring, which is broken at a node that generates slots. It is intended to operate at data rates of about 150 Mbit/s, using optical fiber as the transmission medium.

9.4 Large-scale networks

9.4.1 General

The bus and ring structures used in LANs and WANs are inadequate for large-scale networks having a very large number of terminal stations distributed over a wide geographical area. It is necessary to have several packet-switching exchanges and employ tandem routings. This usually results in a hierarchical network, as shown in Figure 1.6. Since end-to-end connections can involve several queues in tandem, the number of levels in the hierarchy is restricted by the permissible delays. If transmission is at 64 kbit/s and delays should seldom exceed 300 ms, the hierarchy is limited to three levels. Some military networks use a mesh network, as shown in Figure 1.2(a), to minimize the loss of routes resulting from damage to nodes or links.

The technique was pioneered in the 1960s by the ARPANET network, linking together establishments of the US Defense Department's Advanced Research Projects Agency.[9,18] Subsequently, telecommunications operating organizations in a number of countries have installed *packet-switched public data networks* (PSPDN). Examples include British Telecom's Packet-switched Service (PSS), Internet and Tymnet in the USA, Data Pac in Canada, Transpac in France and Datel in Germany. International standards have been established to enable networks to be interconnected.

9.4.2 Datagrams and virtual circuits

The simplest form of packet-switched service uses the *connectionless mode* or *datagram mode* of operation. Every packet is treated as a separate message and the network establishes a connection for each one. Therefore, packets of the same message can be routed through different intermediate switching nodes when there are alternative routes to the destination.

In the *connection-oriented mode* or *virtual-call mode*, a *virtual circuit* is established by the initial exchange of packets between a pair of terminals and subsequent packets use the same route. This is more convenient for a transaction involving a sequence of packets. It provides an apparently circuit-switched connection, although economy is obtained by sharing the circuits with other messages. Consequently, this mode is usually employed in public networks. A telecommunications operator can lease a private data circuit to a customer by setting up a *permanent virtual call* (PVC). This operates permanently in the data-transfer mode.

9.4.3 Routing

The switching centres may use fixed routes to each destination. However, adaptive routing may be employed, in which each exchange may use different routes to the same destination, depending upon traffic conditions. The scope for this is greater for datagram operation, where a route is selected for each packet, than for virtual-call operation, where a single routing decision is made at the time of initial call set-up.

9.4.4 Flow control

Terminals generate calls in ignorance of the traffic already on the network, so congestion can occur. In an ideal network, the throughput would rise with increasing offered traffic until the capacity of the network is reached and then remain constant, as shown in Figure 7.9(a). However, overload can cause the exchanges to waste processing capacity on abortive attempts, with the result that the throughput begins to fall with increases in the traffic offered, as shown in Figure7.9(b). It is therefore necessary to provide *flow control*, as shown in Figure 7.9(c).

To some extent, flow control is provided end-to-end by the terminals. Any increase in cross-network delay caused by congestion increases the time taken for a sending terminal to obtain a 'permission to send' signal from the receiving terminal and thus reduces the rate at which it can send packets. This should be supplemented by network flow control,[19,20,21] in which switches reject incoming packets when the lengths of their queues exceed prescribed limits.

9.4.5 Standards

The international standard[22] for packet-switched networks is provided by CCITT Recommendation X.25. This caters for virtual-circuit services. Recommendation X.25 defines the interface between a subscriber's *data terminal equipment* (DTE) and the termination of the network, known as a *data circuit terminating equipment* (DCE). The DTE may be a computer. Alternatively, a number of low-speed character-mode terminals may be connected to the DCE by a *packet assembler/disassembler* (PAD), which acts as a statistical multiplexer. If the DTE is connected to the PSPDN by an analog customer's line, the DCE is a data modem.

Although Recommendation X.25 pre-dates the OSI seven-layer model described in Section 1.7, it has a three-layer structure which conforms with layers 1 to 3 of the OSI model. The recommendations are too lengthy and complex to describe here, so only a brief outline will be given. Further details can be found in references 22 and 23.

Layer 1, the physical layer, defines the connections between the DTE and DCE and specifies the electrical representation of '0' and '1', timing, etc. The specifications are contained in CCITT Recommendations X.21, X.21 bis and V.24. The specifications for connecting low-speed character-mode terminals via a PAD are contained in CCITT Recommendations X.3, X.28 and X.29. These are often called the 'Triple-X standard'.

[244] Packet switching

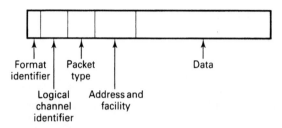

Figure 9.8 X.25 packet format.

In addition, there is Recommendation X.31, which defines how connection should be made to an integrated-services digital network (ISDN).[24]

Corresponding to layer 2 of the OSI model is the *link-layer protocol* or *frame-layer protocol*. This is based on the ISO high-level data-link control (HDLC) protocol described in Section 8.9.2. It uses the HDLC in the form known as *balanced link access protocol* (LAP-B).[2] Each packet is carried within a LAP-B frame, as shown in Figure 8.15. The beginning and end of each frame is indicated by an 8-bit flag consisting of the digits '01111110'. Since this sequence of digits can occur in messages, it is prevented from being interpreted as a flag by using bit stuffing as described in Section 8.9.2.

There are three kinds of frame: *information frames* (I frames), *supervisory frames* and *unnumbered frames*. An information frame contains a message packet. The supervisory frames are the 'receive ready' (RR) frame, 'the receive not ready' (RNR) frame and the 'reject' (REJ) frame. If there is a sequence error, the receiving terminal sends back a REJ frame and the sending terminal retransmits an I frame, starting from that indicated by the sequence number in the REJ frame. The unnumbered frames are used by a terminal to report an error condition which is not recoverable by retransmitting the identical frame, and then to initiate the link-resetting procedure.

The layer-3 protocol is the *packet layer*. It defines the form of the packet that is sent in the information field of the HLDC frame. The format of the packet is shown in Figure 9.8. The *format identifier* is a 4-bit field which indicates the detailed format of the rest of the packet. The *logical channel identifier* corresponds to the particular virtual call, so that packets belonging to the same call can always be identified. The *packet-type identifier* specifies the function of the packet, since packets can be of several different types. For call initiation, the address is contained in the next field. For a virtual call, subsequent packets do not need to contain the address; routing is now determined by the logical channel identifier. The address field is then used for the sequence number or for acknowledgements. Acknowledgement and control packets do not contain a data field. Information packets contain a data field of up to 128 octets in length.

The procedure for a virtual call is shown in Figure 9.9. The sequence of events is as follows:

1. The calling terminal sends a 'call request' packet and the network delivers this to the called terminal as an 'incoming call' packet.

Large-scale networks

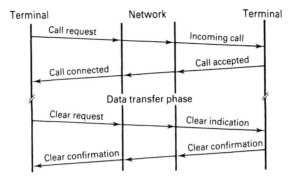

Figure 9.9 Virtual call set-up and clear.

2. The called terminal returns a 'call accepted' packet and the network delivers this to the calling terminal as a 'call connected' packet.
3. The call then enters the data-transfer phase and information packets are transmitted.
4. At the end of the call, either terminal may initiate clearing down of the call, using the packet sequence: 'clear request', 'clear indication' and 'clear confirmation'.

9.4.6 Frame relay

Conventional packet-switched networks use the X.25 protocols at the interfaces between all links and switching centres. Thus, errors are corrected by retransmission on a link-by-link basis and an acknowledgement must always be received by a switching centre before sending the next packet of any message. This is unnecessarily complex if the error rates of the links are very low. End-to-end error control between terminals is then sufficient.

A technique known as *frame relay*[24] has been developed for applications where the links in the network are known to have very low error rates. Each switching centre simply relays frames to the next without checks or acknowledgements. In this way, a greatly increased throughput can be obtained with smaller delays.

The main differences between frame relay and X25 packet switching are:

1. There is no link-by-link flow control or error control. These are the responsibility of the users' terminals.
2. Switching of logical connections takes place at layer 2 instead of layer 3, thus eliminating one layer of processing.
3. Call-control signalling is carried out on a logical connection separate from the data. As a result, intermediate nodes do not need to process call-control messages on a basis of individual connections.

The link-by-link error control provided by X.25 has been lost. However, the very low error rates of modern optical-fiber digital transmission systems make this superfluous. Frame relay is used at data rates of up to 2 Mbit/s. It is the basis of the *frame-mode bearer* service in an ISDN.

9.5 Broadband networks

9.5.1 General

Telephone circuits are used for many other purposes, including the Telex service, telemetry, facsimile (fax) and data transmission. However, the bandwidth provided by a public switched telephone network is limited to about 3 kHz. Basic-rate ISDN access can cater for signals which do not require a digit rate higher than 64 kbit/s. This can accommodate high-speed data and fax transmission and slow-scan television (e.g. for surveillance). Primary-rate access can cater for up to 1.5 Mbit/s or 2 Mbit/s. However, there are signals (e.g. high-definition colour television) whose bandwidth is too great for the existing ISDN. This has therefore been called *narrowband ISDN* (N-ISDN). The demand for these broadband services is expected to increase, so plans are being considered for the introduction of a future *broadband ISDN* (B-ISDN). This is also referred to as the *integrated broadband communications network* (IBCN). Much research has been done[25] and recommendations have been published by the CCITT.[26]

In order to convey a large number of broadband signals between its switching nodes, an IBCN needs very high-capacity transmission links. These can now be provided over optical fibers by transmission systems in the synchronous digital hierarchy (SDH) described in Section 2.6.4. The customers' distribution networks must also be able to provide broadband transmission. Considerable research has already been done on providing broadband customers' distribution over optical fibers.[27]

An IBCN should be able to provide its customers with any required bandwidth on demand. A circuit-switched network is unsuitable for this. It gives its customers connections of fixed bandwidth, so the maximum bandwidth must be provided even when not required. This is clearly uneconomic.

Packet switching is potentially suitable. Digitally, encoded samples of analog signals can be assembled into packets for transmission and switching. Provided that the overall digit rate of the network is high enough, a user can transmit a broadband signal by sending packets frequently and a narrowband signal by transmitting packets less frequently. A disadvantage is that assembling signal samples to form packets at the sending terminal causes delay and further delays are encountered in queues at switching nodes. Moreover, these delays are variable. Thus, for speech transmission, received packets must be stored in a buffer in order to emit speech samples at a constant rate. If the buffer is full when a packet arrives, it is discarded. Fortunately, occasional loss of packets is not noticeable.

These delays make conventional packet switching unsuitable for telephony. As explained in Section 2.7.1, unless echo-control devices are used, the delay across a network should not exceed about 25 ms. It may be noted that the requirements for various types of traffic differ. For data transmission, delays can be tolerated but errors must be minimal. For telephony, quite high error rates can be tolerated, but delay and jitter are severe impairments.[3] These conflicting requirements have been called *semantic transparency* and *time transparency*, respectively.

These factors led to the adoption of a modified form of packet switching for broadband services. This has been called *fast packet switching* (FPS), *asynchronous time division* (ATD) and the *asynchronous transfer mode* (ATM). The term ATM has been adopted by the CCITT.[28] This term is confusing. It does not imply that transmission links or switches operate asynchronously; they do not! The word 'asynchronous' is used because the method allows asynchronous operation of the clocks at the sending and receiving terminals. Any timing difference between them is accommodated by inserting or removing empty packets in the bit stream between them.

9.5.2 The asynchronous transfer mode

When existing packet networks for data were designed, the available circuits in the telephone network had a relatively poor error performance. Complex arrangements were therefore necessary for error control. Packets were of varying length, so complex protocols were needed to handle them. However, this was possible because speeds were slow. Delays were comparatively long, but time transparency was not a requirement.

Subsequent developments, particularly the introduction of optical-fiber transmission, have enabled both much higher operating speeds and much lower error rates to be achieved. This has permitted simplifications to be introduced for ATM, which enable it to obtain sufficient time transparency for telephony and still retain sufficient semantic transparency for data transmission.[29,30] The features of ATM are as follows:

1. Packets are of fixed length, known as *cells*.
2. ATM operates in the virtual-call mode. If the necessary resources for a call are available, they are reserved for it. If not, the call is rejected.
3. No error control or flow control is provided on a link-by-link basis. Error control is omitted for the information in cells (but not for their headers) because of the low error rates of modern transmission systems. End-to-end error control for data connections can be provided by terminal equipments.
4. The header is short. The header only identifies the virtual connection and sequence numbering is not provided. However, error-control bits are included, because an error would cause a cell to be misrouted.
5. The information field is short. This also enables the cell to be short. Use of a short cell minimizes packetizing delay at the sending terminal and queueing delays at switching nodes and so enables adequate time transparency to be provided.

[248] Packet switching

6. A policing function is provided. This enables the network to ensure that the user is not sending and receiving digits (and therefore cells) at a greater rate than is being paid for.

If the information field is too short, the header occupies most of each packet and transmission capacity is used inefficiently. If the information field is too long, delays are excessive. The choice of cell size is a compromise between these conflicting requirements. The CCITT has recommended a 53-octet cell containing a 48-octet information field preceded by a 5-octet header.[28] Transmission is at 149.76 Mbit/s, which corresponds to the payload of the synchronous transport module STM-1 of the SDH hierarchy of transmission systems.

Example 9.5

An ATM network uses transmission links that operate at 150 Mbit/s and have a propagation delay of 5 μs per km. It uses cells of length 53 octets, consisting of a 5-octet header and a 48-bit information field. The maximum delay introduced by a switching centre is 300 cells.

Find the maximum delay encountered by a telephone call over a connection of length 500 km that passes through six switching centres.

The speech encoder produces a PCM signal at 64 kbit/s. The packetization delay is the time taken to fill an information field of 48 octets, i.e. 384 bits.

∴ Packetization delay = 384/64 = 6 ms
Total propagation delay = 500 × 5 μs = 2.5 ms
Delay per switching centre = 300 × 53 × 8/150 μs
= 848 μs
∴ Total switching delay = 6 × 0.848 = 5.1 ms
Delay encountered by telephone call = 6 + 2.5 + 5.1 ms
= *13.6 ms*

This is satisfactory for a national network. It uses just over half of the 25 ms specified by the CCITT for the maximum delay of an international connection without the use of echo suppressors or cancellers.

9.5.3 ATM switches

The basic functions of an ATM switch are shown in Figure 9.10. A space switch is required to route calls from an incoming trunk I_j to an outgoing trunk O_k. The incoming header H_i must be changed to an outgoing header H_o, which is required to operate the next switching node in the connection. Several cells for the same outgoing trunk may arrive simultaneously on different incoming trunks. To prevent loss of cells, queues must be provided. Thus, the functions to be provided by an ATM switch are:

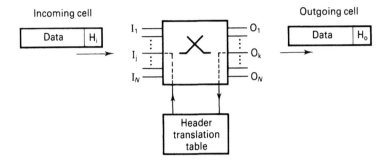

Figure 9.10 Principle of ATM switch.

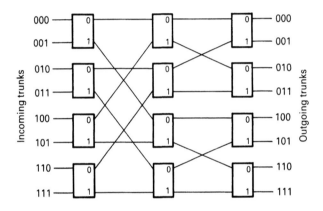

Figure 9.11 Self-routing ATM switch.

- Call routing (by space switching)
- Header translation
- Queuing.

Several different switching architectures have been proposed, based on different queuing methods and different basic switching elements.[29] One multistage switch is shown in Figure 9.11. This is called a *delta switch*[31] and it has a self-routing ability. Its network is called a *banyan network* because its shape resembles the tropical tree of that name.

There is only one path through the network from an incoming trunk to each outgoing trunk. Each switching element has two inlets and two outlets. The outlet chosen at each stage is determined by one digit in the header of a cell. If this digit is '0', the upper outlet is chosen; if it is '1', the lower outlet is chosen. Thus, successive bits in the header route the cell through the switching stages until the required outgoing trunk

[250] Packet switching

is reached. Figure 9.11 shows a three-stage network connecting eight incoming trunks to eight outgoing trunks (numbered 0 to 7). Since cells may arrive simultaneously at the two inlets of any switching element, the contention must be resolved by storing one cell in a buffer until the other has been sent.

A network with switching elements of size 2×2 requires $\log_2 N$ stages to reach N outgoing trunks and uses $\log_2 N$ header digits to route the calls. Larger networks can use larger switching elements. For example, a single element of size 8×8 would use three header digits and would replace the whole of the network of Figure 9.11. Three stages would give access to $8^3 = 512$ outgoing trunks. Since these networks consist of elements that are identical and have regular interconnections, they are suitable to be manufactured as large-scale integrated circuits.

Notes

1. A CRC uses a set of parity bits that cover overlapping sets of data bits.[1] Small numbers of errors are always detected and large numbers of errors are detected with probability 1 in 2^M, where M is the number of bits in the check code. They are generated and checked by using cyclic shift registers.
2. LAP-B caters for links which can be controlled from either end. The *normal response mode* (NRM) caters for the situation where links are always controlled from one station, as in polling networks.
3. The same delay restrictions apply to a video signal if it accompanies a sound signal; e.g. the lip movements of a speaker must coincide with the speech.

Problems

1. (a) Explain the differences between a circuit-switched and a packet-switched network and discuss their relative merits.
(b) x bits of data are to be sent at b bit/s over a switched connection of l links. The propagation time for each link is t_p seconds. If circuit switching is used, the time taken by each exchange to set up its connection is c seconds. If packet switching is used, the packets consist of i information bits and h header bits. The delay through a packet-switched exchange is s seconds. It can be assumed that both networks are error-free.
(i) In each case, find the time taken to convey the message.
(ii) Under what conditions does the packet-switched network convey the message quicker?

2. What is meant by polling? Explain the differences between roll-call polling and hub polling. Under what circumstances would each be used?

3. (a) A pure ALOHA system transmits at 16 kbit/s. Each station, on average, sends a 512-bit packet every 20 seconds. What is the maximum number of stations that the system can accommodate?
(b) How many stations could the system accommodate using a slotted-ALOHA protocol?
(c) In each case, when the system is fully loaded, what is the average number of attempts required to send a packet successfully?

4. The stations of a slotted-ALOHA system

attempt to send 100 packets per second (including initial requests and attempts at retransmission). The slot duration is 20 ms. It may be assumed that the attempts are independent random events with a Poisson distribution.

(a) What is the probability that the first attempt to send a packet will be successful?
(b) What is the probability of exactly three collisions followed by a successful attempt?
(c) What is the average number of attempts per transmitted packet?

5. (a) A LAN bus using the CSMA/CD protocol is of length 2 km. Its propagation delay is 5 μs per km and the digit rate is 5 Mbit/s. Each packet contains 720 bits, including a 176-bit header and 32 check bits. Find the maximum data throughput of the system.
(b) Find the maximum total data throughput if:
  (i) The bit rate of the system is increased to 10 Mbit/s
  (ii) The packet length is increased to 10 000 bits.

6. A slotted-ring LAN operates at 10 Mbit/s. It is filled with a packet of 40 bits overall size containing 16 data bits. Find the longest delay in obtaining access and the throughput of data obtained by a user when:

(a) No other users have data to send
(b) There are 250 simultaneous users.

7. Describe briefly the functions of the layers of the OSI model for open systems interconnection. Show how the X.25 protocol for packet switching conforms with layers 1 to 3 of the OSI model.

8. A packet-switching network uses packets consisting of h header digits and i information digits. The bit error probability is P and it can be assumed that errors are independent random events. Thus, the probability that a packet is received correctly, P_c, is given by

$$P_c = (1 - P)^{i+h}$$

and the probability, P_m, that m transmissions are needed to receive a packet correctly is

$$P_m = (1 - P_c)^{m-1} P_c.$$

(a) Hence, show that the average number of transmitted digits, n, required to convey one packet is

$$n = (i + h)/(1 - P)^{i+h}$$

(b) The optimum packet length, $i + h$, is that which, on average, requires the minimum number of digits to be transmitted per information digit. If h is fixed and P is small, find the optimum packet length.

9. An X.25 packet-switched network has a 64 kbit/s circuit between exchanges A and B. The frames have an average length of 64 octets. It may be assumed that the frame lengths have a negative exponential distribution. (This is an approximation, since HDLC frames have both a maximum length and a non-zero minimum length.)

For frames received at A for onward transmission to B, find the percentage of frames that are delayed, the percentage delayed by at least 10 ms and the mean delay when the mean rate of arrival of frames is:

(a) 25 per second
(b) 100 per second
(c) 120 per second.

10. An ATM network uses cells of length 53 octets, consisting of a 48-octet information field and a 5-octet header. These are transmitted at 150 Mbit/s and the propagation delay through the network is 5 μs per km.
(a) Determine the packetization delay for:
  (i) An 8-bit speech coder sending at 64 kbit/s
  (ii) An 8-bit video coder sending at 2 Mbit/s.
(b) A telephone connection is 1000 km in length. It passes through five switching centres, each introducing a maximum delay of 1 ms. (If this delay would be exceeded, a cell is rejected.) Find the maximum delay encountered by the speech signals.
(c) If the number of information bits per cell is trebled to improve the throughput of the network, what effect does this have on telephone transmission? Comment on the result.

References

[1] Brewster, R.L. (1989), *Communication Systems and Computer Networks*, Ellis Horwood, Chichester.
[2] Halsall, F. (1988), *Data Communication Networks and OSI*, 2nd edn, Addison Wesley, Reading, MA.
[3] West, B.G. and Wright, P.G. (1989), 'The philosophy of the ISO seven-layer model', Chapter 5 in Brewster, R.L. (ed.), *Data Communications and Networks*, 2nd edn, Peter Peregrinus, Stevenage.
[4] Wilbur, S.R. (1989), 'High-speed local area networks', *ibid*, Chapter 9.
[5] Stallings, W. (1987), *Local Networks: an introduction*, Macmillan, New York.
[6] Nouri, M. (1991), 'Satellite communication', Chapter 13 in Flood, J,E. and Cochrane, P. (eds), *Transmission Systems*, Peter Peregrinus, Stevenage.
[7] Martin, J. (1972), *Introduction to Teleprocessing*, Prentice Hall, Englewood Clifs, NJ.
[8] Abrahamson, N. (1970), 'The ALOHA system – another alternative for computer communications', *AFIPS Conf. Proc.*, **37**, 281–5.
[9] Kleinrock, L. (1976), *Queuing Systems*, vol.2: *Computer Applications*, Wiley, New York.
[10] Dunlop, J. and Smith, D.G. (1989), *Telecommunications Engineering*, 2nd edn, Chapman and Hall, London.
[11] ISO, 'CSMA/CD local area networks', IS 8802.3.
[12] 'Ethernet LAN: data link layer and physical layer specifications', version 2, DEC, Intel and Xerox Corporations, 1983.
[13] Wilkes, M.V. and Wheeler, D.J. (1983), 'The Cambridge digital communications ring', *Proc. Local Areas Symp.*, NBS/Mitre, Boston, MA.
[14] Temple, S. (1984), 'The design of the Cambridge fast ring', Chapter 6 in Dallas, I.N. and Spratt, E.B. (eds), *Ring-technology Local Area Networks*, North-Holland, Amsterdam.
[15] ISO, 'Token ring local area networks', DIS 8802/5.
[16] Burr, W.E. (1986), 'The FDDI optical data link', *IEEE Comms. Mag.*, **24**, 18–23.
[17] Mollenauer, J.F. (1988), 'Standards for metropolitan networks', *IEEE Comms. Mag.*, **26**, 15–19.
[18] Roberts, L.C. (1970), 'Computer network development to achieve resource sharing', *AFIPS Conf. Proc.*, **36**, 543.
[19] Schwartz, M. (1977), *Computer-communication Network Design and Analysis*, Prentice-Hall, Englewood Cliffs, NJ.
[20] Tannnenbaum, A.S. (1981), *Computer Networks*, Prentice-Hall, Englewood Cliffs, NJ.
[21] Schwartz, M. (1987), *Telecommunication Networks, Protocols, Modelling and Analysis*, Addison-Wesley, Reading, MA.
[22] CCITT Recommendation X.25: 'Interface between data terminal equipment (DTE) and data-circuit terminating equipment (DCE) for terminals operating in the packet mode and connected to public data networks by dedicated circuits'.
[23] Deasington, R.J. (1985), *X.25 Explained*, 2nd edn, Ellis Horwood, Chichester.
[24] Stallings, W. (1992), *ISDN and Broadband ISDN*, 2nd edn, Macmillan, New York.
[25] *Integrated Broadband Services and Networks*, IEE Conf. Pub. no.329, 1990.
[26] CCITT, Recommendation no. I.121: 'Broadband aspects of ISDN'.
[27] Adams, P.F., Rosher, P.A. and Cochrane, P. (1991), 'Customer access', Chapter 15 in Flood, J.E. and Cochrane, P. (eds), *Transmission Systems*, Peter Peregrinus, Stevenage.
[28] CCITT Recommendation I.150, 'BISDN functional characteristics'.

[29] de Prycker, M. (1991), *Asynchronous Transfer Mode: solution for broadband ATM*, Ellis Horwood, Chichester.
[30] Cuthbert, L.G. and Sapanel, J-C. (1993), *ATM: the Broadband Telecommunications Solution*, Peter Peregrinus, Stevenage.
[31] Ahmadi, H. and Denzel, W. (1989), 'A survey of modern high-performance switching techniques', *IEEE Jour. on Selected Areas in Communications*, **SAC**-7, 1091–103.

CHAPTER 10

Networks

10.1 Introduction

A national telecommunications network is large and complex and different parts of it are planned by different groups of engineers. It is therefore essential that they adhere to common standards in order to obtain a satisfactory performance. National plans are needed to govern the design of a network and its constituent local, junction and trunk networks. These must include the following:

- A transmission plan
- A numbering plan
- A charging plan
- A routing plan
- A signalling plan
- The grades of service
- The capabilities of the switching equipments
- Interconnection with other networks
- Network management.

These considerations are not independent; they are interrelated. For example, charging, routing and numbering are closely related. The directory number of a called customer defines both the route for a call and its charging rate. The flexibility with which these can be handled is determined by the capabilities of the switching equipment and signalling systems employed. Either transmission or signalling standards may limit the size of a local-exchange area or the number of links that may be connected in tandem for a connection. Performance standards must be adequate not only for connections within the network, but also those that extend into other networks: for example, customers' private networks, cellular mobile radio networks and connections into the networks of other countries via international circuits.

A telecommunications network must provide its customers with services of a satisfactory quality at a price that they are willing to pay. Network planners must therefore achieve a compromise between performance and cost. This consideration affects the ways in which permissible impairments are allocated to different parts of the network hierarchy.

Customer-access networks typically account for over one third of the total cost of a national network, so economy in their provision is essential. Consequently, the smallest practicable wire gauges are used. Junction circuits are less numerous, so they can use heavier conductors and have less attenuation. Long-distance circuits must contain amplification, so that they can provide low attenuation. As a result, the largest part of the allowable attenuation for a limiting connection is allocated to the customers' lines and the least part to the trunk network.

The overall grade of service for a connection is the sum of those for its constituent parts. It is economic to provide circuits more generously where they are cheap and less generously where they are expensive. Consequently, specified grades of service may be, for example, 0.001 for trunks between switches in an exchange, 0.01 for junction circuits between local exchanges and 0.1 for very expensive long-distance and international routes. Thus, the latter have high occupancy and the revenue that they earn is maximized.

These considerations result in a set of standards for a national network. They should be mutually consistent and fit the overall strategy of the network operator. Furthermore, to enable international communications to be effective, national plans should conform with international objectives recommended by the ITU.

10.2 Analog networks

The number of levels in the hierarchy of a public switched telephone network (PSTN) depends on the relative costs of transmission and switching. A country that is small and densely populated will have relatively short distances between its primary centres and large amounts of traffic between them. A country that is large or sparsely populated will have longer distances and less traffic between its primary centres. The cost of providing direct routes between primary centres is much less in the former case than in the latter. As a result, the small, densely populated country will have more of its primary centres directly connected and fewer levels in its hierarchy. These differences are illustrated in Figures 10.1 and 10.2, which show the former analog networks of the UK and North America.

The UK trunk network[1] had three levels of switching centre: group switching centres (GSC), district switching centres (DSC) and main switching centres (MSC). The trunk network used four-wire transmission, but circuits between local exchanges (LE) and from LEs to tandem exchanges and GSCs were two-wire unamplified junctions. LEs, tandem exchanges and GSCs employed two-wire switching, but the trunk transit exchanges (DSCs and MSCs) used four-wire switching. Most trunk traffic was handled over direct circuits between GSCs, or over two GSC–GSC links in tandem. Only a small proportion of traffic reached the trunk transit network. However, this network was necessary for the large number of different possible

[256] Networks

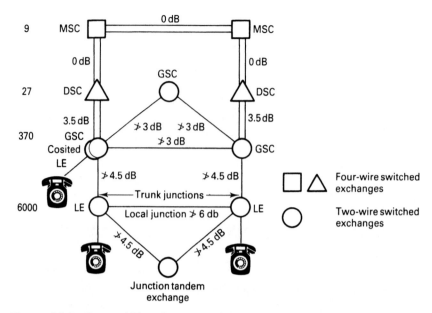

Figure 10.1 Former UK analog network. LE = local exchange, GSC = group switching centre, DSC = district switching centre, MSC = main switching centre. (MSCs and DSCs are collectively known as transit centres.)

connections which could not be made directly between GSCs, although each of these generated relatively little traffic.

The North American network,[2] shown in Figure 10.2, differed from the UK network in several respects. There were four levels of toll switching centre: toll centres (class 4 offices), primary centres (class 3), sectional centres (class 2) and regional centres (class 1). Local exchanges (end offices) are class 5 offices. Four-wire switching was used in class 1 and class 2 offices. Two-wire switching was used elsewhere, but considerable care was taken to control impedances to obtain high balance return losses. Long distances made echo a problem, so echo suppressors were fitted on circuits between regional centres.

During the 1980s, the structure of the PSTN in the USA was altered because of changes in the regulatory environment.[3] Ownership of the regional Bell operating companies was separated from AT&T. Now, AT&T is one of several competing long-distance carriers, known as *inter-exchange carriers* (IXC), and *local-access and transport areas* (LATA) belong to *local-exchange carriers* (LEC). These include the former regional Bell operating companies and independent telephone companies.

Because a LATA includes many local-exchange areas, a LEC completes those toll calls which are within its own LATA. However, it does not convey traffic between LATAs; this is done by an IXC. Each IXC interfaces with a LATA at a single point, as shown in Figure 10.3. This is known as a *point of presence* (POP). End offices may be

Analog networks [257]

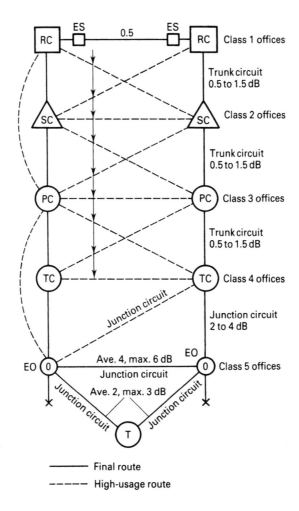

Figure 10.2 Former North American analog network. X = customer's station, EO = end office, T = tandem office, TC = toll centre, PC = primary centre, SC = sectional centre, RC = regional centre, ES = echo suppressor.

directly connected to a POP or routed via an *access tandem* (AT), which takes the place of a class 4 office in the previous Bell hierarchy.

In an analog network, four-wire connections have an overall nominal loss to ensure stability when component links have losses less than their nominal values, as described in Section 2.3.3. When these component losses are more than nominal, the overall loss of a connection is increased. For several links in tandem, it can be very large and the worst-case connection can have an extremely large overall loudness rating (OLR). For example, the limiting connection in the UK network[4] had an OLR of

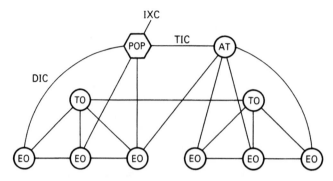

Figure 10.3 Local-access and transport area (LATA) in the USA. EO = end office, TO = tandem office, AT = access tandem, DIC = direct inter-LATA connection, TIC = tandem inter-LATA connection, POP = point of presence, IXC = inter-exchange carrier.

29 dB. This was unsatisfactory; however, it was encountered by only a small proportion of calls.

10.3 Integrated digital networks

Since the analog networks of Figures 10.1 and 10.2 were designed in the 1950s, there have been great reductions in the per-circuit cost of long-distance transmission, particularly because of the introduction of high-capacity digital optical-fiber transmission systems. During the same period, electronic digital time-division switching systems have been developed to replace electromechanical analog space-division switching systems.

If space-division switching is used to interconnect circuits provided over high-capacity transmission systems, it is necessary to install channelling equipment at each end of every link in order to demultiplex every channel to audio-frequencies before switching it and, after switching, to multiplex it again for onward transmission. If both the transmission systems and the switching systems use digital time-division multiplexing, the need for channelling equipment is eliminated, since channels are both transmitted and switched in PCM frames. The cost saving is very large. Further cost savings are obtained by replacing channel-associated signalling with common-channel signalling (although space-division SPC systems also have this capability).

A network having compatible digital transmission and switching is known as an *integrated digital network* (IDN). The above changes in costs also result in changes in network configurations. It is now economical to use more direct routes between trunk exchanges and this results in a hierarchy with fewer levels. It also makes it economic for an exchange to have routes to two or more switching centres at the next higher level, instead of the single backbone route in a traditional analog network. If one of these

Integrated digital networks [259]

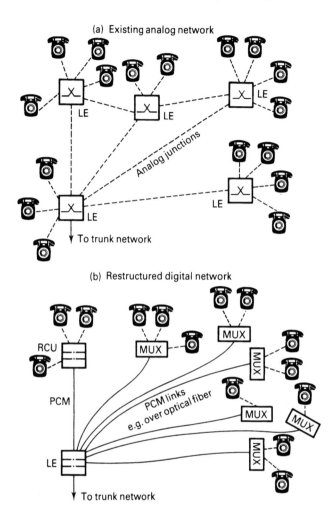

Figure 10.4 Local network restructuring. (a) Previous analog network. (b) Subsequent digital network. LE = local exchange, MUX = multiplexer, RCU = remote concentrator unit.

routes breaks down, the other can still carry traffic. This increases the *resilience* of the network, i.e. its ability to cope with failures and overload.

The equipment for a digital switching network occupies much less floor space than that for a space-division switching network, so an exchange building can serve many more customers' lines. If customers' line units can be located remote from the exchange, then still more customers can be served by the switching equipment and processors. This favours the use of remote switching units, remote concentrators and multiplexers in access networks. Since these are connected to the main exchange by PCM links, the size of an exchange area is no longer limited by the DC resistance and

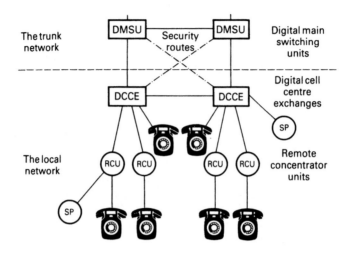

Figure 10.5 British Telecom integrated digital network. RCU = remote concentrator unit, DCCE = digital cell centre exchange, DMSU = digital main switching unit, SP = service provided.

attenuation of customers' lines. As shown in Figure 10.4, many small local exchanges can be replaced by multiplexers and remote concentrators. Furthermore, having fewer local exchanges results in fewer and larger junction routes and so reduces the costs of junction plant and tandem switching. Having fewer local exchanges also reduces the costs of operation and maintenance.

The IDN of British Telecom is shown in Figure 10.5. Local areas are served by *digital cell centre exchanges* (DCCE). Customers obtain access to a DCCE via a remote concentrator unit (RCU) unless they are situated very close to the DCCE. The concentrator is connected to the main exchange via two low-capacity digital links, usually operating at 2 Mbit/s or 8 Mbit/s. These are diversely routed to provide against failures. Concentrators parented on about 400 DCCEs will replace over 6000 local exchanges.

The trunk network is nonhierarchical, containing only one level of switching centre. These are called *digital main switching units* (DMSU) and are fully interconnected. The digital trunk network contains only 53 DMSUs, whereas the analog network of Figure 10.1 had over 300 GSCs. Each DCCE has junction routes to two DMSCs. Thus, it can still connect trunk calls in the event of breakdown of one route or its own DMSU.

The integrated digital long-distance network of AT&T[5] is shown in Figure 10.6. This is a nonhierarchical network, in which one class of tandem switching centre replaces both the sectional and regional centres of the previous analog network. Traffic entering this network continues to be concentrated by a hierarchy of toll centres and primary centres. The nonhierarchical long-distance network uses dynamic alternative routing, as described in Section 10.10.2.

Integrated digital networks [261]

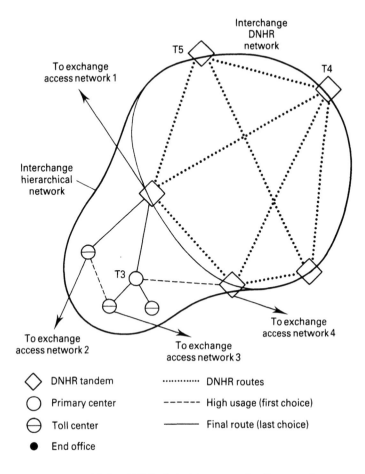

Figure 10.6 AT&T digital integrated toll network.

In an IDN, the data links used for common-channel signalling between digital exchanges form a separate signalling network, as shown in Figure 10.7. However, this signalling network uses channels in the basic transmission bearer network, just as does the PSTN. In order to maintain synchronism of the PCM frames in all the digital transmission links and exchanges, it is necessary to distribute synchronizing signals from a national reference clock source. This results in a synchronizing network linking it to all the digital exchanges. Finally, centres remote from exchanges have been introduced to collect bulk data (e.g. traffic statistics, billing, etc.) from exchanges and to provide human–machine interfaces for maintenance and software changes. Management centres receive real-time information on traffic flows and allow traffic to be rerouted to minimize the effects of congestion. These require an *administration data network* connecting the remote operation and management centres to the nodes of the

[262] *Networks*

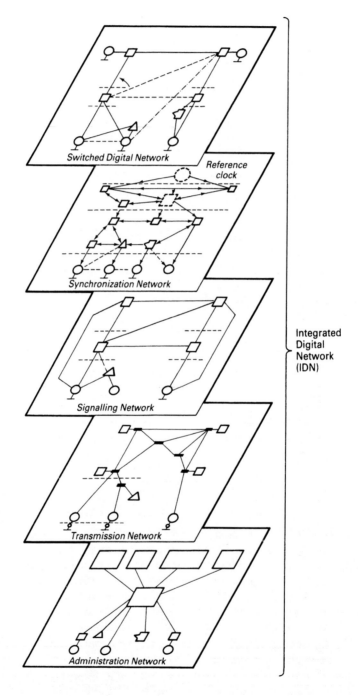

Figure 10.7 Component networks of an IDN.

PSTN. It is thus seen that four separate networks are required. However, as shown in Figure 10.7, all of these use channels in the basic transmission bearer network, making five networks in all.

Digital transmission links provide telephone circuits that are virtually free from noise. Moreover, in an IDN, only one A/D and one D/A conversion occur in a connection, however many links there are in tandem. The only noise introduced into a connection is quantization noise due to the codecs in the customers' line circuits at the two ends of a connection.

It has been shown in Section 10.2 that an analog network cannot provide a satisfactory overall loudness rating for every possible connection. However, digital transmission links have zero attenuation, so this will apply to any connection set up in an IDN. The only variation in overall attenuation between different connections is that due to differences in the lengths of customers' lines and the sensitivities of their telephones. This represents an enormous improvement and can ensure that all calls through the network have overall loudness ratings within the preferred range. For example, all connections in the IDN of British Telecom[4] have an OLR not exceeding 16 dB.

The final stage of this evolution is to extend digital transmission to customers' premises, which enables an IDN to evolve into an *integrated services digital network* (ISDN). This permits the transmission of high-speed data in addition to speech and so can provide many additional services. It also improves telephony performance, because there is no longer any variation in attenuation due to different lengths of customers' lines.

10.4 Integrated services digital networks

Extension of the digital transmission of an IDN over the access network to the customer's premises enables a wide variety of teleservices and bearer services to be provided in addition to telephony. An *integrated services digital network* (ISDN)[6,7] is one that can provide these services over a common network via the local exchange and the customer's line. Thus, the customer has a single access point to the network, instead of a separate interface for each service.

The services available from an ISDN[6] are those already provided by a PSTN, those that obtain a significant benefit from an ISDN and those that could not be carried on a normal PSTN. Telephony is the obvious example of the first category. However, use of an ISDN can enable a telephone to provide a range of supplementary services[6] (e.g. calling line identification, call transfer, conference calls, etc.). An example of the second category is facsimile. Use of 64 kbit/s transmission enables a page of text to be sent in about 3 seconds instead of 3 minutes. Examples of new services that can be provided are high-speed data transmission and video-conferencing over switched connections.

Two forms of access to an ISDN have been standardized by the CCITT:

1. *Basic-rate access (CCITT Recommendation I. 420)* The customer's line carries two 64 kbit/s 'B' channels plus a 16 kbit/s 'D' channel (for a common signalling channel) in each direction.
2. *Primary-rate access (CCITT Recommendation I.421)* Two lines are provided to carry a complete PCM frame in each direction. In countries that use 2 Mbit/s PCM, this provides 30 'B' channels plus a 64 kbit/s 'D' channel (in time-slot 16). In countries that use 1.5 Mbit/s PCM, it provides 23 'B' channels plus a 'D' channel (in time-slot 24)

Basic-rate access requires a bit rate of $2 \times 64 + 16 = 144$ kbit/s. The addition of 'overheads' for frame alignment, etc., gives a total digit rate of about 160 kbit/s. This signal must be sent in each direction over the same two-wire line. Several techniques have been developed for providing this duplex transmission.[6,7] If 'burst-mode' transmission is used, the signal is sent in each direction in a different time interval. Thus, the digit rate must be more than doubled in order to send the signal in less than half the time. An alternative is to use a form of electronic 'hybrid' with an adaptive echo canceller. The latter is now favoured, because it does not increase the digit rate and so can be used over longer lines.

When primary-rate access is used it is also possible to use fewer than 30 or 23 channels in order to obtain channels with greater digit rates for broadband applications such video communications. These are:

- The H_0 channel at 384 kbit/s
- The H_{11} channel at 1536 kbit/s (for use with 1.544 Mbit/s access)
- The H_{12} channel at 1920 kbit/s (for use with 2.048 Mbit/s access).

Both H_{11} and H_{12} may be used as single channels, or they may carry multiplexed H_0 channels. In order to switch these channels, a digital exchange must be able to make connections using n time-slots (where $n = 6$ for H_0, $n = 24$ for H_{11} and $n = 30$ for H_{12}).

The standard interfaces between the user and the network are shown in Figure 10.8. The functional units are as follows:

1. *The exchange termination ET* This connects the access network to the core network at the local exchange.
2. *The line termination LT* This is also in the local exchange and provides the appropriate form of signal for the customer's line (for basic-rate or primary-rate access).
3. *The network termination NT1* This terminates the access line at the customer's end.
4. *Network termination NT2* This enables switching functions to be performed.
5. *Terminal equipment TE1* This provides functions required to handle the layer 1, 2 and 3 protocols for terminals, such as digital telephones, workstations, etc., which conform to CCITT specifications.
6. *Terminal equipment TE2* This provides functions corresponding to the capabilities of existing equipments which do not conform to the CCITT standards.

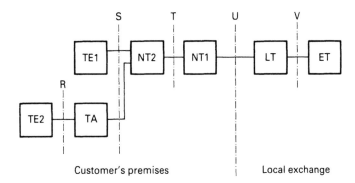

Figure 10.8 The ISDN user access reference configuration.

7. *Terminal adapter TA* This converts the layer 1, 2 and 3 protocols of a TE2 into those of a TE1, to enable equipment on the TE2 to operate over the ISDN.

The CCITT has also defined the reference points R, S, T, U and V at the interfaces between the above functional units. These are physical interfaces if separate pieces of equipment are used, but they are virtual interfaces if different functions (e.g. NT1 and NT2) are combined in the same equipment.

The NT1 at the customer's premises and LT in the exchange cater for digital transmission over the access line and provide line-maintenance functions, such as performance monitoring and loop-back testing. Because of this close interaction, the NT1 is provided by the network operator in most countries. Thus, the customer's interface with the network is the T interface. However, in the USA, the FCC forbids network operators to provide any equipment on customers' premises as part of the network. Consequently, the U reference point is the interface between the network operator and the customer. The latter must obtain an NT1, known in the USA as a *network channel terminating equipment* (NCTE), which is compatible with the digital transmission technique used by the network operator. The same interface is used at the S and T reference points. This gives flexibility, e.g. NT2 may not be required or NT1 and NT2 may be combined in one unit.

In a basic-rate installation, the digit rate at the T interface is increased from 144 kbit/s to 192 kbit/s by the inclusion of 'overheads' and a 16 kbit/s 'echo' channel (E). This interface provides a bus for up to eight terminals to be connected directly. For incoming calls, each is addressed separately from the exchange over the D channel. For outgoing calls, there can be contention, since two TEs may attempt to send simultaneously. A form of CSMA/CD is used. A terminal monitors the D channel and does not send until it is free. Then, the NT echoes back over the E channel the D-channel bits it receives. A sending TE monitors these to see if they match its transmitted D-channel bits. If not, it indicates a collision and the TE ceases sending. In a more complex installation, an NT2 is employed to provide a switching function (for example, a PBX).

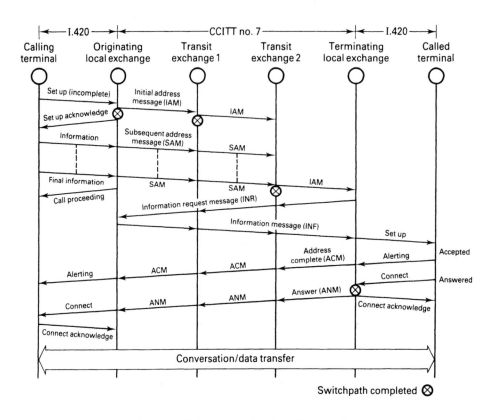

Figure 10.9 Example of an ISDN call.

In an ISDN the switching systems must be able to set up and clear down connections rapidly. They must also be able to cater for circuit switching and packet switching and to provide point-to-point, multipoint and broadcast connections in addition to coping with a range of data rates, message sizes and call holding times.

This necessitates digital switching of 64 kbit/s channels (and $n \times 64$ kbit/s), stored-program control and common-channel signalling. As described in Section 8.9, the CCITT common-channel signalling system no.7 provides the necessary capabilities between exchanges, by means of its ISDN user part. Common-channel signalling is provided between the user and the exchange over the D channel, as described in Section. 8.10. A typical sequence of signals exchanged over these systems for an ISDN call is shown in Figure 10.9.

The possibility exists of merging the transmission, processing and storage of voice, fax, data and video information in a single terminal apparatus, sometimes called a *multimedia terminal*. Thus, there is a demand for the conveyance of signals, such as colour graphics and high-definition television, whose bandwidth is too great for the existing ISDN. This has therefore been called the *narrowband ISDN* (N-ISDN) and plans are being considered for a *broadband ISDN* (B-ISDN). A B-ISDN[7] will require

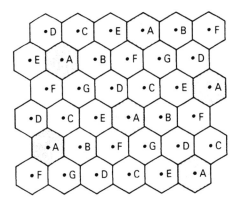

Figure 10.10 Principle of cellular mobile radio system (showing re-use of frequency groups A to G).

customers' distribution over optical-fiber cables and use the asynchronous transfer mode (ATM), as described in Section 9.5.

10.5 Cellular radio networks

The use of radio enables users of a network to move about, instead of being associated with fixed stations. Early mobile radio systems had a capacity for only a small number of users. This was because they required a wide area of coverage, but only a limited number of radio frequencies is ever available. Such systems are still used for private networks, known as *trunked radio mobile systems*.[8] However, public telecommunications operators now use *cellular radio systems*.[8,9] These have provided an enormous increase in capacity and have led to a corresponding growth in traffic. In a cellular network a country is divided into a large number of small areas, known as *cells*. Since the cells are small, low transmitter powers can be used and the same radio frequencies can be used in nonadjacent cells. Thus, for example, 1000 radio channels can accommodate about one million users.

The principle is illustrated in Figure 10.10, where each cell uses one of the available groups of frequencies (groups A to G). All cells have the same number of radio frequencies, but they are unlikely to to have the same number of customers wishing to use them. This problem is solved by having many small cells where there are many customers (e.g. in cities) and few large cells where there are few customers (e.g. in rural areas).

Each cell has a *radio base station* (RBS) for communicating with the customers' *mobile stations* (MS). Its transmitter and receiver cater for both a number of voice channels and for control channels. As shown in Figure 10.11, the RBSs in a group of cells are connected to a *mobile switching centre* (MSC). The MSCs are linked by fixed

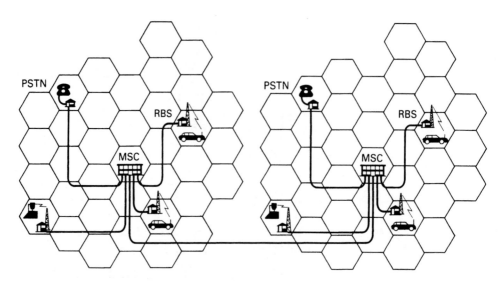

Figure 10.11 Cellular mobile radio network. MS = mobile station, RBS = radio base station, MSC = mobile services switching centre, PSTN = fixed network.

circuits and have interfaces to the PSTN. Thus, calls can be made between mobile users in different areas and between them and the customers of the PSTN.

In order to originate a call, the mobile user accesses the cellular network via a base station. The cell in which the user is located is therefore known. For a call to a mobile user, all that is initially known is the directory number of the user. It is therefore necessary for the network to determine in which cell the user is located. Furthermore, users move about, so the location of a telephone may change from one cell to another while a call is in progress. There must then be a *hand-off* or *hand-over* of the call from one base station to another. It is therefore necessary for the network to keep track of the locations of all its users.

There is a process of *periodic registration*, which notifies the network that a mobile telephone is switched on, and this is also used for locating it. The mobile telephone continuously monitors a control channel and so receives information that identifies the area. If the received signal strength falls below a threshold, the mobile telephone automatically switches to another control channel. If a different area identification is then received, this triggers a new registration to tell the network that the mobile is now active in a different cell.

Each mobile telephone has a home switching centre. This contains a *home location register*, which stores customer data, including the directory number, equipment serial number and class of service. When registration occurs, the switch that has received the registration requests the customer data from the home switch and stores it. At the same time, the home switch records where the request came from. It can therefore route calls for that customer to the correct switch and so to the correct base station.

There are a number of cellular radio standards in use. These include: the Advanced Mobile Phone System (AMPS) in North America, the Nordic Mobile Telephone Service (NMT) in Scandinavia, the Total Access Communication System (TACS) in the UK, Network C used in Germany and Portugal, Radiocom 2000 in France, RTMS in Italy, the Nippon Automatic Mobile Telephone System (NAMTS) in Japan and UNITAX in China and Hong Kong. All these systems use analog transmission of the speech signals.

Since these systems were developed there has been a move towards digital speech transmission. This should minimize inter-channel interference and enable closer channel spacings to be used. In Europe, the technical standards have been specified by a group called the Groupe Spéciale Mobiles (GSM) for a pan-European digital mobile service. The GSM system[8] will enable users to roam anywhere in Europe and originate and receive calls just as in their home country.

In the UK, plans are also being made for *personal communication networks* (PCN) in the 1.7 GHz to 2.5 MHz band. These will use very small cells, known as microcells, and so should be able to accommodate an enormous number of users.

A cheaper alternative to cellular radio is the *telepoint service*. Customers use cordless telephones to communicate via base stations situated in places where people congregate, such as airports, railway stations and shopping malls. Thus, anybody within range of a base station can use a cordless telephone to make calls over the PSTN. Although the telepoint service is much cheaper than cellular radio, it is less useful because it cannot be used to receive incoming calls. In the UK it was not a commercial success.

The combination in a handset of a radio pager and a cordless telephone leads to the idea of a *personal communicator*. If the handset is within range of a base station, a number delivered by the pager can be automatically redialled by the telephone. This enables incoming calls to be received and so gives a service almost equivalent to a cellular radio system. However, the user must not move outside the range of the base station during a call, since there is no hand-off procedure.

10.6 Intelligent networks

In the era of electromechanical switching systems, the introduction of new services, such as subscriber trunk dialling (direct distance dialling), necessitated the design of new hardware and modifications to all exchanges. This was costly and disruptive. When the introduction of SPC was proposed it was thought that new services would require only small modifications to call-processing programs and that these would be easy and cheap to implement. In practice, however, it has been found that making consistent software upgrades to the many interconnected exchanges in a network is as costly and time-consuming as the previous hardware changes. Consequently, such changes are made very infrequently and introducing new services is still a slow process.

[270] *Networks*

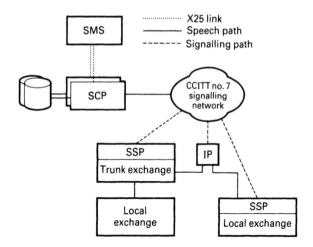

Figure 10.12 Intelligent network architecture. SSP = service switching point, IP = intelligent peripheral, SCP = service control point, SMS = service management system.

A possible solution to this problem is to separate the software that controls basic functions, such as setting up and supervising connections, from the software required for providing more complex services. These include, for example, freephone service, calling-card service and value-added services. The more complex services can be controlled by a centralized processor, called a *service control point* (SCP), which is remote from the exchange setting up the connection. A telecommunications network that has been enhanced in this way is called an *intelligent network* (IN). The exchange that makes the required connection is called a *service switching point* (SSP). In the AT&T network[10] it is called an *action control point* (ACP). The SSP communicates with its controlling SCP via a common-channel signalling network.

The architecture of an IN is shown in Figure 10.12. The SSP may be at any level in the network hierarchy (e.g. a local exchange, a trunk exchange or an exchange in a special overlay network). Its software is modified so that a number of events can trigger it to suspend normal call processing and request the intervention of the SCP. These events may be the caller's class of service, the code digits dialled or some subsequent event (e.g. ring tone no reply). As a result of the event, the SSP sends information to the SCP and resumes call processing when it receives back information on how to proceed.

The SCP is a centralized processor with access to a large database. Its software is organized in three levels:

1. *Node software* This provides common utilities, such as signalling, database access, transaction monitoring and alarm reporting.
2. *Service logic programs (SLP)* These are the programs that control the various services. They are constructed, as far as possible, from modules, known as

service-independent building blocks (*SIB*), which correspond to frequently used functions, such as: translate, verify, compare, charge and queue.

3. *The service logic execution environment* (*SLEE*) This is the program-execution environment that hosts the various SLPs and interworks with the basic call-control and switching operations of the SSP.

Other specialized functions required for IN services are provided by *intelligent peripherals* (IP). Examples are digit-collection units and voice-guidance systems which give instructions to users by means of recorded announcements. These are associated with SSPs, but they can be controlled by the SCP via common-channel signalling.

The common-channel signalling network is the means to pass messages between the SCP and SSPs and IPs. The signalling is nonassociated because, usually, the SSP interacts with the SCP before the route is selected for a call. Thus, messages between the SSP and SSC pass through signal-transfer points. This has necessitated modification of the basic CCITT common-channel signalling system no.7. As explained in Section 8.9.1, the signalling-connection control part (SCCP) and transaction capabilities (TC) have been added. The TC is a message structure designed specifically for the query and response messages used for IN transactions and the SCCP provides the information for routing these messages.

In a network with a number of SCPs, there may also be a *service management system* (SMS), connected to all the SCPs by data links. This manages the addition of new customers, updates of data (from the network operator or customers) and data reloads (if an SCP crashes).

Several telecommunication network operators have installed IN equipment, or are planning to do so. An example is the British Telecom *digital derived services network* (DDSN)[11] shown in Figure 10.13. This is an overlay and contains 13 *digital derived services switching centres* (DDSSC), which are connected to the trunk exchanges (DMSU). The DDSCs perform the SCP function and three *network control points* (NCP) perform the SCP function.

The DDSN network of BT caters for freephone numbers (0800 XXXXXX), for which call charges are levied against the called customer instead of the caller. Use of an IN enhances the freephone service in several ways. For example, if the organization being called has several offices, the call can be routed to whichever is nearest to the caller. However, outside of normal working hours all calls can be routed to a single office and this can be changed from day to day. Also, voice guidance can be used to enable a caller to choose between different departments of the organization being called. The network also caters for premium-rate calls (08xx XXXXXX) to service providers, such as information services and the cellular radio networks. In this case, the caller pays at a higher rate than normal and the income from these calls is shared between BT and the service provider.

At present, intelligent networks use a centralized structure, as shown in Figure 10.12. In future, the growth of IN-based services may lead to decentralization, with the SLEE, SLP and database located at local or trunk switching centres rather that at a remote SCP. This will give faster response times and reduce the load on the

[272] Networks

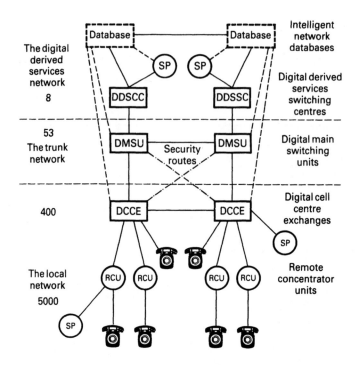

Figure 10.13 British Telecom digital derived services network. DDSSC = digital derived services switching centre, SP = service provider.

common-channel signalling network, and it may be cheaper for widely used services. However, centralized-SCP control will probably be retained to enable new services to be developed rapidly. When these services have proved successful and their traffic has grown, they can then be transferred to distributed control.

10.7 Private networks

Organizations operating in several locations often have sufficient traffic between sites to make it economic to join them with a private network, instead of sending traffic through the PSTN. A private network is usually provided by leasing circuits from a public telecommunications operator (PTO) or even, in the case of a multinational company, the PTOs of several countries.

Most organizations have separate networks for voice and data traffic. The voice network links their PBXs and a WAN links their LANs for data communication. However, ISDN technology is applicable to private networks, as well as to public networks. Use of integrated-services PBXs (ISPBX) enables both voice and data traffic to share the same digital circuits.

Some PTOs offer a *Centrex service*, in which users' telephones are served by switching equipment in a public exchange instead of by a PBX on the customer's premises. Line costs are, of course, greater, but a fee to the PTO replaces the cost of a PBX. There are circumstances in which Centrex can save money, for example if staff are located in several buildings (each of which would otherwise need a separate PBX).

A further development is to route traffic over the PSTN instead of using a private network. This is known as a *virtual private network* (VPN). It enables the customer's staff to form a closed user group within the PSTN. The economic justification is that fewer circuits are needed to carry the private traffic when mixed with public traffic on large circuit groups than if separate small groups of circuits are provided. A VPN must provide the same numbering scheme, signalling protocols and services as a private network. These usually differ from those of the PSTN and require an additional database and different software. It is preferable to have these in a central location, rather than in every local exchange, so an intelligent network (as described in Section 10.6) is required.

Nationwide introduction of a VPN service will encourage more *teleworking*.[12] Members of staff of an organization can work at home and communicate with colleagues as if they were using the same PBX in the same building. They can exchange documents as faxes and have full access to their company's databases and computing facilities.

10.8 Numbering

10.8.1 National schemes

In order to establish a connection through a network it is necessary for the caller to inform the switching centre of the address of the customer being called, i.e. the called customer's directory number. This determines both the route for the call and the charging rate. Thus, a numbering plan is required to allocate a unique number to each customer. At first, each numbering scheme applied only to a single exchange and exchanges were identified by the names of their towns. Later, *linked numbering schemes* were applied to multi-exchange areas.

In a linked numbering scheme the 'local' numbering scheme covers a number of exchanges, so a call from any exchange in the area uses the same number to reach a particular customer. The first part of the directory number is an exchange code and the remainder is the customer's number on that exchange. For example, a six-digit linked number scheme has a theoretical capacity for 100 four-digit exchanges. In practice, this is reduced by the need to allocate codes for access to various services.

The subsequent introduction of direct distance dialling (DDD) or subscriber trunk dialling (STD) required the development of national number plans.[1,13] Later, the introduction of international subscriber dialling (ISD) made it necessary for national number plans to conform to an international number plan. The CCITT recommended[14] that the maximum number of digits for an international call should be eleven. The maximum number of digits in a national number is thus $11 - N$, where

N is the number of digits in the country's code in the world numbering plan.

In general, a national number contains three parts:

- An area code
- An exchange code
- The customer's number on the local exchange.

For a local call, only the exchange code and customer's number are normally dialled, but the full national number is needed for a long-distance call. In order to enable an originating local exchange to differentiate between them, a *trunk prefix* is dialled before a national number. This routes the call to a trunk exchange, where a register translates the area code to determine the required trunk routing. An additional prefix (following the trunk prefix) denotes an international call and this routes the call from the trunk exchange to the international gateway exchange. The international prefix is followed by the international code for the required country and the national number of the called customer in that country.

The numbering plan adopted for the UK is now shared by competing network operators. It mainly uses three-digit area codes and six-digit linked numbering schemes in these areas. However, the largest cities are director areas having seven-digit numbering schemes, as described in Section 3.7. Each of these areas is identified by a two-digit area code. The maximum number of dialled digits, excluding the trunk prefix, is thus nine. This complied with the CCITT recommendation for national numbers. (For the UK, the country code is '44' and $11 - 2 = 9$.) The trunk prefix is '0' and the international prefix is '010'.

The North American numbering plan[13] has uniform ten-digit numbers. A three-digit area code is followed by a three-digit office code and a four-digit customer's station code. Two toll prefixes are used: '1' for customer-dialled station-to-station calls and '0' for calls requiring operator assistance (e.g. person-to-person and reverse-charge calls).

The capacity of national numbering schemes is reduced by the need to avoid use of the trunk prefix as the initial digit of any local-exchange code to avoid ambiguity. Similarly, the international prefix must not form the initial part of any area code. There is also a need to reserve codes for accessing various services. For example, in the UK the code '999' is used for emergency services, '100' to obtain operator assistance, '192' for directory enquiries, and so on. Thus, the digits '0', '1' and '9' are not available as initial digits for local numbers. This reduces the capacity of the numbering schemes of all local areas. The numbering capacity of local areas is also reduced by the need to provide direct dialling in (DDI) to PBXs. This requires each extension to have a number within the scheme for the area. Thus, a PBX uses a large block of numbers instead of a single number. The advent of the integrated services digital network (ISDN) also has an impact. It may be necessary for a basic-rate-access subscriber to have up to eight numbers in order to receive calls through the PSTN.

In countries where the telecommunications service is 'liberalized', codes must be reserved to enable customers to obtain access to different competing networks. In the UK, area codes may not use the initial digit '8', since this gives access to the BT digital

Table 10.1 World numbering zones

Code	Zone
1	North America (including Hawaii and the Caribbean Islands, except Cuba)
2	Africa
3 & 4	Europe
5	South America and Cuba
6	South Pacific (Australasia)
7	CIS (fomerly USSR)
8	North Pacific (Eastern Asia)
9	Far East and Middle East
0	Spare code

derived-services network.[11] This network caters for the freephone service (0800) and for calls charged at a premium rate (e.g. message services). It also provides access to other networks, e.g. cellular radio networks. Thus, '836' might appear to be an area code, but '0836 XXXXXX' would be the number of a mobile telephone which could be anywhere in the UK.

In the USA, each of the competing long-distance carriers is allocated a three-digit *carrier-identification code* (CIC), e.g. '288' for AT&T and '222' for MCI. A customer on a SPC local office may choose a particular carrier in advance for all long-distance calls. This information is stored in the database of the office and causes toll calls to be routed to the chosen carrier. If the customer wishes to use another carrier, the toll prefix '1' is followed by '0' and the three-digit CIC in order to route the call to the appropriate network.

The introduction of mobile services has led to numbers being associated with individual customers, rather than geographical locations as in the fixed PSTN. Ideally, this would apply to everybody. There would be no need for abortive calls when customers are away and no need for customers to have their numbers changed if they move to another town. This would require an intelligent network, in which every local exchange could identify complete directory numbers for distant customers (instead of only area codes) and access a complete national database to determine the routing for every such call.

10.8.2 International numbering

The introduction of international subscriber dialling made it necessary for every subscriber's station in the world to have a unique number. In the CCITT world numbering plan[14] each subscriber's number consists of a country code followed by the subscriber's national number.

For numbering purposes, the world is divided into zones, each given a single-digit code. Each country within a zone has the zone number as the first digit of its country

Table 10.2 Examples of typical international numbers

Zone	Country	Country code	No. of digits in national number	Total no. of digits
1	USA	1	10	11
1	Canada	1	10	11
2	Egypt	20	8	10
2	Liberia	231	6	9
3	France	33	8	10
3	Portugal	351	7 or 8	10 or 11
4	UK	44	8 or 9	10 or 11
4	Switzerland	41	8	10
5	Brazil	55	9	11
5	Ecuador	593	7	9

code. However, the European numbering zone has been allocated two codes because of the large number of country codes required within this zone. The codes for the world numbering zones are listed in Table 10.1.

Within each zone every country has been alloted a single two-digit or three-digit code number. For example, within zone 3 (Europe) Holland has the code '31' and Albania has '355'. The three-digit codes have been allocated to smaller countries, having fewer digits in their national numbering plans, to minimize the total number of digits in customers' international numbers. Exceptions occur when an integrated numbering plan covers an entire zone; countries in these zones require only a single-digit code. Thus, '1' is the country code for all countries in the North American numbering plan.[14] Some examples of international numbers are given in Table 10.2.

10.8.3 Numbering plans for the ISDN era

In 1984 the CCITT made recommendations for an international numbering plan to cover integrated services digital networks (ISDN).[16] This extends the existing numbering plans for PSTNs. Thus, a customer's ISDN access will normally be indistinguishable from PSTN access and will be provided by the same local exchange.

The maximum length of international numbers is extended to fifteen digits. The area code becomes a *network destination code* (NDC). This arises from the present situation whereby some 'area' codes have been used for access to mobile networks and special services (which are nongeographic). An international exchange will analyze six digits to determine a route. These can therefore include digits within the NDC in addition to the country code, in order to route a call to the appropriate network in the destination country.

It is intended that provision will also be made for *sub-addressing*. This will use additional digits following a customer's national number, which will be transmitted to the destination for use on the customer's premises. These can select the appropriate terminal for an ISDN connection or provide DDI on a PBX without using up numbers in the area numbering scheme. The sub-address field can range from four digits for simple applications up to 40 digits for use in open systems interconnection (OSI). For OSI, a global scheme for the identification of network service access points has been developed by the International Standards Organization (ISO).

The CCITT recommended the full implementation of this plan by the end of 1996. It is hoped that, by that date, stored-program-controlled digital exchanges will have penetrated all countries so that they can handle fifteen-digit international numbers and the increased number of digits to be analyzed.

In the UK the changes will be implemented in April 1995. The international prefix will change from '010' to '00', which is recommended by the ITU for use worldwide. The trunk prefix will become '01'. This will be followed by a nine-digit national number, containing a two-digit area code followed by seven digits or a three-digit area code followed by six digits.[17] Nongeographic services (e.g. freephone, premium-rate services and cellular radio) will continue to be accessed by '08' and the code '05' has also been allocated for freephone, etc. Codes beginning with '02', '03', '04', '06', '07' and '09' will be available for future sevices.

In North America it is planned to extend carrier-identification codes from three digits to four digits in 1995 when the present scheme is expected to become exhausted.

10.8.4 Public data networks

For public switched data networks[15] an international number consists of two parts: the *data network identification code* (DNIC) of four digits and the *network terminal number* (NTN) of ten digits. Thus, the complete number has fourteen digits. The DNIC consists of a *data country code* (DCC) of three digits followed by a 'network' digit. By this means a single country can have up to ten different data networks. Furthermore, a country can have more than one DCC. For example, the UK has been allocated codes '234' to '237'; thus it can ultimately have up to forty different public data networks.

The format used for the ten-digit NTN can be determined by the network operator, since only the DNIC needs to be analyzed to route the call into that network from another country. However, the CCITT recommended that the last two digits shall be a sub-address for use by the subscriber. In the British Telecom packet-switched service (PSS), the first three digits of the NTN serve as a routing code to identify the packet-switching exchange and the next five digits identify the network terminating point serving the subscriber.[15] The final two digits correspond to a sub-address for use within the customer's own system.

10.9 Charging

The cost of providing a telecommunications network consists of the capital cost and the current operating expenses. All of these costs must be met by income obtained by the telecommunications operator from its customers. It is equitable that the charges paid by each customer should be related, if possible, to the proportion of these costs incurred in providing services. For this reason, the charges that are made to the customer are levied in the following ways:

- An initial charge for installing the customer's line
- An annual rental or leasing charge
- Call charges.

The customer's share in the capital costs should be covered by the connection charge and part of the rental. Part of the operating cost is incurred even if the network carries no traffic, so this should also be recovered by rental charges. Those parts of the capital cost and operating cost which depend on the traffic carried should be recovered by call charges.

Because the major proportion of the cost of a local network is independent of traffic, some operators do not make a separate charge for local calls but include them in the rental. This is known as a *flat-rate tariff*. Others charge a unit fee per call or a fee proportional to call duration. This is known as *message-rate charging*.

Call charges may be made by either metering or ticketing, as described in Chapter 3. The former results in bulk billing, but the latter permits an itemized bill to be computed. Since the processors in SPC exchanges can perform an automatic ticketing process for the connections that they establish, itemized billing is now being introduced more widely.

The quantity of switching equipment and of junction and trunk transmission plant required depends on the busy-hour traffic. Calls made at off-peak times virtually incur no capital cost, since no plant would be saved if these calls were not made. Because of these relative costs, and to restrict peak demand, it is common to make call charges vary with the time of day.

A charging plan for long-distance calls should satisfy the following criteria:

1. The call revenue should recover the capital and operating costs, since these are almost entirely traffic-dependent.
2. Charges to and from customers who are geographically close should be similar, to give equitable treatment and avoid complaints.
3. The charging plan should be easily understood by customers.
4. The charging plan should be suitable for implementation by automatic equipment. It must therefore be compatible with the numbering and routing plans.

Traditionally, charges for long-distance calls have been proportional to distance × duration. It would be unnecessarily complicated to work out for each call the distance between the originating and terminating local exchanges. Instead, the charge is based on the distance between the originating and terminating primary centres. In order to establish a connection, the originating primary centre translates the address digits dialled by the caller into a different set of digits used to route the call. This translation can include additional digits representing the charging rate. These may be used to operate periodic-pulse metering, whereby the interval between meter operations during the call depends on the charging rate. They can also indicate the rate per

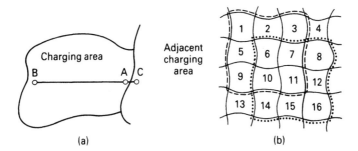

Figure 10.14 Unit-fee boundaries in relation to charging areas. (a) Anomaly when unit-fee range covers only one area (call A-B costs 1 unit; call A-C costs 2 units). (b) Unit-fee range covering adjacent areas, ----- Unit-fee boundary for customers in area 6, xxxxx Unit fee boundary for customers in area 11.

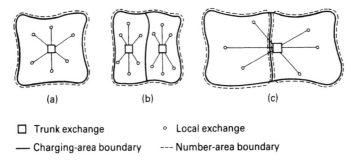

☐ Trunk exchange　　　　○ Local exchange
— Charging-area boundary　--- Number-area boundary

Figure 10.15 Charging areas in regions having different telephone densities. (a) Charging area and number area with same boundaries. (b) Charging area divided into two number areas in region of high telephone density. (c) Two areas served by single trunk exchange in region of low telephone density.

unit time to be used in automatic ticketing.

Boundaries between charging areas must coincide with boundaries between number-plan areas. Charges between all exchanges in two areas are then equal. Anomalies must be introduced between adjacent exchanges on opposite sides of a boundary, as in Figure 10.14(a). The effect of these anomalies can be reduced by charging the same fee for calls within an area and to adjacent areas, as shown in Figure 10.14(b). The boundary of the customer's own area is then less important than the boundaries of adjacent areas. Since they are more distant, anomalies are less significant.

The requirement that charging areas shall be approximately equal in size may conflict with the requirement that their boundaries coincide with numbering areas. Numbering areas are usually small in urban regions where there are many exchanges. This situation can be accommodated by including two or more numbering areas within a single charging area, as shown in Figure 10.15(a). For example, London is divided into two areas: Central London (code 071) and Outer London (code 081). New York is divided into three areas (codes 212, 718 and 917).

Numbering areas are large in rural regions where there are few exchanges. This situation can be accommodated by dividing an area into two or more, as shown in Figure 10.15(b). However, this makes wasteful use of the numbering plan, since two or more area codes are used to route calls to the same trunk exchange.

The ability to make the same charge for calls to different areas enables the size of area covered by the same fee to increase with distance. This is consistent with the economics of long-distance transmission. With high-capacity multiplex systems, such as those operating on optical fibers, the cost per channel-kilometre is extremely low, since the transmission medium is shared by very many channels. Thus, the cost of calls in a trunk network is becoming almost independent of distance. In the UK, the same charging rate is employed for all calls over more than 35 miles (56.4 km). Call charges for the BT packet-switched data network are independent of distance.

10.10 Routing

10.10.1 General

In a hierarchical national network, as shown in Figure 1.6, the minimum tree configuration of exchanges is usually augmented by direct routes interconnecting those exchanges where a high community of interest results in sufficient traffic. There is thus a backbone route joining each switching centre to the highest level via intermediate centres, together with transverse branches between some centres at the same level. There may also be some other direct routes between centres at different levels which violate this pattern.

If it is assumed that a country is already divided into local-exchange areas and the locations of their exchanges have already been decided, a routing plan should be developed to determine:

- Which exchanges should be interconnected by direct circuits, and which connections made indirectly via tandem switching centres
- The number and location of tandem switching centres
- The number of levels of tandem switching to be used in the network
- Whether automatic alternative routing is to be used and, if so, under what conditions.

This routing plan must be consistent with the plans for numbering, charging, transmission and signalling.

Large groups of circuits are more efficient than small groups because of their higher occupancy (i.e. traffic per circuit), as explained in Section 4.6.2. If there is a large amount of traffic between two exchanges, it is economic to provide a direct route between them. If there is little traffic between two exchanges, it is more economic to combine this with traffic to other destinations to produce a large amount of traffic over

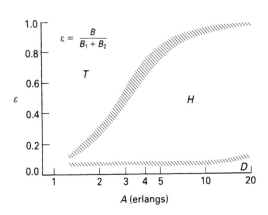

Figure 10.16 Domains for employment of tandem (T), high-usage (H) and direct (D) circuits as a function of traffic (A) and cost ratio (ε). B = marginal cost per circuit on direct route, B_1, B_2 = marginal costs of circuits forming tandem connection. Minimum no. of circuits on high-usage routes = 4. From Rapp[19].

a common route to a tandem switching centre. The correct solution obviously depends on the cost of the circuits as well as the amount of traffic. If circuits are cheap, it is less expensive for them to be lightly loaded than to incur the cost of switching equipment in a tandem exchange. Thus, many direct routes are provided between local exchanges in a small area with a high customer density but not to more distant exchanges.

In some networks, *automatic alternative routing* (AAR) is used. Direct routes are underprovided with circuits; when all circuits on a direct route are busy, traffic overflows to a fully provided tandem route through a switching centre at a higher level in the hierarchy. An underprovided direct route is called a *high-usage route* and the fully provided indirect route to which its traffic finally overflows is called a *final route*. Only a small proportion of calls use the complete backbone of final routes, since transverse routes are used whenever these are free.

The traffic levels at which direct routes, tandem routing and automatic alternative routing should be used depend on the relative costs of direct and tandem routes (including associated switching and signalling equipment). Low traffic and high-cost direct routes indicate tandem working. High traffic and low-cost direct routes indicate direct connection. Intermediate situations favour automatic alternative routing, as shown in Figure 10.16, due to Rapp.[19]

Example 10.1

A simple junction network has four local exchanges (A, B, C and D) and it may also have a tandem exchange (T). During the busy hour, each local exchange originates 2.0 E of traffic to each of the others and the required grade of service is 0.02.

[282] *Networks*

1. Determine the total number of one-way junctions required in the network if:
   (i) Direct junctions only are provided (i.e. no tandem exchange).
   (ii) All junction calls are routed through a tandem exchange.
   (iii) Automatic alternative routing is used and there are four direct junctions between each pair of local exchanges (two in each direction), with overflow traffic routed via the tandem exchange.
2. If each direct junction costs D and each junction to the tandem exchange costs T (including associated switching and signalling equipment in each case), over what range of values of the ratio T/D does each of the above methods provide the most economic routing?

1. (i) From Table 4.1 the number of junctions required from A to B and from B to A is 6.
   The number of routes between exchanges is 6.
   ∴ Total number of junctions is $6 \times 2 \times 6 = 72$.
   (ii) Total traffic from A to T is $3 \times 2 = 6$ E.
   Table 4.1 shows that the number of junctions required is 12.
   ∴ Number of junctions between A and T is 24.
   The number of routes is 4. ∴ Total no. of junctions is 96.
   (iii) Number of direct junctions is $2 \times 2 \times 6 = 24$.
   Outgoing traffic from A requires 12 junctions.
   However, six are direct junctions ∴ six junctions go to T.
   Similarly, incoming traffic to A requires 12 junctions
   ∴ No. of junctions required from T to A is 6.
   ∴ Total number of tandem junctions is $4 \times 2 \times 6 = 48$.
   Thus, total number of junctions in network is $24 + 48 = 72$,
   i.e. both (i) and (iii) require fewer junctions than (ii).
2. Cost of scheme (i) is $C_1 = 72D$
   Cost of scheme (ii) is $C_2 = 96T$
   Cost of scheme (iii) is $C_3 = 24D + 48T$
   Hence:

   $C_3 \leq C_1$ for $48T \leq 48D$ ∴ $T \leq D$
   $C_3 \leq C_2$ for $24D \leq 48T$ ∴ $T \leq T/2$

Thus:
Ratio T/D: < 0.5 $0.5 < T/D < 1.0$ > 1.0
Cheapest scheme: (ii)   (iii)   (i)

Modern networks have SPC exchanges, so their routing translations can readily be changed. Moreover, the use of common-channel signalling enables the routing tables of exchanges to be changed remotely. The common-channel signalling network can link the processors of the exchanges to a central *network management centre* (NMC). This can monitor the traffic on all the routes at frequent intervals to enable its staff to change routings in order to bypass failures and congestion.

The ability to change routing tables in exchanges permits *dynamic routing*, whereby the preferred choices of routes are changed from time to time. Non-coincident busy hours make it advantageous to use indirect routes over links that are lighly loaded by direct traffic. For example, the different time zones of the East and West of the USA result in little traffic between New York and Los Angeles and between Washington and Los Angeles in the morning when the route between New York and Washington gets busy. Thus, circuits on the route New York–Los Angeles–Washington can supplement those on the direct route New York–Washington. It has been estimated[20] that the use of dynamic routing in the AT&T long-distance network provides a cost saving of about 15%.

It is particularly advantageous to make use of time-zone differences in the international network. For example, it is economic to route traffic between the UK and other European countries via the USA in the morning when the time difference ensures that there is little direct traffic between Europe and North America. Similarly, the USA may use the UK as a tandem to route calls to Pacific Rim countries while it is night-time in Europe.

10.10.2 Automatic alternative routing

If an exchange has direct high-usage routes to a number of others and an overflow route to a tandem exchange, the traffic on the tandem route is the sum of the overflows from all the direct routes. The number of circuits provided on each of the direct routes is deliberately chosen to give a grade of service much worse than that required. As a result, congestion exists on these routes for a high proportion of the busy hour and the occupancy of each circuit is a large fraction of an erlang. If the traffic offered to the high-usage routes is random, peak loads will seldom occur simultaneously on all of them. Thus, fewer additional circuits are required to carry the overflow traffic than would be needed if more circuits were added separately to each of the direct routes.

Example 10.2

A small exchange A has a direct route to exchange B and a final route to exchange C, which carries traffic to all other destinations. The total traffic from A is 5 E, of which 1 E is to exchange B. The grade of service is required to be not worse than 0.01.

Find the total number of outgoing trunks from A that is required if:

1. All traffic from A to B is carried on the direct route.
2. There are two trunks on the direct route and traffic overflows from these to the final route.

1. Table 4.1 shows that the direct route needs 5 trunks
   Table 4.1 shows that the final route needs 10 trunks
   Total: 15 trunks.
2. The grade of service of route AB is

$$B = \frac{1/2}{1 + 1 + 1/2} = 0.2$$

Thus, 0.2 E overflows from AB to AC.
∴ Traffic on route AC = 4.2 E and Table 4.1 shows that 10 trunks are required. (This assumes an Erlang distribution on AC, which is not strictly correct. However, it is a reasonable approximation, since the overflow traffic is less than 5% of the whole.)
Total number of trunks needed = 2 + 10 = 12
i.e. method 2 requires three fewer trunks than method 1.

Automatic alternative routing is applicable in private networks as well as in PSTNs. A PBX in a private network may be programmed so that, when all private circuits to another PBX in the network are busy, the traffic overflows onto the PSTN.

Automatic alternative routing will reroute calls away from a high-usage route whenever it is unable to carry traffic. This can happen if there is a breakdown instead of high traffic. This is beneficial, since traffic can still reach its destination, whereas it could not if there were only the direct route. However, the large amount of additional traffic offered to the backbone route can cause severe congestion, resulting in the loss of calls to destinations served only by that route (which could be to most other exchanges in the network). A solution to this difficulty is *trunk reservation*.[21,22] A proportion of the circuits on the final route is reserved for calls which can only take that route. Thus, these calls still obtain a reasonable grade of service when the overflow traffic offered to the route is abnormally large.

The number of circuits required by a final route to carry the overflow traffic should not be calculated from Erlang's loss formula, because this traffic is not Poissonian. Figure 10.17(b) shows that if traffic $A(t)$ is offered to a high-usage group of N trunks, the traffic carried is $A(t)$ when $A(t) \leq N$ and is N when $A(t) \geq N$. As shown by Figure 10.17(c), the traffic which overflows to the final route is zero when $A(t) \leq N$ and is $A(t) - N$ when $A(t) \geq N$. The traffic offered to the final route is thus more peaky than Poissonian traffic. It cannot be described adequately by its mean value alone; its variance must also be taken into account. A theory for determining the number of circuits required for an overflow route has been given by Wilkinson.[23] An outline of this theory is given in Appendix 2.

In a traditional hierarchical network each exchange normally has only one backbone route to others at higher levels. However, as explained in Section 10.3, an integrated digital trunk network is usually nonhierarchical. If the network is fully interconnected, a direct route between two exchanges can have an alternative via any other exchange. Thus, each direct route between two exchanges can also form part of a tandem route between other pairs of exchanges. This gives a very large number of options. Each exchange has a routing table containing all its allowed options and will select from these in some preferred order.

Common-channel signalling is employed between the processors of the exchanges. This enables crank-back to be used. If a connection is required between exchanges A and B and all circuits on the direct route A–B are busy, then exchange A attempts to route the call via exchange C if there are free circuits on route A–C. However, the attempt will be unsuccessful if there is congestion on route C–B.

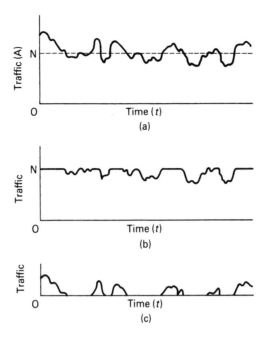

Figure 10.17 Character of traffic for high-usage route with overflow. (a) Traffic offered to high-usage route. (b) Traffic carried by high-usage route. (c) Overflow traffic.

Exchange C signals this information back to A. On receipt of the message, exchange A releases the connection A–C and makes a new attempt, say over route A–D–B.

In the form of dynamic routing described in Section 10.10.1 the routes are predetermined on the basis of traffic forecasts for different times of the day. More sophisticated methods are also possible, in which the choice of routes is automatically adapted according to the traffic conditions actually encountered. Bell Canada[24] has used a method in which a central routing processor periodically receives information from exchanges as to the number of idle trunks on each of their outgoing routes. Routes are then selected centrally and the choices signalled back to the exchanges.

British Telecom has a distributed dynamic alternative routing scheme which uses a learning approach.[25] When all trunks are busy on a direct route, the alternative two-link route which was last used successfully is selected again. If this is busy, a new alternative route is selected at random.

The AT&T toll network uses a distributed dynamic alternative routing scheme[26] known as *real-time network routing* (RTNR). Every switching centre has a table of data containing the load condition of each of its direct routes and the two-link alternatives. These tables are updated by messages sent between the centres over the common-channel signalling network. When a direct route is busy, the switching centre checks the available capacity of each of the alternatives and selects that which is least heavily loaded. In making the selection, account is taken of the class of service of the call, which determines the bandwidth required. An ISDN call may require 64 kbit/s for voice or switched data, 384 kbit/s for an H_0 channel or 1536 kbit/s for an H_{11} channel connection.

Use of AAR increases the traffic capacity of a nonhierarchical network under normal conditions. However, once the total traffic exceeds a certain level, the traffic actually carried begins to decrease with further increase in offered traffic, as shown in Figure 7.9(b). An analysis of this behaviour is complex.[27–29] However, a simple explanation is that a large number of calls are then using alternative routes. Each such connection uses two trunks in tandem; thus one call being carried can prevent two new direct connections from being made. Clearly, this situation would be exacerbated if connections were made over more than two links in tandem. In practice, therefore, AAR in nonhierarchical networks is restricted to routing with only two tandem links.

Measures must be taken to prevent overload reducing network capacity. One method is to use trunk reservation. Calls overflowing from a direct route are blocked whenever there are less than a certain number of circuits free on the alternative route. However, first-choice traffic always has access to all circuits on that route. It has been shown that use of common-channel signalling allows both-way working on routes between exchanges. Thus, an overload of traffic from A to B may prevent calls from B to A. This can be avoided by reserving some of the trunks for use in one direction only.

Another strategy for preventing overload is *call gapping*. This is useful in the case of a *focused overload*, due to failure of an exchange or transmission link. It may also be due to an exceptionally large amount of traffic to one destination. This may be caused, for example, by a local disaster or by a 'phone-in' programme on radio or television inviting people to make calls to a particular telephone number. If there is severe congestion in one part of the network, most attempts to route calls into it will be unsuccessful and will only cause congestion to spread elsewhere. It is therefore better to stop these calls where they originate. The NMC can instruct the processors of originating exchanges to throttle back this traffic by allowing through only one call every t seconds, where t is varied according to the capacity of the destination to receive calls.

10.11 Network management

The management of both public and private telecommunications networks is carried out at a number of levels[30,31] as follows:

1. *The business level* This is the management of the network as a business. It includes sales, customer administration and billing, accounting, inventory control and investment planning.
2. *The service level* This is management of the services provided to the customers. It include both basic services (e.g. telephony) and value-added services.

3. *The network level* This includes route optimization, traffic management, contingency planning to cope with emergencies and the planning of changes and extensions to the network.
4. *The network element level* This includes the installation of equipment, diagnosis of faults and the management of maintenance, repairs and alterations.

Computer tools are used in all these activities. Savings can be made and the quality of service improved by coordinating them, which involves using common databases and data communication between them. Since a telecommunications network is distributed over a wide area, this necessitates the use of a data network to reach the elements to be monitored and controlled. This leads to the evolution of an overlay network for administration of the telecommunications network, as shown in Figure 10.7.

In a complex network many different proprietary systems are involved, so standard interfaces are needed between them to build an effective management network. International cooperation in standards making has resulted in a set of open-systems standards for this purpose.[32]

10.12 Conclusion

During the last few decades, technological progress has reduced the cost of telecommunications and improved both the quality of service obtained from telecommunications networks and the range of services which they can provide. The combination of technology-push and market-pull should ensure that this process continues.

The large investment in present customer-access networks ensures that copper wire will continue to be the principal means of access to most customers. However, ISDN will enable them to obtain a wider range of services. Optical fiber is being increasingly deployed where it is economically justifiable.[33] As costs fall, it will become the standard method of provision for business customers and, eventually, for domestic customers. This will enable broadband services to be provided over a B-ISDN.

In the core transmission-bearer network the present PDH systems will be replaced by SDH transmission. This can be managed automatically to give remote monitoring and greater flexibility for rerouting and reconfiguration. It will eliminate the 'multiplex mountain' and result in fewer transmission nodes. The inner core will remain a mesh network, but the outer core may use SDH rings, as shown in Figure 2.19. Ultimately, SDH-based cross-connect units may give customers control of their access bandwidth.

The process of network management will be increasingly automated. Eventually, automatic collection and correlation of data may enable all but the most unusual events in the operation of a network to be handled without human intervention. Automatic processing of service requests will enable customers to initiate and discontinue services, and businesses will be able to reconfigure private networks under their own control. It will be possible to manage services end-to-end across different networks, for example across fixed and mobile networks, the networks of different operators and both public and private networks[34]

The number of 'intelligent' service-control points may increase to provide a universal intelligent network. The creation and control of services will reside on separate computers and databases connected to the switching centres through defined interfaces. Thus, advanced services should be switch-independent and the network better able to respond quickly to new requirements.

Some personal mobile services already exist. However, present cellular radio services are expensive and lower-priced 'telepoint' services, based on cordless telephones, are less useful. New digital systems are planned, including the pan-European GSM system.[8] Designers are already considering possible third-generation systems that could lead to a universal personal mobile service. In Europe, this is called *Universal Mobile Telecommunications Service* (UMTS); internationally it is called *Future Land Mobile Telecommunications System* (FLMTS). The objective is a universal radio interface. If this is linked to an intelligent fixed network it could give customers identical services whether on a wired or wireless basis. The use of a number associated with a person, rather than a location, should enable a customer to receive a full range of services, whether at home, at work or travelling abroad. Moreover, the services provided may also be expected to develop further as telecommunications moves into the next century.[35]

Problems

1. Compare the hierarchy of a telecommunications network with any other hierarchy (e.g. a transport or electricity supply system or the organization of a company, a university, the Army or the Church). Illustrate your answer with suitable diagrams and discuss the similarities and differences that you find.

2. Explain why a large town has several telephone exchanges, instead of a single large one.

3. A fictitious city has a square boundary of side B km. The city is divided into m square areas, with a telephone exhange at the centre of each. Customers' stations are uniformly distributed throughout the city with a density of s customers per km².

The cost of a customer's line is $k_1 l$ (where l is the length of the line in km). The cost of a local exchange is $k_2 + k_3 n$ (where n is the number of customers' lines).

If points are uniformly distributed over a square of side b, the mean distance from a point to the centre of the square is $\bar{l} = k_4 b$. (It can be shown that $k_4 = 0.382$.)

(a) Sketch curves showing how the cost of customers' lines, the cost of local exchanges and the total cost of local-area plant vary with the number of exchanges (m).

(b) Show that the total cost of local-area plant is a minimum when

$$m = B^2 \left[\frac{k_1 k_4 s}{2k_2} \right]^{2/3}$$

i.e. the optimum number of local exchanges is proportional to the two thirds power of the customer density.

4. (a) In the square city, each customer has a total both-way traffic of a erlangs during the busy hour. Each customer is equally likely to call any other, so all junction routes carry equal traffic.

(i) Show that the total junction traffic is $A = Na(1 - 1/m)$, whether there are direct junctions between all local exchanges and no tandem switching centre or a tandem switching centre and no direct junctions.

(ii) Why, then, are different numbers of junctions needed when there is direct routing and when there is a tandem switching centre?

(b) The cost of a junction group is $k_5 + k_6 A_j$, where A_j is its traffic in erlangs.

The cost of a tandem switching centre is $k_7 + k_8 A_t$, where A_t is its total traffic in erlangs.

(i) Sketch curves showing how the cost of the junction-network plant varies with m, when all junction traffic is carried over direct routes and when all junction traffic is routed via a tandem switching centre. Hence, modify the curve for local-plant cost obtained in part (a) to provide a curve showing how the total cost of the network varies with the number of exchanges (m).

(ii) If $m \gg 1$, show that the cost for tandem routing is less that the cost for direct routing if

$$m^2 > \frac{2 k_7 + N a k_8}{k_5}$$

i.e. the number of local exchanges for which tandem routing is economic increases with the cost of exchange plant (k_7, k_8) and with traffic (Na) and it decreases with the cost of line plant (k_5).

5. (a) Explain the principles determining a national numbering plan. How is this influenced by the need for an international numbering plan?

(b) For equity in charging, number groups should cover approximately equal areas. To avoid wasting number capacity they should contain approximately equal numbers of exchanges and so be of different sizes. How is this conflict resolved?

6. Direct cable routes exist between switching centres P and Q and between Q and R. There are no direct circuits between P and R. Two alternative plans are proposed to cater for traffic between P and R:

(a) Q provides transit switching for traffic between P and R in addition to switching its own originating and terminating traffic.

(b) Some of the cable pairs from P to Q are directly strapped at Q to pairs in the cable from Q to R. Thus, no transit switching is needed at Q.

It is given that:

(i) Distances PQ and QR are 8 km and 7 km, respectively.
(ii) Busy-hour traffic between P and Q is 10 E and between Q and R is 16 E.
(iii) Cost of signalling equipment is £40 per circuit.
(iv) Cost of transit or terminal switching equipment is £50 per circuit.
(v) Cable cost is £30 per circuit kilometre.
(vi) The traffic table for the required grade of service can be approximated as follows:

$$N = 1.42A + 2.9$$

where N is the number of circuits required to carry A erlangs of traffic.

At what traffic load will costs begin to favour through strapping instead of transit switching?

7. A simple network contains only three local exchanges, A, B and C, which are 20 km apart. A tandem exchange, T, is located at the exact centre of the three local exchanges. If exchanges A and B each originate 0.9 E of traffic to C in the busy hour, calculate the total length of junction pairs required under each of the routing schemes below:

(a) Direct routing of all traffic from A to C and from B to C.
(b) Indirect routing of all traffic from A to C and from B to C via the tandem exchange T.
(c) One direct junction from A to C and from B to C, with tandem routing of the overflow traffic.

The following table should be used to determine the number of junctions required:

Traffic offered (E)	0.25	0.53	0.90	1.32	1.8	2.31
No. of junctions	3	4	5	6	7	8

8. Exchanges A_1 to A_5 and D_1 to D_5 are primary centres in a hierarchical trunk network. Exchanges A1 to A5 are connected to secondary centre B, and exchanges D1 to D5 are connected to secondary centre C. The

[290] Networks

switching centres B and C are joined by a high-capacity multiplex transmission system.

For traffic from B to exchanges D_1 to D_5 via exchange C, the following alternatives are to be considered:

(i) All traffic routed by direct circuits, strapped through at C and loaded at 0.6 E.
(ii) High-usage direct circuits loaded to 0.75 E carrying 90% of the traffic, together with overflow circuits switched at C and loaded to 0.6 E.

The average cost of a complete transmission path is £1100 per circuit and the cost of switching equipment at C is £800 per erlang. Find which method is cheaper.

9. The traffic from exchange A to exchange B is 2 E. This is first offered to a direct route and any overflow is routed via a third exchange C. Routes AC and CB are large routes and may be assumed to have an occupancy of 0.8 E. A direct circuit AB costs 0.7 of the total cost of a pair of circuits AC and CB. Losses on route ACB are negligible.

Prove that the maximum number of direct circuits worth providing is only one. (Fractional numbers of trunks may be used in the calculation. Thus, if all the traffic were routed indirectly, it would require $2/0.8 = 2.5$ trunks on each route.)

10. A nonhierarchical network has only three exchanges: X, Y and Z. The total traffic capacity of each route is $2A$ erlangs if it is bidirectional and A erlangs in each direction if one-way working is employed. It may be assumed that the loss probability is negligible if the offered traffic is less than this and that all offered traffic beyond this level is lost.

The following routing strategies may be employed:

(i) Unidirectional working and no automatic alternative routing (AAR)
(ii) Bidirectional working and no AAR
(iii) Unidirectional working with AAR
(iv) Bidirectional working with AAR.

(a) The traffic offered to routes XY, YX, ZX and ZY is zero, but the traffic offered to each of the routes XZ and YZ is $2A$. Find, for each routing strategy, the traffic carried from X to Z and from Y to Z.
(b) Traffic of A erlangs is offered to each of routes XY, XZ, ZX, YX and ZY, but traffic $2A$ is offered to route YZ. For each routing strategy, obtain a rough estimate for the traffic carried from Y to Z and from Y to X. To simplify the approximate calculations for AAR, assume that the congested route between Y and Z can carry no overflow traffic. (If required, this first estimate can then be improved by an iterative process to include the overflow traffic on YZ.)

In each case, what is the total traffic between originating and terminating exchanges carried by the network?

11. Explain the meanings of the following terms:

(a) Bearer service
(b) Teleservice
(c) Intelligent network
(d) Managed network
(e) Dynamic routing.

12. You have been asked to advise a small travel agency on its telecommunications requirements. At present, the office is served by two PSTN lines, which are each rented at a cost of £10 per month. One is used as a telephone line and the other is connected via a 1200 baud modem to a videotex (viewdata) terminal. The terminal is used to access tour operators' databases to select and book holidays for the agency's clients.

The videotex terminal is used, on average, twenty times in each of the 25 days in a month that the agency is open. Each videotex call lasts for 10 minutes, of which 2 minutes is spent waiting for the screen to be updated whenever a new page is requested. The database response is near-instantaneous; the delay is due to transmission across the modem link. The charge to connect to a database is 10p per minute.

The databases can also be accessed via ISDN links at data rates of up to 9.6 kbauds. You are aware that ISDN is now available in the locality and suspect that basic-rate access could provide a more cost-effective solution to the travel agent's needs. Basic-rate ISDN can be obtained by paying an installation charge of £120 and a rental of £18 per month, call charges being the same as for PSTN. However, a digital telephone with a compatible 9.6 kbaud terminal adaptor would cost £900.

(a) How many months will it take for the travel agent to recover the investment if the change is made to ISDN access?
(b) Would you recommend that the travel agent should change to ISDN?
(c) Although you might recommend a change to ISDN, the travel agent is not fully convinced by the financial argument and would like to know what other advantages would be gained. Outline the other benefits (both immediate and long-term) that the change would bring to the business.

References

[1] Francis, H.E. (1959), 'The general plan for subscriber trunk dialling', *PO Elec. Engrs. Jour.*, **51**, 258–67.
[2] Clarke, A.B. (1952), 'Automatic switching for nation-wide telephone service', *Bell Syst. Tech. Jour.*, **31**, 823–31.
[3] Andrews, F.T. (1985), 'Divestiture: a record of technical achievement', *IEEE Comms. Magazine*, **23**, December, 54–8.
[4] McLintock, R.W. (1991), 'Transmission performance', Chapter 4 in Flood, J.E. and Cochrane, P. (eds), *Transmission Systems*, Peter Peregrinus, Stevenage.
[5] Ash, G.R. and Mummert, V.S. (1984), 'AT&T carves new routes in its nationwide network', *AT&T Bell Labs. Rec.*, **62**, August, 18–22.
[6] Griffiths, J. (ed.) (1992), *ISDN explained*, 2nd edn, Wiley, Chichester.
[7] Stallings, W. (1992), *ISDN and Broadband ISDN*, 2nd edn, Macmillan, New York.
[8] Macario, R.C.V. (ed.) (1991), *Personal and Mobile Radio Systems*, Peter Peregrinus, Stevenage.
[9] Parsons, J.D. and Gardiner, J.G. (1989), *Mobile Communication Systems*, Blackie, Glasgow.
[10] 'Stored program controlled network', Special Issue of *Bell Syst. Tech. Jour.*, 1982, **61**, No.7, Pt 3.
[11] Webster, S. (1989), 'The digital derived services intelligent network', *Brit. Telecom. Eng. Jour.*, 8, 144–9.
[12] Huws, U., Kurte, W.B. and Robinson, S. (1990), *'Teleworking: towards the elusive office*, Wiley, Chichester.
[13] Nunn, W.H. (1952), 'Nationwide numbering plan', *Bell Syst. Tech. Jour.*, **31**, 851–9.
[14] Munday, S. (1967), 'New international switching and transmission plan recommended by the CCITT for public telephony', *Proc. IEE*, **114**, 619–27.
[15] McLeod, N.A.C. (1990), 'Numbering in telecommunications', *Brit. Telecom. Eng. Jour.*, **8**, 225–31.
[16] CCITT Recommendation E.164, 'Numbering for the ISDN era'.
[17] McLeod, N.A.C. (1992), 'Development of the national code change', *Brit. Telecom, Eng. Jour.*, **10**, 303–5.
[18] CCITT Recommendation X.121, 'International numbering plan for public data networks'.
[19] Rapp, Y. (1964), 'Planning a junction network in a multiexchange area', *Ericsson Tech.*, **20**, 77–130.
[20] Ash, G.R., Cardwell, R.H. and Murray, R.P. (1981), 'Design and optimization of networks with dynamic routing', *Bell Syst. Tech. Jour.*, **60**, 1787–820.

[21] Katzschener, L. (1973), 'Service protection for direct final traffic in DDD networks', *3rd Int. Teletraffic Congress*, Stockholm.
[22] Ingham, A.R. and Elvidge, A.M. (1989), 'Trunk reservation with automatic alternative routing', *6th UK Teletraffic Symposium*.
[23] Wilkinson, R.L. (1956), 'Theories for toll traffic engineering in the USA', *Bell. Syst. Tech. Jour.*, **35**, 421–514.
[24] Cameron, W.H. et al. (1983), 'Dynamic routing for intercity telephone networks', *10th Int. Teletraffic Congress*, Montreal.
[25] Stacey, R.R. and Songhurst, D.J. (1989), 'Dynamic routing in British Telecom Network', *12th Int. Switching Symposium*.
[26] Ash, G.R. et al. (1992), 'Real-time network routing in the AT&T network: improved service quality at lower cost', *Proc. IEEE Global Telecom. Conf.*, pp. 802–9.
[27] Akinpelu, J.M. (1983), 'The overload performance of engineered networks with hierarchical and nonhierarchical routing', *10th Int. Teletraffic Congress*, Montreal.
[28] Schwartz, M. (1987), *Telecommunication Networks: protocols, modeling and analysis*, Addison-Wesley, Reading, MA.
[29] Girard, A. (1990), *Routing and Dimensioning of circuit-switched Networks*, Addison-Wesley, Reading, MA.
[30] Held, G. (1992), *Network Management: techniques, tools and systems*, Wiley, New York.
[31] 'Network management systems', Special Issue of Brit. Telecoms. Eng., 1991, 10, Pt 3.
[32] Black, V. (1992), *Network Management Standards: the OSI, SNMP and CMOL standards*, McGraw-Hill, New York.
[33] Adams, P.F., Rosher, P.A. and Cochrane, P. (1991), 'Customer access', Chapter 15, in Flood, J.E. and Cochrane, P. (eds), *Transmission Systems*, Peter Peregrinus, Stevenage.
[34] Valdar, A., Newman, D., Wood, R. and Greenop, D, (1992), 'A vision of the future network', *Brit. Telecom. Eng. Jour.*, **11**, 142–52.
[35] Davies D.E.N., Hilsum, C. and Rudge, A.W. (eds) (1993), *Communications after AD2000*, Chapman & Hall, London.

APPENDIX 1

Basic probability theory

A1.1 Definitions

A1.1.1 Probability

Probability can be defined as the relative frequency of occurrence of a random event.[1] If event A occurs n_A times among a large number of events, n, the probability, $P(A)$, of A occurring is:

$$P(A) = n_A/n$$

If A never occurs, then $P(A) = 0$. If it always occurs, then $P(A) = 1$.

$$\therefore 0 \leq P(A) \leq 1$$

A1.1.2 Conditional probability

The probability of A conditional on B, $P(A|B)$, is the probability that A occurs given the occurrence of B. If A and B are *independent*, the probability of A occurring does not depend on whether B occurs.

$$\therefore P(A|B) = P(A), \quad P(B|A) = P(B)$$

A1.1.3 Joint probability

The joint probability of A and B, $P(A.B)$, is the probability that both occur.

$$\therefore P(A.B) = P(A|B)\,P(B) = P(B|A)\,P(A)$$

If A and B are *independent*, then:
$P(A.B) = P(A)\,P(B)$.

A1.1.4 Mutually exclusive events

Events are mutually exclusive if they cannot occur together, i.e. $P(A.B) = 0$. Then the probability of *either* occuring, $P(A + B)$, is:

$$P(A + B) = P(A) + P(B)$$

If A and B are *not* mutually exclusive, then $P(A)$ and $P(B)$ both include $P(A.B)$.

$$\therefore P(A + B) = P(A) + P(B) - P(A.B)$$

A1.1.5 Complementary events

Events A and B are complementary if either A or B always occurs, but never both together.

$$\therefore P(A) + P(B) = 1, \quad P(A.B) = 0$$

A1.2 Discrete probability distributions

A1.2.1 Mean and variance

The values of $P(x_1)$, $P(x_2)$,...., $P(x_n)$ for a discrete random process $X = x_1, x_2,...., x_n$ may be plotted as a bar chart or a histogram, as shown in Figure A1.1. For this:

$$\sum_{j=-\infty}^{\infty} P(x_j) = 1$$

The value of x corresponding to the centre of gravity of this diagram is called the average or

Appendix 1

mean usually denoted by $\bar{x}$ or μ, or the expectation $E(X)$.

$$E(X) = \sum_{j=-\infty}^{\infty} x_j \, P(x_j)$$

i.e. the mean is given by the first moment of the distribution.

The variance, var (X) or σ^2, is a measure of the dispersion or spread of the histogram. It is defined as the mean squared deviation of x from the mean.

$$\therefore \quad \text{var}(X) = \sum_{j=1}^{n} (x_j - \bar{x})^2 \, P(x_j)$$

$$= \sum_{j=1}^{n} x_j^2 \, P(x_j) - \bar{x}^2$$

The *standard deviation*, σ, is defined as the square root of the variance, i.e. $\sigma_X = \sqrt{\text{var}(X)}$.

A1.2.2 Mean and variance of sum or difference of independent random variables

If X and Y are independent random variables, then:[2]

$E(X + Y) = E(X) + E(Y)$ and
var $(X + Y) = $ var $(X) + $ var (Y)
$E(X - Y) = E(X) - E(Y)$ but
var $(X - Y) = $ var $(X) + $ var (Y)

A1.2.3 Bernouilli or binomial distribution

Consider a series of trials which satisfies the following conditions:

1. Each trial can have two possible outcomes, e.g. success or failure, with probabilities p and $(1 - p)$.
2. The outcome of each trial is an independent random event.
3. Statistical equilibrium (i.e. the probabilities do not change).

The probability of one particular combination of x successes and $n - x$ failures is $p^x (1 - p)^{n-x}$.

However, there can be $\binom{n}{x}$ such arrangements of successes and failures, where $\binom{n}{x}$ is the binomial coefficient given by:

$$\binom{n}{x} = \frac{n!}{x! \, (n-x)!}$$

$$= \frac{n(n-1)(n-2)\ldots(n-x+1)}{x!}$$

$$\therefore \quad P(x) = \binom{n}{x} p^x (1-p)^{n-x} \qquad (A1.1)$$

The mean is $\mu = n\,p$
The variance is $\sigma^2 = n\,p\,(1 - p)$.

A1.2.4 Poisson distribution

The Poisson distribution is the limiting case of the Bernouilli distribution when n is very large. From equation (A1.1):

$$P(x) = \frac{n(n-1)(n-2)\ldots(n-x+1)}{x!} p^x (1-p)^{n-x}$$

$$= 1\left(1 - \frac{1}{n}\right)\left(1 - \frac{2}{n}\right)\ldots$$
$$\left(1 - \frac{x+1}{n}\right) \frac{n^x}{x!} p^x (1-p)^{n-x}$$

$$\therefore \text{ if } n \text{ is large, } P(x) \to \frac{n^x}{x!} p^x (1-p)^{n-x}$$

But
$\mu = np \therefore p = \mu/n$.

$$\therefore \quad P(x) = \frac{\mu^x}{x!} \left[1 - \frac{\mu}{n}\right]^n \left[1 - \frac{\mu}{n}\right]^{-x}$$

$$\to \frac{\mu^x}{x!} \left[1 - \frac{\mu}{n}\right]^n \qquad \text{(since } \mu \ll n\text{)}$$

$$\therefore \quad P(x) = \frac{\mu^x}{x!} \left[1 - n\frac{\mu}{n} + \frac{n(n-1)}{2!}\left(\frac{\mu}{n}\right)^2 \right.$$
$$\left. - \frac{n(n-1)(n-2)}{3!}\left(\frac{\mu}{n}\right)^3 + \ldots\right]$$

$$= \frac{\mu^x}{x!}\left[1 - \frac{\mu}{1} + \frac{\mu^2}{2!} + \frac{\mu^3}{3!} + \ldots\right]$$

$$\therefore \quad P(x) = \frac{\mu^x}{x!} e^{-\mu} \qquad (A1.2)$$

The mean is still μ.
The variance for the binomial distribution is:

$$\sigma^2 = np(1-p) = \frac{n\mu}{n}\left[1 - \frac{\mu}{n}\right]$$

$$\therefore \lim_{n \to \infty} \sigma^2 = \mu$$

Discrete probability distributions [295]

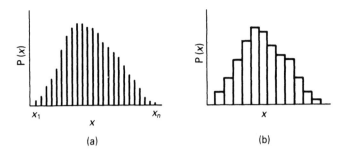

Figure A1.1 Discrete probability distributions. (a) Bar chart. (b) Histogram.

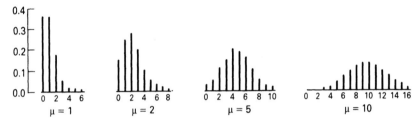

Figure A1.2 Examples of the Poisson distribution.

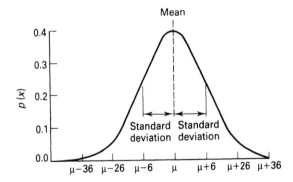

Figure A1.3 The Gaussian or normal distribution.

[296] Appendix 1

It is thus a property of the Poisson distribution that the variance equals the mean.

Some examples of Poisson distributions for different mean values are shown in Figure A1.2.

A1.3 Continuous probability distributions

A1.3.1 General

If the interval δx between all values of x_j and x_{j+1} tends to zero, X becomes a continuous function of x and the histogram becomes a continuous curve. However, $P(x)$ also tends to zero, so the *probability density*, $p(x)$, is plotted against x, e.g. as shown in Figure A1.3. The probability density, $p(x)$, is defined by:

$$p(x_j)\delta x = P(x_j \leq x \leq x_j + \delta x)$$

$$\therefore \quad P(a \leq x \leq b) = \int_a^b p(x)\, dx$$

and

$$\int_{-\infty}^{\infty} p(x)\, dx = 1$$

For a discrete distribution: $\mu = \sum_{-\infty}^{\infty} x_j\, P(x_j)$

$\therefore$ for a continuous distribution:

$$\mu = \int_{-\infty}^{\infty} x\, p(x)\, dx$$

For a discrete distribution:

$$\sigma^2 = \sum_{-\infty}^{\infty} x_j^2\, P(x_j) - \mu^2$$

$\therefore$ for a continous distribution:

$$\sigma^2 = \int_{-\infty}^{\infty} x^2\, p(x) - \mu^2$$

Since a continuous distribution is the limiting case of a discrete distribution, the formulae for the mean and variance of the sum and difference of independent random variables still apply.
i.e. $E(X \pm Y) = E(X) \pm E(Y)$
$\text{var}(X \pm Y) = \text{var}(X) + \text{var}(Y)$

A1.3.2 Negative exponential distribution

A sequence of random events is a discrete random process. However, the intervals between these events can have any value and so constitute a continuous random process.

Let the interval between two successive events be t and its probability density be $p(t)$. Let one event occur at $t = 0$ and the next between t and $t + \delta t$. If δt is very small, the probability of two or more events during it is negligible. Let the probability of one event during δt be $P_{\delta t}(1)$.

If the event between t and $t + \delta t$ is the next in the sequence, there must be no event during t. Let the probability of this be $P_t(0)$.

Thus, the probability that the next event occurs between t and $t + \delta t$ is:

$$P = P_t(0)\, P_{\delta t}(1) = p(t)\, \delta t.$$

If the random events are independent, the number of events during a time t has a Poisson distribution, given by equation (A1.2). If the mean rate of occurrence of the events is λ per unit time, the mean number of events during a time t is $\mu = \lambda t$.
Hence, from equation (A1.2):

$$P_t(0) = e^{-\lambda t}, \quad P_{\delta t}(1) = \lambda \delta t\, e^{-\lambda \delta t}$$
$$\therefore \quad P = p(t)\, \delta t = \lambda \delta t\, e^{-\lambda(t + \delta t)}$$
$$= \lambda e^{-\lambda t} \delta t \quad \text{(since } \delta t \text{ is very small)}$$

$$\therefore \quad p(t) = \lambda\, e^{-\lambda t} \qquad (A.1.3)$$

The mean of the negative exponential distribution is:

$$\mu = \int_{-\infty}^{\infty} t\, p(t)\, dt = 1/\lambda$$

The variance is:

$$\sigma^2 = \int_{-\infty}^{\infty} t^2\, p(t)\, dt - \mu^2 = 1/\lambda^2$$
$$\therefore \quad \sigma = 1/\lambda$$

Thus, the standard deviation is equal to the mean.

The probability that the interval between events, T, will exceed time t is given by:

$$P(T \geq t) = \int_t^{\infty} p(t)\, dt = \lambda \int_t^{\infty} e^{-\lambda t}\, dt = e^{-\lambda t}$$
$$= e^{-t/\bar{T}}$$

where $\bar{T}$ is the mean interval between events.

A1.3.3 Gaussian or normal distribution

In many practical cases[3] it is found that the probability density function has the well-known bell shape shown in Figure A1.3. It is therefore called the *normal distribution*. The following equation was fitted to it by Gauss:

$$n(\mu, \sigma; x) = p(x) = \frac{1}{\sigma\sqrt{2\pi}} e^{-(x-\mu)^2/2\sigma^2} \quad (A.1.4)$$

where μ = mean
σ = standard deviation

The *central limit theorem* states that the probability density function of the sum of a large number of independent variables tends towards $n(\mu,\sigma;x)$ as n increases, provided that the variances of all the components are small compared with the variance of their sum.[4,5] This is usually the case if, for example, the results of a large number of similar measurements are plotted.

Substituting $t = (x - \mu)/\sigma$ in equation (A1.4) gives:

$$p(t) = \frac{1}{\sqrt{2\pi}} e^{-t^2/2}$$

This is the *standard normal distribution*.
The probability that $0 \leq t \leq z$ is given by:

$$\phi(z) = \frac{1}{\sqrt{2\pi}} \int_0^z e^{-t^2/2} \, dt$$

Tables of $\phi(z)$ have been published[2,6] usually for $0 \leq z < 4$. Since

$$\int_{-z}^0 p(t)dt = \int_0^z p(t)\,dt$$

and $\int_{-\infty}^0 p(t)\,dt = \int_0^\infty p(t)\,dt = 1/2,$

these tables enable $P(a \leq x \leq b)$ to be determined for any values of a and b, given μ and σ.

As a result of the central limit theorem, most distributions tend towards $n(\mu,\sigma;x)$ when they result from a large number of events. This applies to the envelope of the Poisson distribution, as can be seen happening in Figure A1.2. Thus, provided that μ is fairly large, a normal distribution with $\sigma = \sqrt{\mu}$ can be used as an approximation for the Poisson distribution. Then:

$$P(x) = \int_{x-\frac{1}{2}}^{x+\frac{1}{2}} n(\mu, \sigma; x) \, dx$$

$$= \phi(z_2) - \phi(z_1)$$

where $z_1 = (x - \tfrac{1}{2} - \mu)/\sqrt{\mu}$
$z_2 = (x + \tfrac{1}{2} - \mu)/\sqrt{\mu}$

References

[1] Aitken, A.C. (1957), *Statistical Mathematics*, 8th edn, Oliver & Boyd, Edinburgh.
[2] Walpole, R.E. and Myers, R.H. (1972), *Probability for Engineers and Scientists*, Collier-Macmillan, New York.
[3] Fry, T.C. (1965), *Probability and its Engineering Uses*, 2nd edn, Van Nostrand, New York.
[4] Beckmann, P. (1967), *Probability in Communication Engineering*, Harcourt, Brace and World, New York.
[5] Freund, J.E. (1962), *Mathematical Statistics*, Prentice Hall, Englewood Cliffs, NJ.
[6] Speigel, M.R. (1972), *Statistics*, Schaum, McGraw-Hill, New York.

APPENDIX 2

Non-Poissonian traffic

A2.1 Smooth and peaky traffic

For pure-chance traffic, as defined in Section 4.5, call arrivals have the Poisson distribution given by equation (A1.2). It is shown in Section 4.5 that the number of calls in progress also has a Poisson distribution. It is a property of this probability distribution that the mean, μ, and variance, σ^2, are equal.

If call arrivals are less irregular than for pure-chance traffic, then the smaller variation of inter-arrival times causes a smaller variation in the number of calls in progress than for pure-chance traffic, i.e. $\sigma^2 < \mu$. The traffic is then called *smooth traffic*.

If there is a greater variation of inter-arrival times than for pure-chance traffic, then there is a greater variation in the number of calls in progress. i.e. $\sigma^2 > \mu$. In this case, the traffic is called *peaky traffic* and the ratio $K = \sigma^2/\mu$ is called its *peakedness*.

These three different types of traffic can be defined as follows:

- Smooth traffic: $\sigma^2 < \mu$, $K < 1$
- Pure-chance traffic: $\sigma^2 = \mu$, $K = 1$
- Peaky traffic: $\sigma^2 > \mu$, $K > 1$.

An example of smooth traffic is that generated by a small number of sources. In pure-chance traffic, the number of sources is theoretically infinite, so the mean call-arrival rate is independent of the number of calls in progress. However, if the number of sources is finite, the arrival rate decreases with the number of calls in progress (since a source that is already busy cannot generate a new call). In pure-chance traffic there is theoretically no limit to the number of calls that can be in progress. However, if the number of sources is finite, the calls in progress cannot exceed this number.

When congestion occurs, the traffic carried by a lost-call system is smoother than the traffic offered to it because the peaks of the offered traffic are rejected. When smooth traffic is offered to a lost-call system, the loss is smaller than for pure-chance traffic, because the offered traffic has smaller peaks to be lost.

In a step-by-step switching system, the traffic offered to the first selectors can be assumed to be Poissonian. However, the traffic offered to the second selectors is smooth, because peaks of traffic are lost by the first selectors; thus, their loss probability is smaller. Traffic tables have been compiled which take this into account.[1]

When a system uses switches that make a sequential search, the first-choice trunk has a high occupancy as discussed in Section 4.6.2. Thus, it carries smooth traffic and traffic overflowing to subsequent choices is peaky.

In a progressive grading the traffic offered to the commoned late choices is peaky, because it has overflowed from the earlier choices. Indeed, the efficiency of gradings depends on this. The traffic capacity is increased because trunks receive a mixture of traffic from several sources whose peaks rarely coincide.

These effects are present when automatic alternative routing (AAR) is employed. The direct (high-usage) routes are dimensioned to have a high occupancy and a high loss probability. They therefore carry smooth traffic and the traffic which overflows to the alternative route is peaky, as shown in Figure 10.17. A theory that enables tandem routes to be dimensioned to carry such overflow traffic was developed by Wilkinson.[2] This is known as the *equivalent random method*. It assumes that the circuits on the tandem route behave in the same way as if they formed part of a larger group of circuits offered an equivalent amount of random traffic.

Let an exchange have k high-usage routes to other exchanges. The overflow traffic from these routes is carried by a common group of m circuits to a tandem exchange.

If the rth high-usage route has n_r trunks and is offered traffic A_r, the mean traffic, A'_r, lost from this route is $A'_r = B_r A_r$, where $B_r = E_{1,n_r}(A_r)$. It can be shown[2,3] that the variance of this traffic is

$$V_r = A'_r \{1 - A'_r/(1 + n_r + A'_r - A_r)\}$$

The mean, M, and variance, V, of the total traffic offered to the tandem route are:

$$M = \sum_{r=1}^{k} A'_r, \qquad V = \sum_{r=1}^{k} V_r$$

The equivalent random traffic is given by

$$A^* = V + 3K(K - 1), \text{ where } K = V/M$$

The traffic A^* is considered to be offered to a full-availability group of $m + n^*$ trunks, where m is the number of tandem trunks and n^* is a fictitious number of trunks carrying traffic equivalent to that on the high-usage routes. It can be shown[2,3] that

$$n^* = \frac{A^*(M + K)}{M + K + 1} - M - 1$$

The total traffic lost from the AAR system is then given by the Erlang loss formula for $m + n^*$ trunks offered traffic A^*. Hence, if the traffic offered to each high-usage group and its number of trunks are known, the number of trunks required on the tandem route to provide a specified grade of service can be found from a standard traffic table.

The same grade of service can be obtained with different combinations of numbers of high-usage and tandem trunks. The minimum-cost solution depends on the relative costs of direct and tandem trunks. A method of finding this from the cost ratios was published by Rapp.[4]

A2.2 Time congestion and call congestion

Time congestion is defined as the proportion of the busy hour for which all trunks are busy. *Call congestion* is defined as the proportion of calls which find all trunks busy, i.e. the probability of a call encountering congestion. In Section 4.3 it is assumed that these are equal. However, this is not always the case.

When pure-chance traffic is offered to a full-availability group of trunks, the mean arrival rate of calls is constant, so the proportion of calls encountering congestion equals the proportion of time for which there is congestion.

If the number of sources generating traffic, S, is less than the number of trunks provided, N, there can be no congestion. Both time congestion and call congestion are zero. The probability of finding x trunks busy is given by the Bernouilli distribution

$$P(x) = \binom{S}{x} a^x (1 - a)^{S-x}$$

where a = trunk occupancy
= total traffic/no. of trunks.

If the number of trunks equals the number of sources, it is possible for all trunks to be busy. The probability of this is $P(N) = a^N$, which is the time congestion. However, there must always be at least one free trunk when a source generates a call, so the call congestion is zero.

If the number of sources exceeds the number of trunks, but is not very much larger, the rate at which calls arrive decreases with the number of sources that are busy, because a source which is already busy cannot generate a call. Thus, the traffic is smooth and the conditions for the Erlang distribution are not satisfied. The call congestion and time congestion are unequal and the distribution which applies, for a lost-call system, is called the Engset distribution. It can be shown[5,6] that the call congestion, i.e. the grade of service, is given by:

$$B_N = \binom{S-1}{N} a^N \bigg/ \sum_{k=0}^{N} \binom{S-1}{k} a^k$$

where S = number of sources
N = number of trunks
a = calling rate per free source.

Tables of the Engset loss formula have been

published.[7] However, it can be shown[6] that:

$$B_N = \frac{(S-N)a B_{N-1}}{N + (S-N)a B_{N-1}}$$

Since $B_0 = 1$, this iterative formula enables B_N to be computed for all values of N.

The quantity a is inconvenient to use, since it depends on both A and S. It can be shown[6] that

$$a = \frac{A}{S - A(1-B)}$$

Thus, B depends on a, which in turn depends on B. Consequently, iteration is required, starting from $a_0 = A/S$, to determine B accurately from A.

The Engset loss formula is applicable, for example, to PBXs having a small number of extension telephones and exchange lines.[8] Since the offered traffic is smooth, the loss probability is less than for the Erlang distribution, but it tends towards it as the number of sources increases. Thus, if $S > 10N$, there is little error if the grade of service is taken to be $B = E_{1,N}(A)$.

References

[1] Atkinson, J. (1950), *Telephony*, Vol. 2, Appendix 1, Pitman, London.
[2] Wilkinson, R.L. (1956), 'Theories for toll traffic engineering in the USA', *Bell Syst. Tech. Jour.*, **35**, 421–514.
[3] Girard, A. (1990), *Routing and Dimensioning of Circuit-switched Networks*, Addison-Wesley, Reading, MA.
[4] Rapp, Y. (1964), 'Planning a junction network in a multiexchange area', *Ericsson Tech.*, **20**, 77–130.
[5] Dunlop, J. and Smith, D.G. (1989), *Telecommunications Engineering*, Chapter 10, 2nd edn, Chapman and Hall, London.
[6] Bear, D. (1988), *Principles of Telecommunication Traffic Engineering*, 3rd edn, Peter Peregrinus, Stevenage.
[7] *Telephone Traffic Theory, Tables and Charts* (1970), Siemens Aktiengellschaft, Munich.
[8] Hunter, J.M., Lawrie, N. and Peterson, M. (1988), *Tariffs, Traffic and Performance*, Com Ed Publishing, London.

APPENDIX 3

Method of Jacobaeus for determining grade of service of a link system

To illustrate this method[1,2] consider the two-stage switching network shown in Figure 5.11 serving outgoing routes having one trunk connected to each secondary switch. Let

A = traffic in erlangs offered to an outgoing route
b = occupancy of links
c = occupancy of outgoing trunks
g = no. of secondary switches.

A call is lost if x of the g trunks on the selected outgoing route are busy and all the $(g - x)$ links between the incoming trunk and free outgoing trunks on the route are also busy. If these events are independent and their probabilities are $P(x)$ and $Q(g - x)$ respectively, then the loss probability is:

$$B_2 = \sum_{x=0}^{g} P(x) Q(g - x) \quad (A3.1)$$

If the link occupancies are independent, then

$$Q(g - x) = b^{g-x}.$$

If the number of trunks on the outgoing route is small, this has a smoothing effect on the traffic carried, so it may be assumed that the number of busy trunks has a Bernouilli distribution.

$$\therefore P(x) = \binom{g}{x} c^x (1 - c)^{g-x} \text{ and}$$

$$B_2 = \sum_{x=0}^{g} \binom{g}{x} c^x (1 - c)^{g-x} b^{g-x}$$

$$= \sum_{x=0}^{g} \binom{g}{x} c^x \left[b(1 - c) \right]^{g-x}$$

$$\therefore B_2 = (c + b - bc)^g \quad (A3.2)$$

This is the same result as that obtained by Lee's method[3] and given by equation (5.25).

If the number of trunks on the outgoing route is large (e.g. $g \geq 30$), it can be assumed that the number of busy trunks has an Erlang distribution. Hence, from equation (4.8):

$$P(x) = \frac{A^x}{x!} \Bigg/ \sum_{k=0}^{g} \frac{A^k}{k!}$$

and

$$B_2 = \sum_{x=0}^{g} \frac{A^x}{x!} b^{g-x} \Bigg/ \sum_{k=0}^{g} \frac{A^k}{k!}$$

$$= b^g \sum_{x=0}^{g} \frac{(A/b)^x}{x!} \Bigg/ \sum_{k=0}^{g} \frac{A^k}{k!}$$

$$= \left[\frac{A^g}{g!} \Bigg/ \sum_{k=0}^{g} \frac{A^k}{k!} \right] \left[\sum_{x=0}^{g} \frac{(A/b)^x}{x!} \Bigg/ \frac{(A/b)^g}{g!} \right]$$

$$= \frac{E_{1,g}(A)}{E_{1,g}(A/b)} \quad (A3.3)$$

Hence, tables based on the Erlang loss formula can be used to determine the grade of service.

The method can be extended to more complex networks, with either a Bernouilli distribution or an Erlang distribution applied to the outgoing trunks and the links at each stage, as appropriate.[1,2]

References

[1] Jacobaeus, C. (1950), 'A study of congestion in link systems', Ericsson Tech., **48**, 1–70.
[2] Bear, D. (1988), Principles of Telecommunication Traffic Engineering, 3rd edn, Peter Peregrinus, Stevenage.
[3] Lee, C.Y. (1955), 'Analysis of switching networks', Bell Syst. Tech. Jour., **34**, 1287–1315.

Answers to problems

Chapter 2

Q2. (a) 15 dB, 22 ms, 6 dB, 20 ms.

Q3. 3.3, 4.0, 4.7, 5.3, 5.8, 6.4, 6.9, 7.3 dB.

Q4. 85 dB.

Q5. (a) 9 h, (b) 1 h 20 min, (c) 2 min 35 s (d) 2.2 s.

Q6. (a) 112×10^3 (b) 3.58×10^6.

Q7. (a) 900, 2700, 10 800.
(b) 30, 120, 480, 1920, 7680.
(c) 24, 96, 672, 2016, 4032, 8064.
(c) 1920, 7680, 30 720.

Q8. (a) 4×10^{-4}, 4×10^{-2}, 4×10^{-3}, 0.4.
(b) 0.04%, 4.2%, 0.4%, 67%.

Q9. (a) (i) 5.5 dB, (ii) 19.5 dB, 3.5 dB.

Q10. (a) 22 dB, (b) 16 dB, 30 ms.

Chapter 3

Q5. (c) 1, 5.

Q7. (a) 40 000, (b) 10.

Q8. (i) 8×8, (ii) 4×4; 1024.

Q9. (a) 58%, (b) 0.53 U, (c) 0.45 U.

Chapter 4

Q1. 5 E, 4.95 E.

Q2. 13 E.

Q3. 0.008.

Q4. 598 calls, 3 min duration.

Q5. 0.91 E, 0.89 E, 0.88 E.

Q6. No. of trunks: 5, 7, 10, 18, 53, 96.
Occupancy: 0.2, 0.29, 0.4, 0.56, 0.75, 0.83 E.
Trunks per erlang: 5, 3.5, 2.5, 1.8, 1.33, 1.2.

Q7. (i) 19, 55, 66, 77, 98.
(ii) 18, 54, 66, 78, 100.
Normal-load criterion applies for $A \leq 50$ E
Overload criterion applies for $A \geq 50$ E.

Q8. (a) $2 + 1$ spare $= 3$.
(b) System crashes.

Q9. 12 messages.

Q10. (a) 12 000 calls per hour
(b) 40 ms.

Chapter 5

Q2. (a) $s = 2$, $d = 4$, $t = 2$, $c = 2$.

Q3. (a) 13.1 E, 17.1 E.
(b) 0.92 E, 0.91 E; 0.4 E.

Q5. (a) 84, (b) 148.
(Note: a single stage needs only 144 crosspoints.)

Q6. (a) 60, (b) 0.003.

Q7. (a) ii, (b) (i) 5000, (ii) 4000.
(c) 0.012, 0.03.

Answers to problems [303]

Q8. $m = 30$, $n = 5$, 220.

Q11. (a) 8000, (b) 0.012, (c) 9600, (d) 15 200.

Q12. (a) (i) 30 000 instead of 90 000, (b) (i) 0.018, (ii) 0.072, (iii) 0.038.

Chapter 6

Q1. (b) (ii) 0.39, (iii) 0.001 55.

Q2. (b) (ii) 0.0038, (iii) 0.0038.

Q4. S–T–S switch requires 2048 crosspoints and 4096 bytes of storage.
T–S–T switch requires 1024 crosspoints and 5120 bytes of storage.

Q5. S–T–S: 267, T–S–T: 196.

Q6. (b) S–T: 0.4, T–S: 1.5×10^{-6}

Q7. 500 extensions and 12 exchange lines.

Q8. $B < 3.4 \times 10^{-5}$.

Q10. (b) 3.5×10^{-7}, (c) 4%.

Chapter 7

Q3. 2×10^{-4}.

Q4. (b) (i) 2, (ii) 4, (iii) 9.4, (iv) 18.2.

Q5. (a) 11.4 years, (b) 114 years.

Q6. (a) 1800 calls/hour, 5 E. (b) 7%, (c) 16%.

Q7. (a) 24.5%.
(b) 21 ms.
(c) 7.2×10^{-4}, (d) 499 ms,
(e) 36%.

Q8. (a) 38.2 ms.
(b) 5.5×10^{-3}.
(c) 800 ms.

Q9. (a) 95%, 77%
(b) 33% more.

Q10. (b) 42%.

Chapter 8

Q1. (a) (i) 69.8: 30.2, (ii) 72.3: 27.7.

Q10. 21%, 3.2 ms.

Chapter 9

Q1. (b) (i) Circuit switching: $cl + pl + x/b$
(ii) Packet switching: $pl + slx/i + (1 + h/i) x/b$
(c) Packet switching is quicker if $x < cli/(sl + h/b)$

Q3. (a) 115.
(b) 230.
(c) 2.72.

Q4. (a) 0.135
(b) 0.087.
(c) 7.4.

Q5. (a) 2.49 Mbit/s.
(b) (i) 4.54 Mbit/s, (ii) 4.75 Mbit/s.

Q6. (a) 4µs, (b) 1 ms.

Q8. (b) $i + h = \sqrt{h/P}$

Q9. (a) 20%, 7.4%, 2 ms,
(b) 80%, 62%, 192 ms,
(c) 96%, 91%, 192 ms.

Q10. (a) (i) 6 ms, (ii) 192 µs.
(b) 16 ms.
(c) Total delay = 28 ms.

Chapter 10

Q6. 12.7 E.

Q7. (a) 200 km, (b) 196 km, (c) 190 km.

Q8. Method (ii).

Q10. (a) (i) A, A, (ii) 2A, 2A, (iii) A, A.
(iv) 2A, 2A.
(b) (i) A, A, 6A, (ii) 1.33A, A, 6A.
(iii) 1.33A, 0.67A, 5.67A,
(iv) 1.67A, 0.75A, 5.5A.

Q12. (a) 10.25 months.

Index

Access network, 7, 10, 43, 255, 259–60, 287
Access tandem, 257
Acknowledged signalling, 181–2, 204
Action control point, 270
AC9 signalling system, 213
Adaptive differential PCM, 32
Add/drop multiplexer, 37, 39
Address signal, 56, 178, 181, 214, 225
Administration data network, 261
Administration software, 198
A law, 32, 33
Alerting, 57, 179, 181
Allotter, 59, 187
ALOHA system, 236–7
Analog network, 40, 255–8
Analog-to-digital conversion, 17, 31–2, 41, 246, 263
Analog transmission, 16, 41
Answer signal, 53, 179, 181, 207
Application layer, 14
Arbiter, 187
Area code, 182, 274
ARPANET, 242
Associated common-channel signalling, 219
Asynchronous time division, 247
Asynchronous transfer mode (ATM), 39, 247–8, 267
 ATM switching, 248–50
AT&T network, 256, 260, 270, 285
Attending, 56
Attenuation, 17, 44, 255, 257
Audio transmission, 43–4, 207
Automatic alternative routing, 6, 126, 281–6, 298–9
Automatic number indication, 57, 177, 182
Auxiliary switching network, 74, 186–7, 196
Availability, 119, 191–2, 200

Backbone route, 6, 258, 281, 284, 285
Backward signals, 181, 182, 183, 207
Balanced link access protocol, 244
Balance impedance, 21
Balance return loss, 22
Bandwidth, 16, 25
Banyan network, 249
Baseband, 17
Basic group, 28
Basic-rate access, 43, 84, 226, 264
Baud, 16
B channel, 84, 264

Bearer channel, 17
Bearer services, 9, 84, 263
Bell, A.G, 1
Benes, V.E., 117, 147
Bernouilli distribution, 294, 299, 301
Betulander, G.A., 66
Binary code, 16, 31
Binary signal, 16, 25
Binomial distribution, 294, 299, 301
Birth and death process, 74
Bit interleaving, 35, 37
Bit stealing, 34, 213
Bit stuffing, 223
Blocking, 70, 128, 138–42, 144
BORSCHT, 78, 82–3, 156
Bothway working, 205, 218, 286
Bridge, 231
British Telecom network, 11, 260, 263, 271, 277, 280, 285
Broadband channel, 17
Broadband ISDN, 246–8, 266, 287
Burst-mode transmission, 264
Bursty traffic, 233
Bus, 4, 187, 189, 196, 235–8, 241
Busy hour, 88, 103
Busy-hour call attempts, 199
Busy test, 53, 56, 68, 178
Byte, 31
Byte interleaving, 35, 37

Call-back *see* Crank-back
Call barring, 55, 76
Call charging, 54–5, 182, 277–80
Call congestion, 114, 299
Call diversion, 76, 263
Called subscriber held, 180
Call gapping, 286
Calling-card service, 270
Calling-line identification, 57, 177, 182, 263
Call packing, 146, 152
Call processing, 56–7, 177–81, 197–8
Call-progress signal, 179, 181, 188
Call record, 197
Call-request signal, 53, 56, 177, 181, 207
Cambridge Ring, 239
Carrier-identification code, 275
Carrier-sense multiple access with collision avoidance, 237–8

Index [305]

Carrier systems, 27–9, 43, 44, 210–13
CCIR, 11
CCITT, 11
CCITT No. 6 signalling system, 220
CCITT No. 7 signalling system, 14, 221–6, 266, 271
CCS, 89
Cell, 247, 267
Cellular radio network, 44, 267–9, 288
Central-battery operation, 53
Central office, 1, 9
   Classes 1 to 5, 5, 9, 256
Centralized control, 68, 76, 176, 191
Central processor, 56, 57, 66, 74, 76, 92, 113, 176, 193–5, 285
Centre de transit, 10
Centrex, 272
Channel, 16
Channel-associated signalling, 182, 205–18
Channel graph, 142
Charging area, 279–80
CHILL, 198
Circuit, 16
Circuit-identity code, 219, 226
Circuit switching, 51
Class mark, 177
Class of service, 55, 76, 177, 178
Clear signal, 53, 180, 182, 207
Clos, C., 148, 169
Codec, 82, 83
Coded call indicator, 208
Coder, 31
Code selector, 63
Cold standby, 193
Command/response bit, 227
Common-channel signalling, 56, 84, 183, 205, 214, 218–27, 258, 285
Common control, 55, 62, 76, 113, 176, 186–90, 193
Companding, 31
Compelled signalling, 204, 219
Complementary events, 293
Concentration, 54, 144
Concentrator, 60, 70, 74, 81, 83, 99, 128, 130–2, 139, 156, 164
Conditional probability, 293
Conditional selection, 70, 128
Conference call, 77, 263
Congestion, 90–1, 114, 200, 243, 284–6
Connection, 52, 57, 178–9
Connectionless mode, 242
Connection-oriented mode, 242
Connection store, 158, 160
Connectivity, 142, 144, 148, 169
Consolidation, 166
Container, 39
Contention, 187, 205, 236, 265
Continuity test, 70, 128, 219
Continuous probability distribution, 296–7
Continuous signal, 204
Control field, 223, 227
Cord circuit, 52, 54
Core network, 7, 44, 287
Country code, 275
Crank-back, 57, 74, 187, 285
Crossbar switch, 66–8, 77, 118, 138
Crossbar system
   ARF System, 76

Bell No. 5 System, 73–4, 76, 187
5005/TXK1 System, 71–2, 126
Crosspoint, 51, 66, 77, 117, 158
Crosstalk, 27, 31
Customer, 2
Customer-line signalling, 53, 58, 181–2, 205–7
Customer loop *see* Access network
Customer node, 8
Cut-off relay, 53, 180
Cyclic redundancy check, 223, 224, 230, 251

Data-circuit terminating equipment, 243
Data communications, 4, 41, 83, 230–46, 263
Data country code, 277
Datagram mode, 242
Data-link level, 222
Data network, 4, 230–46, 277
Data-network identification code, 277
Data terminal equipment, 243
Data-user part, 222
D channel, 84, 264, 265
Decibel
   dB, dBm, dBmO, dBr, DBW, 17–18
Degraded minutes, 42
Delay probability, 107–8, 112
Delay system, 30, 90, 105–13, 186
Delta network, 249
Destination point code, 219, 225
Diagnostic program, 194, 195
Dialling, 58, 206, 208
Dial tone, 59, 178
Digital cell centre exchange, 260
Digital cross-connect system, 166, 287
Digital derived-services network, 271
Digital derived-services switching centre, 271
Digital distribution frame, 66
Digital main switching unit, 260
Digital private-network signalling system, 227
Digital subscriber signalling system, 156, 226–7
Digital subscriber switching subsystem, 156
Digital switching subsystem, 156, 194
Digital switching systems, 80–4, 156–76, 188, 190
   AXE-10 system, 83, 126, 156, 164, 194, 196
   DMS-10 system, 83, 157
   E-10 system, 81, 156, 157
   E-12 system, 83
   EWS-D system, 83, 157, 196
   HDX-10 system, 198
   NEAX system, 83, 157
   System X, 83, 156–7, 164, 193, 194, 196
   System 12, 189, 194
   No. 4 ESS, 81, 156, 162
   No. 5 ESS, 83, 157, 196
Digital-to-analog conversion, 31, 41, 246, 263
Digital transmission, 16, 25–7, 41, 44
Digit analysis, 62, 178
Dimensioning, 91, 113
Direct dialling in, 274
Direct distance dialling, 64, 273
Director area, 63, 274
Director system, 63, 75
Directory number, 63, 274, 273–5, 276–7
Discrete probability distribution, 293–6
Distortion, 16, 31, 208
Distributed control, 61, 75, 176, 189, 196–7
Distributed queue dual bus network   242

Index

Distribution frame, 64–6
Distribution network *see* Access network
Distribution stage, 137, 138, 143
District switching centre, 255
Double-current working, 208
Drop-and-insert multiplexer, 37, 39
Dual-tone multifrequency signalling, 206
Dynamic routing, 260, 282–3, 285–6

E and M wire signalling, 210–11
E channel, 265
Echo, 21–3, 41, 161, 247
Echo suppressors and cancellers, 23, 41, 256, 264
Electronic switching, 76–8
En-block signalling, 214
End office *see* Local exchange
End-to-end signalling, 214–16, 217
Engset distribution, 299–300
Equalization, 25
Equipment number, 65, 177
Equivalent random method, 125, 126, 285, 298
Ericsson 500-line system, 63
Erlang (unit of traffic), 89
Erlang, A.K., 89, 96, 105
  Erlang's delay formula, 108
  Erlang's first (B) distribution, 97, 98, 114, 301
  Erlang's ideal grading formula, 125
  Erlang's lost-call formula, 98, 300, 301
  Erlang's second (C) distribution, 107, 114
Error control, 13, 27, 216, 221, 223, 224–5, 230, 247
Errored seconds, 42
Ethernet, 238
European Telecommunication Standards Institute, 12
Exchange code, 62, 63, 273, 274
Exchange termination (ISDN), 264
Expander, 61, 69, 74, 129, 132, 134
Expansion, 144–5
Exponential distribution, 93, 110, 296

Facilities, 49, 76, 206
Facsimile, 263
Fast packet switching, 247
Federal Communications Commission, 11
Ferreed, 78
Fiber-optic distributed data interface, 241
Final route, 281, 285
Final selector, 60, 65, 75
First-in first-out queue, 110
Five-stage switching network, 149, 162, 164
Flag, 223, 224, 231, 244
Flat-rate tariff, 55, 278
Flow control, 243
Focused overload, 286
Forced release, 180
Format identifier, 244
Forward signal, 181, 182, 183, 207
Four-stage network, 70, 137, 141–2, 148
Four-wire circuit, 20–4, 40, 41, 255
Four-wire switching, 24, 78, 162–3, 255
Frame, 30, 223, 226, 244
Frame alignment, 30, 33, 34, 36, 170–1
Frame-layer protocol, 244
Frame-mode bearer service, 246
Frame relay, 245–6
Freephone service, 270, 271, 275
Frequency-division multiplexing, 17, 27–9, 44, 72

Full availability, 97, 105, 118
Functional division, 186
Future Land Mobile Telecommunications System, 288

Gain, 17
Gateway, 7, 10, 231
Gaussian distribution, 297
Graded group, 121
Grade of service, 91, 98, 99, 103, 112, 124–5, 128, 138–42, 167, 255
Gradings, 119–26
Graph theory, 142–4
Grooming, 166
Ground start, 206
Groupe Spéciale Mobiles, 269, 288
Group selector, 60, 75
Group switching centre, 255
Guard circuit, 212
Guarding, 54

H_0, H_{11}, H_{12} channels, 264, 286
Hand off/hand over, 268
Header, 230, 238, 247
Hierarchical network, 7, 242, 255, 280–1, 285
Higher-order multiplexing, 28, 34–9, 44
High-level data-link control, 223, 244
High-level language, 198
High-usage route, 281, 283
Highway, 158
Holding time, 89
Home location register, 268
Homogeneous grading, 124
Hot standby, 193
Hub polling, 235
Hybrid transformer, 21, 76, 78, 83, 264

I 420 interface, 265
Ideal grading, 124–5
Inband signalling, 211–13, 214, 216–19
Information field, 223, 230, 247
Information frame, 244
Initial address message, 225
Integrated broadband communications network, 246
Integrated digital network, 14, 34, 41, 43, 81, 171, 258–63, 285
Integrated-services digital network, 14, 84, 219, 244, 263–7, 272
Integrated-services PBX, 272
Intelligent multiplexer, 222–3
Intelligent network, 269–72, 273, 275, 288
Intelligent peripheral, 271
Inter-exchange carrier, 256
Intermediate distribution frame, 65
Intermediate reference system, 39
Intermediate service part, 222
International exchange, 7, 10, 144, 276
International Frequency Registrations Board, 11
International network, 7, 283
International numbering, 275–6
International prefix, 274, 277
International Standards Organization, 12
  ISO 7-layer model for OSI, 12–14, 221, 231
International subscriber dialling, 64, 273, 275
International Telecommunications Union, 11
Inter-register signalling, 214–18
Interrupt handling, 197

ISDN call, 266, 276
ISDN numbering plan, 276-7
ISDN user part, 222, 266
ITE-Radiocommunication, 11
ITU-Telecommunications, 11

Jacobaeus, C., 138, 301
Jitter, 27, 41
Joint probability, 293
Junction call, 56, 61-2, 182
Junction circuit, 5, 9
Junction network, 5, 7, 44, 255
Junction signalling, 207-9
Junctor, 10
Justification, 36

Kendall, D.G., 114
Keyphone, 206
Keypulsing, 214

Lee, C.Y., 138, 301
Length indicator, 224, 231
Limited availability, 119
Linear predictive coding, 32
Line balance, 21
Line circuit 54, 78, 81, 82-3, 180, 192
Line finder, 59, 75
Line relay, 53, 180
Line signalling, 207-14
Line store, 177, 181, 197
Line testing, 65, 78, 83, 219, 265
Link-access protocol
   LAP-B, 244, 251
   LAP-D, 226
Link-by-link signalling, 214-16, 219
Linked numbering area, 62, 273
Link frame, 69-70
Link-layer protocol, 13, 222, 244
Link system, 52, 68, 126-53
Listener echo, 22
Load-sharing processors, 193, 194
Local-access and transport area, 256
Local-area network (LAN), 4, 8, 12, 84, 231, 235-41, 272
Local exchange, 9, 10, 74, 144, 156, 256, 260
Local-exchange carrier, 256
Local network see Access network
Logical-channel identifier, 244
Long-distance DC signalling, 208-9
Loop/disconnect signalling, 53, 54, 58, 83, 205, 208
Loss see Attenuation or Grade of service
Lost calls held, 114
Lost-call system, 51, 91, 96-105
Loudness rating, 39-41, 257, 263

Main distribution frame, 64-5
Majority-decision logic, 194
Manual system, 49, 51-6, 57, 75, 208, 214
Map in memory, 78, 197
Marker, 67, 69-70, 128
Markov process, 94, 104
Mean delay, 110, 111, 112, 233
Mean of probability distribution, 293-4, 296
Mean time between failures, 191
Mean time to failure, 191
Mean time to repair, 191
Memoryless traffic, 93

Mesh network, 4
Mesosynchronous network, 171
Message-based signalling, 219
Message indicator, 225
Message-rate tariff, 278
Message register see Metering
Message signal unit, 223-5
Message switching, 49-51, 113, 230
Message-transfer part, 222
Metering, 54-5, 182, 278
Metropolitan-area network, 242
Microcell, 269
Microprocessor, 176, 196
Mismatch see Blocking
Mixing stage, 136, 138, 143
Mobile switching centre, 267
Mobile telephone, 267-9, 275, 288
Modulation, 27, 29, 31
Molina, E.C., 114
Mu law, 32, 33
Multicomputer architecture, 194
Multiframing, 213-14
Multifrequency signalling, 206, 214, 216-18
   DTMF, 206
   MF2 system, 218
   R1 system, 216
   R2 system, 217-18
Multimedia terminal, 266
Multiple, 53, 54, 59, 65, 66, 70, 77
Multiplexer, 33-4, 37, 259
Multiplexing, 17, 27-39, 43
Multiprocessor architecture, 194
Mutually exclusive events, 293

Narrowband ISDN, 84, 266
National network, 7, 9, 255-63
Negative exponential distribution, 93, 110, 296
Network channel terminating equipment, 265
Network configurations, 2-8
Network control point, 271
Network destination code, 276
Network layer protocol, 13, 222, 244
Network management, 37, 39, 218, 261, 286-7, 287-8
Network management centre, 200, 261, 282, 286-7
Network planning, 254-5
Network terminal number, 277
Network termination (ISDN), 264
Node, 8, 142
Node latency, 240
Node, software, 270
Noise, 25-6, 31, 41, 263
Nonassociated signalling, 219
Nonblocking networks, 146-51, 169
Nonhierarchical network, 260, 285, 286
Nonpoissonian traffic, 285, 298-300
Normal distribution, 297
Normal-response mode, 251
North American network, 256-7, 260
Numbering schemes, 62, 273-7
   International numbering, 273, 275-6
   ISDN numbering, 276-7
   Linked numbering   62, 273
   National numbering   273-5
   North American numbering, 274, 275, 277
   Public data networks, 277
   UK numbering   274-5, 277

Index

Numerical selector, 63, 75
Nyquist, H., 25

Occupancy, 90, 99
Octet, 31
O'Dell, G.F., 121
Office of Telecommunications, 11
One-only selector, 187
One-in-n sparing, 193
Open network access, 11
Open network provision, 11
Open systems interconnection, 12–14, 277, 287
Operating system, 197–8
Optical fiber, 37, 44, 241, 267, 287
Order wire, 56
Originating-point code, 220, 225
Outband signalling, 210–11
Overall loudness rating, 39–41, 257, 263
Overflow traffic, 283, 298–9
Overhead digits, 36, 38, 265
Overlap signalling, 214
Overlay network, 271
Overload control, 199–200, 286
Overload criterion, 100, 103

Packet, 50, 230, 244, 246
Packet assembler/disassembler, 243
Packetizing delay, 246
Packet layer, 244
Packet switching, 50, 113, 230–53
Packet-type identifier, 244
Pair gain, 43, 259
Palm–Jacobaeus formula, 125
Panel call indicator system, 208
Panel system, 63, 208
Pan-European Digital Mobile Service, 269, 288
Parking, 180–1
Party line, 43, 178, 179
Path overhead, 39
Payload, 38
Payphone, 55, 206
Peaky traffic, 102, 298–9
Percentage difficulty, 40
Periodic pulse metering, 55, 208, 279
Permanent loop (permanent glow), 180
Personal communication network, 269
Phantom circuit, 207
Photonic switch, 118
Physical layer, 13, 222, 243
Plesiochronous digital hierarchy, 34–7, 287
Plesiochronous digital network, 171
Pointer, 39
Point of presence, 256
Poisson distribution, 93, 96, 236, 294
Polling, 235–6
Post-dialling delay, 63, 214, 215
Power-drive system, 63, 208
Power levels, 17–18
Preamble, 238
Prefix digit, 274, 277
Premium-rate services, 271, 275
Presentation layer, 13
Primary centre, 10, 255
Primary multiplex group, 33–4, 190, 213
Primary-rate access, 43, 84, 226, 227, 264
Private branch exchange, 8, 56, 83, 84, 166, 178, 206, 227, 272, 274, 284, 300
Private circuit, 8
Private network, 8, 9, 84, 227, 272–3, 284
Probability theory, 293–7
Proceed-to-send signal, 59, 178, 181, 207, 243
Processor architecture, 193–5, 196–7
Progressive control, 60, 76, 194
Progressive grading, 121, 298
Propagation delay, 22–3, 41, 44, 247
Protection, 65, 78, 83
Protocols, 12–14, 181, 221–7, 231, 244, 245–7
Provision period, 103
Public data network, 9, 242–5, 277
Public switched telephone network, 7, 9, 255–63
Public telecommunications operator, 9
Public Utilities Commission, 11
Pulse-code modulation, 17, 31–2, 44, 81
Pulse signalling, 204, 207, 212
Pure-chance traffic, 93, 95, 96, 105, 298
Push-button telephone, 206
P wire, 54, 59, 67, 78, 197

Q Sig system, 227
Quantization, 31
Quantization noise, 31, 41, 263
Quasi-associated signalling, 219, 220
Quasi-nonblocking network, 145, 169
Queue capacity, 109
Queuing system, 50, 90, 105–13, 186

R1 signalling system, 216
R2 signalling system, 217–18
Radio base station, 267
Radio link, 44
Random access, 235, 236–41
Rapp, Y., 281, 299
Real-time network routing, 285–6
Rearrangeable network, 146–7
Reconfiguration, 195
Reed–electronic switching system, 78–80
   AXE System, 80
   No. 1 ESS, 78, 126, 130, 187, 190, 194
   Metaconta System, 80
   TXE2 System, 80, 130, 136
   TXE4 System, 80, 130, 136, 153, 187–8, 190, 194
Reed relay, 78, 81
Reference level, 18
Regeneration, 17, 27, 208
Regional centre, 257
Regional processor, 196
Register, 62–3, 74, 207
Registration, 268
Regulation, 10–11
Relative level, 18
Release of connection, 52, 53, 57, 180–1
Reliability, 191
Remote concentrator unit, 164, 259
Remote switching unit, 164, 259
Remreed, 78
Renewal process, 94
Repeat attempts, 97
Repeater, 16, 17, 27, 45, 207
Repeater station, 16, 210
Resilience, 6, 259, 284
Revertive pulsing, 63, 208
Ringing, 52, 78, 83, 179

Index [309]

Ringing tone, 179
Ring network, 4, 39, 239-41, 287
Roll-back, 195
Roll-call polling, 235
Rotary system, 63
Round-trip delay, 23
Route switch, 74, 81, 128, 129, 156
Routing, 243, 280-6

Sampling, 29, 35
Satellite communication links, 23, 41, 44, 205
Satellite exchange, 62
Scanning, 55, 187, 189, 190
Scheduling, 197
Second-attempt feature, 70, 128, 146
Sectional centre, 257
Sectionalized switching network, 151-3
Section overhead, 38
Security, 164, 192-3
Seize signal, 53, 56, 177, 181, 207
Self-healing ring, 39, 240
Self-routing switch, 249
Semantic transparency, 247
Sender, 63, 74
Sequence number, 224, 230, 244
Sequential search, 100-2, 121, 298
Serial trunking, 187-8
Server, 105
Service-access-point identifier, 226
Service control point, 270, 271
Service-independent building block, 271
Service-information octet, 224
Service logic execution environment, 271
Service logic programs, 270
Service management system, 271
Service node, 8
Services, 2, 8-9, 263, 269
Service switching point, 270
Session layer, 13
Severely errored seconds, 42
Signal imitation, 206, 211
Signalling, 8, 33, 181-3, 204-29, 245, 266
Signalling-connection control part, 222
Signalling control subsystem, 221
Signalling-information field, 224, 225
Signalling network, 219-20, 261, 271
Signalling-network level protocol, 221, 222
Signalling termination subsystem, 221
Signal point code, 222
Signal-transfer point, 219
Signal unit, 220, 221, 223-6
Simulation, 113-14, 138, 146
Singing, 21, 23
Singing point, 24
Single-server system, 111, 235, 239
Skipped grading, 124
Slip, 41, 170-1
Slotted ALOHA, 237
Slotted ring, 239
Smooth traffic, 105, 115, 298-9
Software, 184, 186, 195, 197-9, 269
Software build, 198
Space division, 78, 186
Space switch (time-divided), 158-9, 162
Space-time-space switching network, 161, 162, 163, 167, 169

Specification and description language, 183, 198
Speech store, 159-60
Stability margin, 24
Standard deviation, 294, 296
Standards, 11-12
Star network, 4, 56, 241
State probability, 95
State transition diagram, 183-6, 198
Statistical equilibrium, 93, 95, 106
Statistical multiplexer, 232-3, 243
Status signal, 179, 181, 188
Step-by-step system, 57-62, 74, 126, 128, 146, 191, 298
Stored-program control, 49, 50, 55, 56, 57, 76, 80, 92, 113, 176, 177, 186, 193-200, 205, 269
Strict-sense nonblocking network, 147-51, 169
Strowger, A.B., 57
Strowger switch, 57-8, 59, 66
Strowger system see Step-by-step system
Sub-addressing, 276
Subscriber, 2
Subscriber's line interface circuit see Line circuit
Subscriber trunk dialling, 64, 273
Supervision, 52, 54, 57, 63, 119, 179, 182
Supervisory circuit, 54, 74, 188
Supervisory frame, 244
Supplementary services, 227, 263
Switching networks, 117-55, 161-70, 248-50
Switching node, 8
Switching systems, 49-86, 156-229, 230-53
Synchronization, 30, 33, 34, 36, 170-3, 208-38
Synchronization network, 173, 248
Synchronous digital hierarchy, 37-9, 248, 287
Synchronous dual processors, 194
Synchronous optical network, 37-9
Synchronous transport module, 37, 248
Syski, R., 117

Takagi, K., 142-3
Talker echo, 22
Tandem exchange, 5, 76, 81, 144, 255
Tariff, 55, 84, 277-8
Telegraphy, 1, 9, 49
Telephone set, 45, 58, 206, 263
Telephone-user part, 222
Telephony, 1, 14, 39-41, 263
Telepoint service, 269, 288
Teleservices, 9, 84, 263
Television, 246, 251
Teleworking, 273
Terminal adapter (ISDN), 264
Terminal end-point identifier, 226
Terminal equipment (ISDN), 264
Three-stage switching network, 132-7, 140-1, 148-9, 161-4, 167
Ticketing, 54, 182, 278
Time congestion, 78, 186, 299-300
Time division, 78, 186
Time-division multiplexing, 17, 29-39, 77
Time-division switching, 77, 80-4, 147, 153, 156-75
Time out feature, 180
Time-slot, 30
Time-slot interchange, 158
Time-space-time switching network, 161-4, 167, 169, 193
Time switch, 158, 159-61, 163, 164
Time transparency, 247
Tip, ring and sleeve 53

Index

Token, 238
Token bus, 238
Token ring, 239
Toll centre *see* Trunk exchange
Toll circuit, 5
Toll network *see* Trunk network
Tone-on-idle signalling, 212
Torn-tape relay, 50
Touchtone dialling, 206
Traffic, 87–115, 298–300
Traffic forecast, 92, 103
Traffic intensity, 88
Traffic measurement, 92
Traffic tables, 103–5, 112–13, 125, 197
Transaction capabilities, 222, 271
Transaction-capabilities application part, 223
Transit exchange, 6, 144, 255, 256
Transition probability, 95
Translation
   DN-to-EN, 65, 72, 197
   EN-to-DN, 65, 66, 177, 197
   Routing, 62–3, 197, 249, 282
Translator, 62–3
Transmission bearer network, 9, 261, 287
Transmission bridge, 54, 57
Transmission node, 8, 287
Transmission performance, 39–41, 257, 263
Transmission systems, 43–4
Transport layer, 13
Tree network, 6, 186, 187
Triangular switch, 118
Tributary, 34
Tributary unit, 38
Triple-X standard, 243
Triplication, 194
Trunk, 9, 81
Trunk circuit, 5
Trunk distribution frame, 66, 126
Trunked mobile radio, 267
Trunk exchange, 5, 7, 9, 76, 81, 144, 156, 255–6, 260

Trunking diagram, 60, 74
Trunk network, 6, 7, 10, 44, 255–6, 260
Trunk prefix, 274
Trunk reservation, 284, 286
Two-motion selector *see* Strowger switch
Two-stage switching network, 69–70, 129–32, 147

UK network, 255–6, 260, 271
Unavailability, 192
Uniselector, 59, 75
Universal Mobile Telecommunications Service, 288
Unnumbered frame, 244
User part, 222

Value-added service, 8, 270
Variance of probability distribution, 294, 296
Video conferencing, 263
Virtual call, 242, 244–5, 247
Virtual circuit, 242
Virtual container, 39
Virtual private network, 273
Virtual tributary, 38
Virtual tributary synchronous payload envelope, 39
Voice-frequency receiver, 212, 215, 216
Voice-frequency signalling, 206, 211–13
Voice guidance, 178, 271

Wander, 27, 41
Wide-area network, 231, 235–42, 272
Wide-sense nonblocking network, 147
Wilkinson, R.L., 125, 126, 285, 299
Wired logic, 61, 186
Word, 31
Word interleaving, 32
Worker and standby, 193

X.25 network, 243–5

Zero-bit insertion and deletion, 223
Zero reference point, 18